Discrete
Dynamical
Modeling

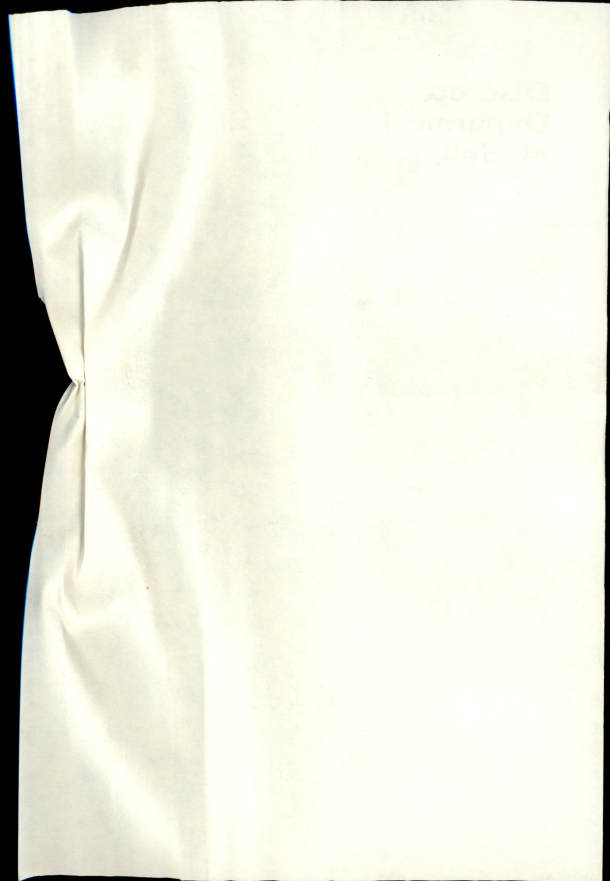

Discrete Dynamical Modeling

JAMES T. SANDEFUR

Georgetown University, Washington DC

NEW YORK · OXFORD
1993

Oxford University Press

Oxford   New York   Toronto
Delhi   Bombay   Calcutta   Madras   Karachi
Kuala Lumpur   Singapore   Hong Kong   Tokyo
Nairobi   Dar es Salaam   Cape Town
Melbourne   Auckland   Madrid

and associated companies in
Berlin   Ibadan

Published in the United States by Oxford University Press, Inc.
198 Madison Avenue, New York, New York 10016-4314

Oxford is a registered trademark of Oxford University Press

British Library Cataloguing in Publication Data
Sandefur, James T.
Discrete dynamical modeling
1. Dynamical systems. Mathematical models
I. Title.
003
ISBN 0-19-508438-1

Library of Congress Cataloging-in-Publication Data
Sandefur, James T.
Discrete dynamical modeling / James T. Sandefur.
Includes bibliographical references
p.   cm.
1. Differentiable dynamical systems—Mathematical models. I. Title.
QA614.8.S25   1993   003'.85—dc20   93-7843
ISBN 0-19-508438-1

Text typeset by the author using TEX.

Design and TEXnical assistance by Frederick H. Bartlett

9 8 7 6 5 4 3 2

Printed in the United States of America
on acid-free paper

To my mother, Evelyn Harris

PREFACE

The study of dynamics is the study of how things change over time. Discrete dynamics is the study of quantities that change at discrete points in time, such as the size of a population from one year to the next, or the change in the genetic make-up of a population from one generation to the next. In this text we tend to concurrently develop a model of some situation and the mathematical theory necessary to analyze that model. As we develop our mathematical theory, we will be able to add more components to our model.

The means for studying change is to find a relationship between what is happening now and what will happen in the "near" future; that is, cause and effect. By analyzing this relationship, we can often predict what will happen in the distant future. The distant future is sometimes a given point in time, but more often is a limit as time goes to infinity. In doing our analysis, we will be using many algebraic topics, such as factoring, exponentials and logarithms, solving systems of equations, manipulating complex numbers, and matrix algebra. We will also use some topics from probability. Most of the probability and matrix algebra used is developed within this text.

After reading this text, you should be able to apply discrete dynamics to many fields in which things change, which is most fields. The goal, then, is not only to learn mathematics, but to develop a different way of thinking about the world.

My own interest in this material is somewhat backwards. Several years ago I became interested in a topic of current mathematical research, chaos. One result of the theory of chaos is that there are certain situations that change over time in an apparently random manner, and no amount of analysis will enable us to make accurate predictions for more than a short period of time. As I studied these chaotic models, I came to learn more and more about situations in which we **can** make accurate long-term predictions. In fact, I discovered that discrete dynamics has a long and useful history in many fields of study, but has been largely ignored by mathematicians until the recent interest in chaos.

My idea was to write a text that demonstrates how discrete dynamics can be used to study applications in many fields, but which can be understood by those with only a knowledge of high school algebra. The reader will also see connections between many different areas of mathematics.

This text will give the reader an understanding of when dynamics can be used to make accurate predictions, and as a result, an appreciation of situations in which math sometimes "fails", that is, chaos. The theory of chaos is not discussed in any detail in this text. The interested reader is referred to other books that deal exclusively with that topic.

Acknowledgements

I express my gratitude to the many people who have helped me in the development of this text. Foremost, I am indebted to E. Ray Bobo and Henry Pollak for their careful reading of a draft of this book. Their comments on the content and form of that manuscript greatly improved the final text. The United States Military Academy used an early version of this text in their classes. The results of that class testing were crucial in shaping this text. Bruce Torrence produced the fractal figures in Chapter 4. My editor, Don Jackson, was invaluable for his support and suggestions. Finally, I would like to thank my wife, Helen, and my son, Scott, for tolerating me during the writing of this text.

Notes to the instructor

While it is not necessary to have a calculator or computer to understand this material, it would be helpful. With the aid of technology, students can study and experiment with many complex and interesting applications, right from the start. Once a simple mathematical relationship is found, students can easily run their own math experiments and make their own hypotheses about what will happen. The verification of these hypotheses will have to wait until the appropriate theoretical model is developed. I have found graphing calculators and computer spreadsheets to be particularly useful. Both the calculations and the graphs in this text can be done on graphing calculators and with spreadsheets.

Chapters 1 and 2 are essential for the rest of the text. Chapters 4, 5, 6, and 8 could be studied in any order. Chapters 3 and 7 introduce needed mathematical concepts with which the reader may be unfamiliar. Many applications in this text use concepts from probability; chapter 3 introduces the basic probability concepts that are needed in these applications. Chapter 7 introduces elementary concepts of matrix algebra, which are then used in Chapter 8. Applications are for the most part independent of one another. Thus, particular applications can be omitted.

The easiest method for compiling tests is often to pick what you want as an answer, then work backwards to get a question with that answer. I must admit that it is difficult to develop good modeling questions.

CONTENTS

Discrete
Dynamical
Modeling

CHAPTER 1

Introduction to dynamic modeling

1.1 Modeling drugs in the bloodstream

Dynamics is the study of how quantities change over time. Dynamic modeling is the process of developing a mathematical relationship that in some sense describes how a real world quantity changes over time. We do not claim that the mathematics gives an exact description of reality. But we do hope the analysis will give us some insight into the phenomenon being studied.

Let's begin by developing a simple model of how medicine is eliminated from the body. An empty container will represent the body. Add 4 cups of water to the container to model the blood in the body. Suppose we are going to take some medicine internally. Let's assume that the proper dose of this medicine, say cough syrup, is 16 ml. To model this, add 16 ml of food coloring to the container. See figure 1.1. (To be precise, the 16 ml of food coloring should be added first, with enough water added to make a total of 4 cups of liquid.)

How do the kidneys remove impurities from the blood? Loosely speaking, during any 4-hour period, the kidneys take in a fixed percentage of the blood and remove the medicine from that blood. Let's assume the kidneys purify one-fourth of the blood during any 4-hour period. This can be modeled by removing one cup of water from the

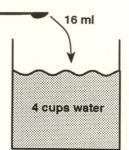

FIGURE 1.1. Simple model of blood system with drug added.

bowl and replacing it with a cup of clear water as shown in figure 1.2. Thus, at the end of the first 4-hour period, one-fourth of the medicine (4 ml) has been removed and thus 12 ml remain in the blood system.

For simplicity, assume for now that no more medicine is taken. How much medicine is left after the next 4-hour period? This is demonstrated by removing another cup of colored water and then adding a cup of clear water. Note that one-fourth of the water is removed so one-fourth or 3 ml of the remaining medicine has been removed. So at the end of 8 hours there are 9, **not 8**, ml of medicine remaining.

ASSUMPTION: Each time period, a certain fraction p of the medicine is removed from the blood. In this example, $p = 0.25$. □

REMARK: Most modeling requires us to make simplifying assumptions such as the one above. Although there are other things affecting the amount of medicine in the blood system, we are trying to determine how this one factor affects the body. As we understand this

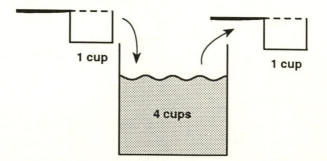

FIGURE 1.2. One fourth of colored water is replaced with clear water.

effect and as we develop our mathematical techniques, we will be able to develop more realistic models.   □

Some notation will be helpful here. Let $a(n)$ represent the amount of drug in the body at the end of n 4-hour periods. Thus, initially (when $n = 0$) the amount of drug in the body is $a(0) = 16$ ml. At the end of the first 4-hour period there is $a(1) = 12$ ml left. Similarly,

$$a(2) = 9, \quad a(3) = 6.75, \quad \ldots$$

For any 4-hour period n, the amount of medicine in the body at the beginning of that period, $a(n - 1)$, will be called the **input** amount and the amount, $a(n)$, in the body at the end of that period will be called the **output** amount. For example, for the first ($n = 1$) 4-hour period, $a(0) = 16$ is the input and $a(1) = 12$ is the output. The relationship between input and output for any 4-hour period n translates into

$$a(n) = 0.75a(n - 1) \qquad \text{for} \quad n = 1, 2, \ldots . \qquad (1.1)$$

Relation (1.1) is called a **discrete dynamical system**. Note that this relationship does not give the amount of drug in the body in any time period n, but instead describes how to find the amount of drug at some point in time **if we know the amount of drug at a previous point in time**.

From the above relationship, we know that $a(3) = 0.75a(2) = 0.75 \times 9 = 6.75$ ml. We can easily compute $a(4)$, $a(5)$, ... using a calculator.

REMARK: On some calculators, such as the Texas Instruments TI-81 and the Casio fx7000, there is an [Ans] button. To use this button, first enter 16, the initial amount of medicine. The calculator stores this number in position [Ans]. Then type the right side of relationship (1.1) using [Ans] in place of $a(n-1)$, that is, 0.75 [Ans] and [ENTER] ([EXE] on the Casio). This gives the new answer of 12, which is $a(1)$. The calculator has the formula stored, and also stores this new answer in [Ans]. Thus, if you press [ENTER] again, it will use 12 for [Ans], giving $a(2)$ or 9 as the answer. Each time you press [ENTER] you get the next $a(n)$ value.   □

Most drug tests can detect small amounts of the drug in the bloodstream. Suppose a certain test can detect 1 ml of this medicine in the bloodstream; that is, 1 ml of food coloring in the container. How long after taking the initial dose of 16 ml will the drug test still be effective? By the recursive process of taking three-fourths of each successive number, you get that the amount of drug after 0, 1, 2, ... time periods is (to one decimal place)

16,   12,   9,   6.8,   5.1,   3.8,   2.8,   2.1,   1.6,   1.2,   0.9.

Since $a(9) = 1.2$, the test is effective for 9 time periods or 36 hours. While this is not difficult, we would not want to work the problem using this method if the test could detect 0.0001 ml of the drug. Let's develop a quicker method to solve this problem.

We know that $a(1) = 0.75a(0) = (0.75)16$ and $a(2) = 0.75a(1)$. By substitution, we get that $a(2) = (0.75)^2 a(0)$. Likewise by substitution, we get that

$$a(3) = 0.75a(2) = 0.75(0.75)^2 a(0) = (0.75)^3 a(0).$$

Continuing, it is easy to see that after k time periods, the amount of dye in the container will be

$$a(k) = (0.75)^k a(0). \tag{1.2}$$

Thus we again get that $a(9) = (0.75)^9 16 = 1.20$ ml. Equation (1.2) is called the **particular solution** to the discrete dynamical system (1.1)

$$a(n) = 0.75a(n-1) \qquad \text{for} \quad n = 1, 2, \ldots.$$

REMARK: In relation (1.1), n is considered a variable since the equation reads **the amount at any one point in time is 75 per cent of the previous amount**. Solution (1.2) gives the amount of drug at some fixed point in time, k. This **convention** will be used throughout this text.   □

With a little trial and error, we can answer the question concerning the length of time necessary for the amount of drug in the bloodstream to be less than 0.0001 ml. This is done by computing $(0.75)^k(16)$ on a calculator using several different values for k, say $k = 10, 20, 30$, and so forth, until we get a result in which $(0.75)^k(16)$ is less than 0.0001. For example, $(0.75)^{40}16 = 0.00016$ and $(0.75)^{50}16 = 0.000009$. Experimenting with $40 \le k \le 50$ gives $(0.75)^{41}16 = 0.00012$ and $(0.75)^{42}16 = 0.000091$. Thus, the test is effective for $k = 41$ 4-hour periods, but not 42.

An even quicker method for solving this problem involves logarithms. In particular, the test is effective for any time period k for which the amount of drug $a(k)$ left in the bloodstream is more that

0.0001 ml, that is,

$$(0.75)^k 16 > 0.0001 \qquad \text{or} \qquad 0.75^k > \frac{0.0001}{16}$$

Taking logarithms of both sides and dividing both sides by $\log 0.75$, remembering to reverse the inequality since $\log 0.75$ is negative, gives that the test is effective for

$$k < \frac{\log 0.0001 - \log 16}{\log 0.75} = 41.653 \qquad \text{or} \qquad k \leq 41.$$

Let's try a different approach to this drug problem. Let the amount of drug in the body at the beginning of a 4-hour period be the input or x, 16 ml initially. Let the amount of the drug at the end of the same 4-hour period be the output or y, 12 ml initially. Dynamical system (1.1) states that the output is **proportional** to the input. Specifically,

$$y = 0.75x. \tag{1.3}$$

This is the equation of a line with slope $m = 0.75$. Note that the slope of the line equals the body's metabolism rate in the sense that the slope equals the fraction of the drug not removed during the time period. Thus, slope is not just a number computed by some formula, but something concrete. In fact 0.75 is the most important number in this problem.

The corresponding inputs and outputs for the above problem correspond to the points $(16, 12)$, $(12, 9)$, $(9, 6.75)$, ..., which all lie on the line $y = 0.75x$. Line (1.3) and these points are given on figure 1.3.

How do we get from one point to the next one? In particular, how could the point $(a(1), a(2)) = (12, 9)$ be used to find the next point $(a(2), a(3)) = (9, 6.75)$? Make a copy of figure 1.3 and put a pencil point at $(12, 9)$. Note that the output for this point is $a(2) = 9$ and that the input for the next point is also 9. The x-coordinate must be changed from 12 to 9. Graphically, the way to change the x-coordinate is to move horizontally. Thus, move horizontally from $(12, 9)$ to $(9, 9)$. This point has the correct input, $a(2) = 9$, but the wrong output. Changing the output means changing the y-coordinate, that is, moving vertically. So now move vertically to the line $y = 0.75x$, ending at the point $(a(2), a(3)) = (9, 6.75)$.

To summarize, suppose we are at a point $(a(n - 1), a(n))$ on $y = 0.75x$. Convert the old output into the new input by moving horizontally to the point $(a(n), a(n))$, which happens to be on the line $y = x$. Now compute the new output from this new input by

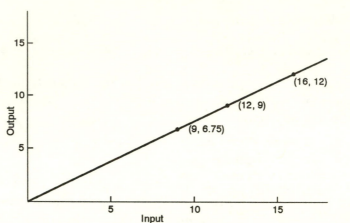

FIGURE 1.3.   Input/output graph, $y = 0.75x$, for drug problem. Input plotted on horizontal axis, output plotted on vertical axis.

moving vertically to $y = 0.75x$. This process is called **cobweb analysis** and can be seen in detail in figure 1.4. Note that we could just as easily have started at the point $(16, 16)$.

In cobweb analysis, the points $(a(0), a(1))$, $(a(1), a(2))$, ..., which correspond to the input/output of the dynamical system, lie on the line $y = 0.75x$. The points $(a(1), a(1))$, $(a(2), a(2))$, ... lie on the line $y = x$. Note that these points are converging toward the point $(0, 0)$ meaning that the numbers $a(0)$, $a(1)$, ... are going toward

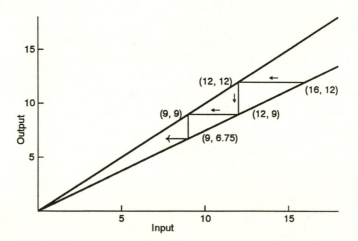

FIGURE 1.4.   Cobweb analysis for drug problem uses the input/output relation $y = 0.75x$, and the line $y = x$ to convert one output to next input.

zero. Also note that the point $(0,0)$ is the point of intersection of the two lines. The x-coordinate of the point of intersection is called the **equilibrium value** or **fixed point** for the dynamical system. This will be discussed in more detail in section 1.3.

To repeat, $x = 0$ is called the equilibrium value for dynamical system (1.1). The reason for this is that if $a(0) = 0$ then $a(1) = 0.75(0) = 0$, $a(2) = 0, \ldots$, that is, if the amount of drug starts at the equilibrium value, then the amount of drug in the body does not change. The body is in equilibrium.

One problem with our model of elimination of drugs is that normally people take another dose of the drug at the end of every time period, while our model considered the case in which only one dose was taken. Let's make our model more realistic.

Set up a container as before, with 4 cups of water and 16 ml of food coloring. Replace one cup of colored water with a cup of clear water. Recall that this is what the body does during the 4-hour period following the original dose of medicine. This leaves 12 ml of food coloring in the container. Now add an additional 16 ml of food coloring to the container to model a person taking a second dose of medicine. (Technically, the 16 ml of food coloring should be added to the cup of clear water before adding it to the bowl, so that the total amount of liquid in the bowl remains at 4 cups.) The amount of medicine/food coloring in the container is now 28 ml. Keeping with the previous notation, $a(0) = 16$ and $a(1) = 28$. Note that $a(n)$ is the amount of medicine/food coloring in the bloodstream/bowl immediately following administration of the $n+1$th dose of medicine (n time periods following the initial dose). It is assumed that the medicine enters the bloodstream instantly.

Repeating the above process gives that in 4 more hours, one cup of colored water is removed, leaving 21 ml of food coloring in the container. Then one cup of clear water with a dose of 16 ml is added, giving a total of $a(2) = 37$ ml of drug in the bloodstream.

Should we be concerned since the amount of drug has risen from 16 to 28 to 37 ml in just 8 hours? Will the blood turn into medicine if we keep taking it long enough?

Let's restate the above question: If 16 ml of medicine is taken every 4 hours for a long period of time, how much drug will eventually be in the container? To answer this question, let's develop a dynamical system to model this situation. The trick to developing dynamical systems is to **assume we know the amount of drug at some point in time, say after n periods, that is, we know $a(n)$.** How much medicine will be in the container one time period later, that is, what is $a(n+1)$? During the time period, one-fourth of the water, and consequently, one-fourth of the medicine is removed from the container. Thus there is $0.75a(n)$ ml remaining. At this point another dose of 16 ml is added, making a total of

$$a(n+1) = 0.75a(n) + 16 \qquad \text{for} \quad n = 0, 1, \ldots. \tag{1.4}$$

Thus, $a(1) = 0.75a(0) + 16 = 12 + 16 = 28$, $a(2) = 0.75a(1) + 16 = 37$, ... This process can be repeated easily on a calculator.

REMARK: Using a calculator with an [Ans] button, press 16, [ENTER], and 0.75[Ans] + 16. Now when you press [ENTER], you will get 28 which is $a(1)$. Each time you press [ENTER], you will get the next $a(k)$ value. If you do this many times, the amount will be slightly less than 64, indicating that after a long period of time, there will be about 64 ml of medicine in the bloodstream. □

REMARK: Dynamical system (1.4) could have been written as

$$a(n) = 0.75a(n-1) + 16 \qquad \text{for} \quad n = 1, 2, \ldots.$$

In either form, the equation is read as **the amount at the end of any time period is three-fourths of the amount at the end of the previous time period plus 16**. □

Let's use cobweb analysis to try to understand why the total amount of medicine approaches 64 ml. If we let the input amount be $x = a(n)$ and the output amount be $y = a(n+1)$, then dynamical system (1.4) becomes

$$y = 0.75x + 16 \tag{1.5}$$

Notice that this relationship means that the points given by plotting input versus output lie on a line with slope 0.75, the metabolism rate of the body, and with y-intercept equal to 16, the size of the dose of the drug.

Figure 1.5 gives the cobweb analysis for dynamical system (1.4). Starting at $(a(0), a(0)) = (16, 16)$ on $y = x$, go vertically to $y = 0.75x + 16$ to find the first output $a(1)$. This is the point $(16, 28)$. Repeat the process of going horizontally to $y = x$ to convert output into next input, then going vertically to get next output from that input. Observe that this process leads to a series of points which converge to the point of intersection of the two lines, $(64, 64)$. We are actually

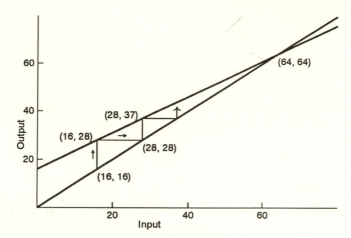

FIGURE I.5. Cobweb analysis for drug model in which 16 ml of the drug is taken in each time period. The input/output relationship is $y = 0.75x + 16$. The line $y = x$ is used to convert one output to the next input.

seeing that

$$\lim_{n \to \infty} a(n) = 64.$$

We could have started at the point $(16, 28)$ instead of $(16, 16)$. What is important is that the sequence of going horizontally and vertically be done in the correct order and that the process goes toward the point $(64, 64)$.

Thus, for dynamical system (1.4), $x = 64$ is the equilibrium value. Note that if the container starts with $a(0) = 64$ ml in it, and 16 ml are added every time period, then

$$a(1) = 0.75a(0) + 16 = 0.75(64) + 16 = 64,$$

and so $a(0) = a(1) = a(2) = \cdots = 64$ and the body is in equilibrium.

There is still the problem of finding the particular solution to dynamical system (1.4). This will not be done until section 2.4.

REMARK: Cobweb analysis can be done easily with a graphing calculator such as the TI-81 or Casio fx7000. For the TI-81 calculator, define functions $y_1 = 0.75x + 16$ and $y_2 = x$. Give $\boxed{\text{RANGE}}$ as $\boxed{\text{Xmin}} = 0$, $\boxed{\text{Xmax}} = 70$, $\boxed{\text{Xscl}} = 10$, $\boxed{\text{Ymin}} = 0$, $\boxed{\text{Ymax}} = 70$, and $\boxed{\text{Yscl}} = 10$. Then press $\boxed{\text{GRAPH}}$. Next press $\boxed{\text{2nd}}$ $\boxed{\text{DRAW}}$ to get a drawing menu and select $\boxed{\text{Line(}}$. The graph will appear with a cursor flashing in the middle. Use arrow keys to move the cursor as close to $(16, 16)$ as possible. Press $\boxed{\text{ENTER}}$ to start the line. Next move vertically using the up arrow to move to the line $y = 0.75x + 16$ and press ENTER to end the line. This point corresponds to $(a(0), a(1))$. While the coordinates will not be exactly correct, the picture gives the qualitative behavior of what is happening. Press ENTER again to start the next line. Move horizontally to the line $y = x$ and press ENTER twice, once to end this line and once to start the next line. Now keep repeating this process and the cobweb will develop.   □

PROBLEMS

1. Suppose a bowl contains five cups of water. Initially, add 20 ml of dye to the bowl. Then every 4 hours, replace one cup of water with a cup of water containing 10 ml of dye.
    a. Develop a dynamical system for $a(n)$, the amount of dye in the bowl after n four-hour time periods.
    b. How much dye will be in the bowl at the end of 12 hours?
    c. What is the equilibrium value for this dynamical system?
    d. Sketch a cobweb diagram for this problem.

2. Suppose a lake contains 20 thousand gallons of water. On the average, each week two thousand gallons of water flows out of the lake, to be replaced by rain water. Originally, the lake contains 20 pounds of insecticides. Each thousand gallons of rain water flowing into the lake contains 2 pounds of insecticide washed in from nearby farmland.
    a. Let $a(n)$ be the number of pounds of insecticide in the lake after n weeks, with $a(0) = 20$. Develop a dynamical system for $a(n)$.
    b. How many pounds of insecticide will be in the lake at the end of the fourth week?
    c. After a long period of time, how many pounds of insecticide will be in the lake?
    d. Draw a cobweb diagram for this dynamical system.

3. Suppose the body eliminates 20 per cent of the medicine every 6 hours. A drug company determines that the optimal level of drug in the bloodstream is 40 ml, which should be the equilibrium value. How many ml of drug should be taken every 6 hours so that the body reaches an equilibrium of 40 ml?

4. Suppose you come home to discover that a friend has been smoking cigarettes in your room. To improve the quality of the air in your room, you put an exhaust fan in the window. The fan

replaces 10 per cent of the air in the room every minute. How long will it take to get rid of 99 per cent of the smoke? What assumptions are you making in working this problem?

1.2 Jargon

The dynamical systems we will consider come in many different forms but, as we will see, seemingly different types of equations can be handled similarly. Therefore we will divide these equations into large classes and study each class separately.

Informally, a discrete dynamical system is a sequence of numbers that are defined recursively, that is, there is a rule relating each number in the sequence to previous numbers in the sequence. One example is the sequence 2, 4, 8, 16, ... The rule relating these numbers is $a(n+1) = 2a(n)$, meaning that each number is twice the previous number. Another example is the sequence of numbers 0, 3, 6, 9, 12, ... The rule relating these numbers is $a(n+1) = a(n)+3$ meaning that each number is three more than the previous number.

It is usually easier to give the rule and the first number, and then compute the sequence. Consider the rule

$$a(n+1) = 2a(n)(1 - a(n)),$$

with the first number being $a(0) = 0.1$. We then get the sequence

$$a(1) = 2a(0)(1-a(0)) = 0.18, \quad a(2) = 0.2952, \quad a(3) = 0.416, \quad \dots.$$

DEFINITION 1.1 *Suppose we have a function $y = f(x)$. A **first-order discrete dynamical system** is a sequence of numbers $a(n)$ for $n = 0, 1 \dots$ such that each number after the first one is related to the previous number by the relation*

$$a(n+1) = f(a(n)).$$

REMARK: Many books and articles discuss sequences of numbers given by the relationship

$$a(n+1) - a(n) = g(a(n)).$$

This relationship is called a **first-order difference equation**. Note that by letting $f(x) = g(x) + x$, these two concepts are seen to be equivalent. $\square$

We will often omit the term "discrete" and just call such sequences **dynamical systems**. We will also equate a dynamical system with the rule that defines it. Three examples of dynamical systems are the relationships

$$a(n+1) = 3a(n), \qquad a(n+1) = 2a(n) + 5, \qquad a(n+1) = \frac{a(n)}{1 + a(n)}.$$

The functions are $f(x) = 3x$, $f(x) = 2x + 5$, and $f(x) = x/(1 + x)$, respectively.

When given a dynamical system, we are usually given what the values of the n variable can be (although sometimes it is obvious from the context of the problem). For example

$$a(n+1) = 1 + 2a(n), \qquad \text{for} \quad n = 0, 1, 2, \ldots.$$

This dynamical system is in reality **an infinite collection of equations**. The above relationship actually represents the equations

$$a(1) = 1 + 2a(0), \quad a(2) = 1 + 2a(1), \quad a(3) = 1 + 2a(2), \quad \ldots$$

Thus, if we are given the value of $a(0)$, we can substitute its value into the first of the above equations to find $a(1)$. We can then substitute the value of $a(1)$ into the second equation to find the value of $a(2)$, and so forth. Given that $a(0) = 5$, you should try finding $a(1)$ through $a(6)$.

When the function $f(x)$ in a dynamical system is of the form $f(x) = mx$ for some constant m, that is, the graph of $y = f(x)$ is a straight line through the origin, the dynamical system is called **linear**. Thus, $a(n+1) = 3a(n)$ is a linear dynamical system. When $f(x) = mx+b$, that is, the graph of $y = f(x)$ is a straight line with y-intercept not equal to zero, the dynamical system is called **affine**. Thus, $a(n+1) = 2a(n)+5$ is an affine dynamical system. We make a distinction between these similar types of dynamical systems because the method for finding solutions is slightly different, and because the term "linear" means something other than "straight line" to mathematicians. Note that the dynamical system $a(n+1) = 0.75a(n)$ which modeled the first drug problem is linear. The dynamical system $a(n+1) = 0.75a(n) + 16$, which modeled the repeated doses of a drug, is affine.

If the function $y = f(x)$ is not a straight line, the dynamical system is called **nonlinear**. Thus

$$a(n+1) = \frac{a(n)}{1 + a(n)} \qquad \text{and} \qquad a(n+1) = 2a(n)(1 - a(n))$$

are nonlinear dynamical systems. These dynamical systems relate to selection in genetics and to population growth, respectively, and will be encountered later in this text.

Linear equations are easy to work with, but nonlinear equations usually model real life better. The way we will resolve this dilemma is to learn as much about linear equations as possible, and then use as much of our linear analysis as possible on the nonlinear equations. We shall see that in many cases this works remarkably well, and in some cases, this approach is useless.

Our definition is not as general as it could be. The function f can also depend on n, that is,

$$a(n+1) = f(n, a(n)),$$

as in $a(n+1) = na(n)$.

When the coefficients of $a(n)$ depend on n, as in the previous example, the dynamical system is called **nonautonomous**. Although nonautonomous dynamical systems come up in many applications, they will rarely be considered in this text. See section 3.1 for one example.

A dynamical system could also be defined as

$$a(n+1) = f(a(n)) + g(n),$$

where g is a function that only depends on time n. Such a system is called **nonhomogeneous**. An affine dynamical system is really a nonhomogeneous dynamical system in which $g(n) = a$, that is, the function g is constant.

Nonhomogeneous dynamical systems in which $f(x) = rx$, such as

$$a(n+1) = 1.1a(n) + 1000n^2 \quad \text{or} \quad a(n+1) = -2a(n) + 3^n$$

are technically called **linear nonhomogeneous**. Thus, nonlinear really means not linear nonhomogeneous. Since the only nonhomogeneous dynamical systems studied in this text are linear, the term "linear" will be omitted.

The dynamical system $a(n+1) = f(a(n))$ is called **first-order** since each number depends only on the previous number.

DEFINITION 1.2 *A dynamical system of the form*

$$a(n+2) = f(a(n+1), a(n))$$

*is called a **second-order dynamical system** since each number depends on the previous two numbers.*

For example,

$$a(n+2) = 2a(n+1) - a(n)$$

is a second-order linear dynamical system and

$$a(n+2) = 2a(n+1) - a(n) + 2n + 1$$

is a second-order nonhomogeneous dynamical system

Consider the second-order dynamical system

$$a(n+2) = 2a(n+1) - a(n) - (n-1)^2, \quad n = 0, 1, 2, \ldots$$

which is equivalent to the equations

$$a(2) = 2a(1) - a(0) - (-1)^2, \quad a(3) = 2a(2) - a(1) - (0)^2, \quad \ldots.$$

Notice that we need to be given both $a(1)$ and $a(0)$ to find $a(2)$ in the first equation. Then we can use $a(1)$ and $a(2)$ to find $a(3)$ in the second equation, and so forth. Given that $a(0) = 2$ and $a(1) = 3$, you should try to find $a(2)$ through $a(6)$.

The values $a(1) = 3$ and $a(0) = 2$ are called the **initial values** for the corresponding dynamical system. To find the values $a(n)$ for all n, you need some initial values to get things started. This is because the first equation always has more than one unknown, so you must be given all but one of these unknowns. Thus, for a first-order dynamical system, one initial value is needed, while two initial values are needed for a second-order dynamical system. This generalizes to third-order, fourth-order, and so forth.

Suppose there are two sequences, say $a(n)$ and $b(n)$, $n = 0, 1, \ldots$, and each number in each sequence is related to the previous number in both sequences, such as

$$a(n+1) = f(a(n), b(n)) \qquad \text{and} \qquad b(n+1) = g(a(n), b(n)),$$

This is called a **dynamical system of 2 equations**. An example would be the system

$$a(n+1) = 0.7a(n) + 0.2b(n)$$
$$b(n+1) = 0.3a(n) + 0.8b(n).$$

For a dynamical system of m equations, there must be one equation or relationship for each sequence, that is, if there are 2 sequences, there must be 2 equations relating those sequences.

It must be observed that quite often two seemingly different dynamical systems are actually the same. Consider the two dynamical systems

$$a(n+1) = -2a(n), \quad \text{for} \quad n = 0, 1, 2, \ldots,$$

$$a(n) = -2a(n-1), \quad \text{for} \quad n = 1, 2, 3, \ldots,$$

If you substitute the corresponding integers for n into these two dynamical systems, you will get the equations

$$a(1) = -2a(0), \quad a(2) = -2a(1), \quad \ldots.$$

Since these two dynamical systems represent the **same set of equations**, we will say that they are the same.

There is a simple way to see if two dynamical systems are the same. The trick is to use substitution, an idea that will be used often. Consider

$$a(n) = -2a(n-1) + n, \quad \text{for} \quad n = 1, 2, \ldots, \text{ and} \quad (1.6)$$

$$b(m+1) = -2b(m) + m + 1, \quad \text{for} \quad m = 0, 1, 2, \ldots. \quad (1.7)$$

Set the highest value of time (terms in parenthesis) in each equation equal to each other, that is,

$$n = m + 1.$$

Now, in equation (1.6), replace each n with $m+1$ (or in equation (1.7) replace m with $n-1$). Dynamical system (1.6) becomes

$$a(m+1) = -2a(m+1-1) + m + 1,$$

which is the same as equation (1.7) after simplifying. The conditions $n = 1, 2, \ldots$ become $m + 1 = 1, 2, \ldots$. Subtracting 1 from both sides gives $m = 1 - 1, 2 - 1, \ldots$ or $m = 0, 1, \ldots$. Thus equation (1.6) represents the same set of equations as equation (1.7) and so they are the same dynamical system.

Some authors write their dynamical systems so that the highest value of time is n, such as in equation (1.6). In this text, dynamical systems will generally be written so that the lowest value of time is n. It is important to be able to convert from one type to the other, so that you can read other articles involving dynamical systems.

In the rest of this text, important applications of each of the types of systems discussed in this section will be seen. We will learn to analyze each type of dynamical system as well as possible and then apply this analysis to the applications. It will be seen that the behavior of the solution to each of these types of dynamical systems can be quite strange and interesting, even for the simpler types.

PROBLEMS

1. Which of the following are dynamical systems? For those that are dynamical systems, classify as linear (affine, nonautonomous, or nonhomogeneous if applicable) or nonlinear.
   a. $a(n) = 2/a(n-1)$
   b. $a(n) = a(n-1) + 1$
   c. $a(n) = 4a(n) - 2$
   d. $b(n) = 2a(n-1)$
   e. $a(n+2) = a(n+1)/(a(n+1)+1)$
   f. $a(n+5) = a(n+6) - 5$
   g. $a(n+1) = a(m) - n$
   h. $a(n+2) = a(n+1) + n^2 - 3$

2. Rewrite the following dynamical systems so that the lowest value of time is n. For example, rewrite

$$a(n+3) = 2a(n+2), \quad \text{for} \quad n = 1, 2, \ldots$$

   as

$$a(n+1) = 2a(n), \quad \text{for} \quad n = 3, 4, \ldots.$$

   a. $a(n+5) = 4a(n+4)$, for $n = -2, -1, \ldots$
   b. $a(n-3) = a(n-4) + 2n$, for $n = 4, 5, \ldots$
   c. $a(n-1) = na(n-2)$, for $n = 2, 3, \ldots$

3. Rewrite the dynamical systems in Problem 2 so that the highest value of time is n.

4. Find $a(6)$ for the dynamical system

$$a(n+2) = a(n+1) + a(n) \quad \text{when} \quad a(0) = 1 \text{ and } a(1) = 1.$$

5. Find $a(6)$ for the dynamical system

$$a(n+2) = 2a(n+1) - 3a(n) \quad \text{when} \quad a(0) = 0 \text{ and } a(1) = 1.$$

6. Find $a(3)$ and $b(3)$ for the dynamical system

$$a(n+1) = 0.4a(n) + 0.8b(n)$$
$$b(n+1) = 0.6a(n) + 0.2b(n)$$

   when $a(0) = 0.3$ and $b(0) = 0.7$.

7. Find $a(3)$ and $b(3)$ for the dynamical system

$$a(n+1) = 2a(n) + 3b(n)$$
$$b(n+1) = 0.5a(n)$$

when $a(0) = 10$ and $b(0) = 5$.

1.3　Equilibrium values

Suppose we borrow $a(0)$ dollars from a friend at 1 per cent per month interest, and that we pay our friend 20 dollars per month. The dynamical system that describes what we owe our friend each month is

$$a(n+1) = 1.01\,a(n) - 20.$$

Here, $a(n)$ represents the amount of money we owe our friend just after we make our nth payment. Notice that if the original debt was $a(0) = 1000$, then $a(1) = 990$, $a(2) = 979.90$, and so forth. When we reach a point where $a(k)$ is negative, we will have paid off our loan. We are then said to have amortized our loan.

On the other hand, suppose that $a(0) = 3000$. Then $a(1) = 3010$, $a(2) = 3020.10$, and so forth. We now observe that we will never pay off our loan and in fact we will owe our friend more and more money as time goes on.

The problem is that if we restrict our payments to 20 dollars a month, then there is an upper limit to what we can borrow if we ever want to pay it back. This limit is 2000 dollars, the point at which our monthly payments equal our one per cent monthly interest charge. Notice that if $a(0) = 2000$, then $a(1) = 2000$, $a(2) = 2000$, and in fact

$$a(k) = 2000 \qquad \text{for} \quad k = 0, 1, 2, \dots.$$

This is a constant solution and the number 2000 is called **an equilibrium value** or **a fixed point** for this dynamical system.

Constant solutions to a dynamical system are of extreme importance in that they often tell us what will eventually happen to our system. Recall the drug problem in section 1.1 in which a container held 4 cups of water. Each time period, one cup of water was replaced with a cup of water containing 16 ml of medicine giving the dynamical system

$$a(n+1) = 0.75a(n) + 16.$$

Notice that if $a(0) = 64$, then $a(1) = a(2) = \cdots = 64$, and $a(k) = 64$ is a constant solution. If you pick any other starting value, such as $a(0) = 10$, and compute $a(1)$, $a(2)$, and so forth, you will observe that your values get close to 64. We say that

$$\lim_{k \to \infty} a(k) = 64$$

for this dynamical system.

In financial examples, constant solutions often tell us what we don't want to happen. In the example above in which we pay off a loan to our friend, the constant solution, $a(k) = 2000$ means we can never pay off this loan if we owe 2000 dollars.

DEFINITION 1.3   *Consider a first-order dynamical system*

$$a(n+1) = f(a(n)).$$

A number a is called an **equilibrium value** *or* **fixed point** *for this dynamical system if $a(k) = a$ for all values of k when the initial value $a(0) = a$, that is,*

$$a(k) = a$$

is a constant solution to the dynamical system.

It is relatively easy to find equilibrium values a. If $a(k) = a$ for all k, it follows that when $k = n$ and when $k = n+1$, that $a(n) = a$ and $a(n+1) = a$, respectively. Thus, substitution of a into the dynamical system for $a(n)$ and $a(n+1)$ should give equality. This proves the following theorem.

THEOREM 1.4   *The number a is an equilibrium value for the dynamical system*

$$a(n+1) = f(a(n))$$

if and only if a satisfies the equation

$$a = f(a).$$

The following examples should help you understand this definition and theorem.

EXAMPLE 1.5   Consider the equation

$$a(n+1) = 2a(n) - 3.$$

Note that $f(a(n)) = 2a(n) - 3$. The solution to

$$a = 2a - 3$$

is $a = 3$. Note that if $a(0) = 3$, then $a(1) = 2a(0) - 3 = 6 - 3 = 3$. By repeating this argument we see that $a(2) = 3$, $a(3) = 3$, and so on. Thus $a = 3$ is an equilibrium value for this dynamical system, and $a(k) = 3$ is a constant solution to this equation.   ∎

THEOREM 1.6   *Consider the first-order affine (or linear) dynamical system*

$$a(n + 1) = ra(n) + b.$$

The equilibrium value for this system is

$$a = \frac{b}{1 - r} \qquad if \quad r \neq 1.$$

If $r = 1$ and $b \neq 0$, there is no equilibrium value. If $r = 1$ and $b = 0$ then every number is an equilibrium value.

To prove this theorem, you only need to solve the equation $a = ra + b$ for a. When $r = 1$ and $b \neq 0$, the equation $a = a + b$ has no solutions. When $r = 1$ and $b = 0$, every number satisfies the equation.

A dynamical system may have many equilibrium values. In general, the more nonlinear the dynamical system is, the more equilibrium values that system may have. For example, if there is an $a(n)a(n) = a^2(n)$ term then there may be up to two equilibrium values, if there is an $a^3(n)$ term then there may be up to three equilibrium values, and so on. The simple dynamical system $a(n + 1) = a(n)$ is an exception in that it has an infinite number of equilibrium values.

EXAMPLE 1.7   Consider the nonlinear dynamical system

$$a(n + 1) = (a(n) + 4)a(n) + 2.$$

In this example, $f(a(n)) = (a(n) + 4)a(n) + 2$. Substituting a for $a(n)$ and $a(n + 1)$ gives

$$a = (a + 4)a + 2 \qquad \text{or} \qquad a^2 + 3a + 2 = 0$$

after simplification. Factoring gives

$$0 = (a + 2)(a + 1).$$

Thus, the equilibrium values are the roots $a = -2$ and $a = -1$. Note that if $a(0) = -2$ then $a(1) = [(-2) + 4](-2) + 2 = -2$, $a(2) = -2$, ... Also, if $a(0) = -1$ then $a(1) = [(-1) + 4](-1) + 2 = -1$. Likewise $a(2) = a(3) = \cdots = -1$. In this example we have **two** equilibrium values.

Let's see what happens to the solution for several values of $a(0)$. If $a(0) = -2.4$, then $a(1) = -1.84$, $a(3) = -1.9744$, ... If $a(0) = -1.01$, then $a(1) = -1.0199$, $a(2) = -1.039404$, ... $a(10) = -1.999966$. If $a(0) = -0.99$, then $a(1) = -0.9798999$, ..., $a(10) = 26611.12$, ...

This is shown graphically in figure 1.6. In this figure, the points $(k, a(k))$ are plotted and successive points connected with straight lines. The lines between the points have no meaning, but make it easier to see the behavior of the $a(k)$ values. Thus, the curves in that figure correspond to particular solutions to this dynamical system.

Notice that when $a(0)$ is "close" to -1, either -1.01 or -0.99, that $a(1)$ is further from -1, $a(2)$ is even further than $a(1)$, etc. But when $a(0)$ is "close" to -2, say -1.01 or -2.4, that $a(1)$ is closer to -2. In fact if $a(0)$ is "close" to -2, then

$$\lim_{k \to \infty} a(k) = -2.$$

Intuitively, $a(k)$ gets "close" to -2 as k gets large.

The equilibrium value -1 is called an **unstable** equilibrium value or a **repelling** fixed point, and -2 is called a **stable** equilibrium value or **attracting** fixed point. Note that when $a(0) = -0.99$, then $a(k)$ goes to infinity. The problem is that -0.99 is not "close enough" to -2. ∎

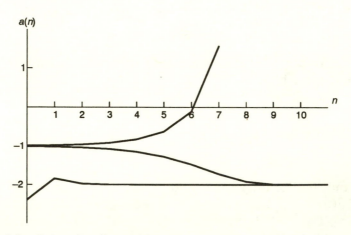

FIGURE 1.6.   Points $(k, a(k))$ satisfying the dynamical system $a(n + 1) = (a(n) + 4) a(n) + 2$ with different initial values.

We will define stable and unstable later in this section. But first we discuss these concepts informally. In the interest problem

$$a(n+1) = 1.01a(n) - 20$$

the equilibrium value, $a = 2000$, is unstable. This is because, if the initial loan is less than 2000, say $a(0) = 1000$, then the amount owed goes toward negative infinity (although we will stop paying when the amount becomes zero), while if the initial loan is greater than 2000, say $a(0) = 3000$, then the amount owed goes to infinity.

For the drug problem

$$a(n+1) = 0.75a(n) + 16$$

the equilibrium value is stable. This is seen in that if the initial drug level is either less than 64, say $a(0) = 16$, or greater than 64, say $a(0) = 100$, the solution goes to 64.

The idea is to first find an equilibrium value a. Then pick several $a(0)$ values close to a. For each of these $a(0)$ values, find $a(1)$, $a(2)$, and so forth. If the values $a(1)$, $a(2)$, ... appear to be getting closer to a for **each** $a(0)$, then a is probably stable. If the values $a(1)$, $a(2)$, ... appear to be going away from a for **each** $a(0)$, then a is probably unstable.

An equilibrium value may not be stable or unstable. Consider the dynamical system

$$a(n+1) = -a(n) + 4.$$

The equilibrium value is seen to be $a = 2$. Pick $a(0)$ close to 2, say $a(0) = 3$. Note that $a(0)$ is one unit from $a = 2$. Then $a(1) = -a(0) + 4 = -3 + 4 = 1$. Note that $a(1)$ is also one unit from $a = 2$. Continuing, $a(2) = -a(1) + 4 = -1 + 4 = 3$ which is one unit from equilibrium. Thus, the values $a(1)$, $a(2)$, ... are not getting further or closer to equilibrium so $a = 2$ is not stable or unstable. This same phenomenon occurs if you pick other values for $a(0)$.

In example 1.16 is another type of behavior in which the equilibrium value is neither stable nor unstable.

Consider the dynamical system

$$a(n+1) = 3.7a(n) - 3.7a^2(n).$$

The two equilibrium values are found to be

$$a = 0 \qquad \text{and} \qquad a = \frac{27}{37} = 0.730$$

to 3 decimal places. Let's pick $a(0)$ "close" to 0.730, say $a(0) = 0.740$. Computations give that

$$a(1) = 0.712, \qquad a(2) = 0.759, \qquad a(3) = 0.677, \qquad a(4) = 0.809.$$

These values are slowly getting further from $a = 0.730$, indicating that this equilibrium value is unstable. Further computations give that

$$a(14) = 0.921, \qquad a(15) = 0.269, \qquad a(16) = 0.728.$$

Note that $a(16)$ is closer to $a = 0.730$ than was $a(0)$, that is, we eventually got closer to a. But this equilibrium value is still unstable. The reason this equilibrium value is unstable is that **once $a(k)$ gets close to $a = 0.730$, the next values start getting further away.** Even though $a(16)$ is close to a, the values $a(17) = 0.732$, $a(18) = 0.725$, $a(19) = 0.737$ are getting further away again.

What is happening here is that when $a(k)$ is "close" to a, the equilibrium value appears to "push" the next values away from it, but once the values get far enough away, something else "pushes" the values "close" again. For some unstable equilibrium values, such as this one, the process of slowly moving away and then jumping back close again repeats indefinitely.

Cobweb analysis may help in understanding stable and unstable by giving a visualization of these concepts.

DEFINITION 1.8   *Suppose we have a dynamical system*

$$a(n+1) = f(a(n)),$$

with $a(0)$ given. Draw a graph of the curve

$$y = f(x) \tag{1.8}$$

and the line

$$y = x. \tag{1.9}$$

Plot the point $(a(0), a(0))$ on $y = x$. The x and y-value of this point correspond to the initial input. Go vertically to a point on $y = f(x)$. This point will be $(x, y) = (a(0), a(1))$, the x-value corresponding to the initial input and the y-value, the corresponding output. Then go horizontally to a point on $y = x$. This point will be $(x, y) = (a(1), a(1))$. The previous output has been converted to the next input. Repeat these steps to get the points

$$(a(1), a(2)) \quad (a(2), a(2)) \quad (a(2), a(3)) \quad (a(3), a(3)) \quad \ldots.$$

*The resulting figure is a **cobweb** for the given dynamical system.*

Cobweb analysis can help in determining the stability of an equilibrium value. In figures 1.4 and 1.5, the cobwebs drawn go **toward** the point of intersection which corresponds to the solution going toward the equilibrium value. This indicates that the equilibrium value is stable.

EXAMPLE 1.9   Consider the dynamical system

$$a(n+1) = 1.5a(n) - 1.$$

The equilibrium is $a = 2$. The lines $y = 1.5x - 1$ and $y = x$ are graphed in figure 1.7. The point of intersection of these two lines is $(2, 2) = (a, a)$, that is, the equilibrium value equals the x and y-coordinate of the point of intersection.

To study the stability of $a = 2$, pick $a(0)$ "close" to 2, say $a(0) = 3$. Plot $(3, 3)$ on $y = x$. Move vertically to the line $y = 1.5x - 1$. Since we move vertically, this point also has $x = 3$, so the y-coordinate must be $y = 1.5 \times 3 - 1 = 3.5$. Thus, we are at the point $(3, 3.5) = (a(0), a(1))$. Move horizontally to the point $(3.5, 3.5) = (a(1), a(1))$. In figure 1.7, the cobweb seen to be moving away from the point of intersection, implying that the values $a(k)$ are moving away from $a = 2$. Thus we see that $a = 2$ is unstable.

If we had started with $a(0) < 2$, such as $(a(0), a(0)) = (1, 1)$, then the "steps" would have gone to the left indicating that $a(k)$ goes to negative infinity.   ∎

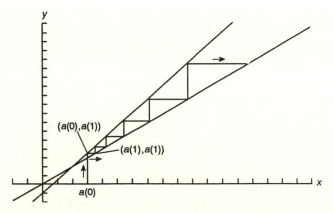

FIGURE 1.7.   Cobweb graph for $a(n+1) = 1.5a(n) - 1$. Points are $(a(k), a(k))$ on the line $y = x$, and $(a(k), a(k+1))$ on line $y = 1.5x - 1$, with $k = 0, 1, \ldots$.

REMARK: In some figures, the scale of the horizontal axis differs from the scale of the vertical axis, as in figure 1.7. This allows more of the important features of the figure to be seen. One problem this causes is that the slopes of lines in the figure will appear to be different from their actual slopes. For example, $y = x$ "appears" to have a slope less than one in figure 1.7. We hope this causes no inconvenience to the reader.   ☐

What is the graphical difference between the equilibrium being stable for the drug problem where $y = 0.75x + 16$ and the equilibrium being unstable in example 1.9 where $y = 1.5x - 1$? One possible answer lies in the looking at how these two lines intersect $y = x$. In the first case, $y = 0.75x + 16$ goes from above $y = x$ to the left of equilibrium to below $y = x$ to the right of equilibrium. In the case of $y = 1.5x - 1$, the reverse is true, that is, the line goes from below $y = x$ to above as we move from left to right.

How can we use cobweb analysis to analyze an affine dynamical system

$$a(n+1) = ra(n) + b \qquad \text{with} \quad r > 0? \qquad (1.10)$$

We graph the lines

$$y = x \qquad \text{and} \qquad y = rx + b,$$

both lines having positive slope. Note that the point of intersection of these two lines is (a, a) where $a = b/(1 - r)$ is the equilibrium value for the dynamical system.

Suppose the slope of the second line is greater than the slope of $y = x$, that is, $r > 1$. Then this line intersects $y = x$ the same way as $y = 1.5x - 1$, that is, it goes from below the curve to the left of equilibrium to above the curve to the right of equilibrium. Thus, the behavior of the cobweb must be the same, that is, the cobweb moves away from the point of intersection. Therefore the equilibrium value must be unstable.

Suppose the slope of the second line is less than the slope of $y = x$, that is, $0 < r < 1$. Then this line intersects $y = x$ the same way as $y = 0.75x + 16$, that is, it goes from above the curve to the left of equilibrium to below the curve to the right of equilibrium. Thus, the behavior of the cobweb must be the same, that is, the cobweb moves toward the point of intersection. Therefore the equilibrium value must be stable.

This indicates that for the dynamical system (1.10), the equilibrium value is stable if $0 < r < 1$, and is unstable if $1 < r$. But what happens if the slope is negative, that is, if $r < 0$? Let's consider a few more examples.

EXAMPLE 1.10   Consider the dynamical system

$$a(n+1) = -0.8a(n) + 3.6$$

with equilibrium value, $a = 2$. Let's draw a cobweb using $a(0) = -4$. We graph the line

$$y = -0.8x + 3.6$$

which has negative slope $r = -0.8$.   The cobweb is given in figure 1.8.   Notice that the solution oscillates towards the stable equilibrium $a = 2$ in that $(a(k), a(k))$ is getting closer to $(2, 2)$, but is to the right $(a(k) > 2)$ when k is odd, and is to the left $(a(k) < 2)$ when k is even.   ■

EXAMPLE 1.11   Consider the dynamical system

$$a(n+1) = -a(n) + 4,$$

also with equilibrium value $a = 2$. Letting $a(0) = 6$, the cobweb gives a square, as seen in figure 1.9.   (As was remarked earlier in this section, because the scales are different for the x and y axes, the rectangle in figure 1.9 does not "appear" to be square.) Notice that the cobweb oscillates between $(-2, -2)$ and $(6, 6)$ on the line $y = x$

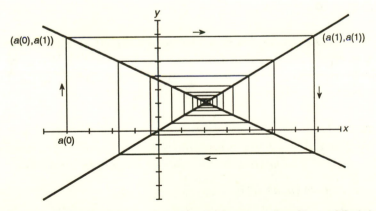

FIGURE 1.8.   Cobweb graph for the dynamical system $a(n+1) = -0.8a(n) + 3.6$. With $a(0) = -4$, it is seen that the solution oscillates to the fixed point $a = 2$.

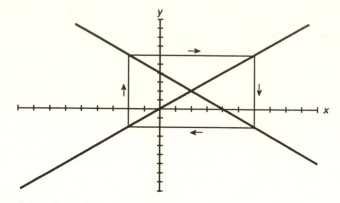

FIGURE I.9.   Cobweb graph for the dynamical system $a(n+1) = -a(n) + 4$.

indicating that $a(k) = 6$ for even k and $a(k) = -2$ for odd k. When a solution $a(k)$ repeats the same two values, the solution is said to form a **2-cycle**. The fixed point $a = 2$ is said to be **neutral** since solutions go neither toward equilibrium nor away from equilibrium.

For this dynamical system, any $a(0)$ value would lead to a 2-cycle about the equilibrium value $a = 2$.   ■

EXAMPLE I.12   Consider the dynamical system

$$a(n+1) = -1.5a(n) + 5,$$

also with equilibrium value $a = 2$. Letting $a(0) = 1.5$, the cobweb cycles outward, as seen in figure 1.10. This indicates that $a = 2$ is unstable and that $a(k)$ oscillates with increasing amplitude. In fact $|a(k)|$ goes to infinity.   ■

The previous examples seem to indicate that as r decreases past -1, the equilibrium value for dynamical system (1.10) goes from stable to unstable. Example 1.11 implies that $r = -1$ is the cut-off point between stability and instability.

THEOREM I.13   *Suppose a cobweb is drawn using the "curve" $y = rx + b$ and the line $y = x$. These curves intersect at the point (a, a) where $a = b/(1 - r)$, $r \neq 1$. The sequence of points determined by the cobweb:*

- *converges to (a, a) if $|r| < 1$,*
- *goes to positive and/or negative infinity if $|r| > 1$, and*
- *oscillates around a square centered at (a, a) if $r = -1$.*

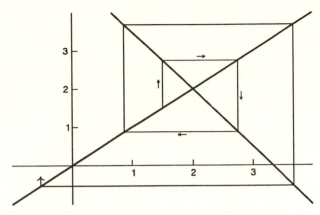

FIGURE 1.10.   Cobweb graph for the dynamical system $a(n+1) = -1.5a(n) + 5$.

Thus, the equilibrium value for dynamical system

$$a(n+1) = ra(n) + b$$

- *is stable if $|r| < 1$,*
- *is unstable if $|r| > 1$, and*
- *is neutral if $r = -1$.*

There are no equilibrium values if $r = 1$ and $b \neq 0$.

This theorem will be proved in section 2.4 when the solution to first-order affine dynamical systems is developed.

Let's now apply the concept of cobwebs to nonlinear dynamical systems to develop an intuitive idea for when a fixed point is attracting or repelling. We will see that many unusual things can happen when the dynamical system is nonlinear.

EXAMPLE 1.14   Consider the dynamical system

$$a(n+1) = 1.5a(n) - 0.5a^2(n).$$

The solutions to the equation

$$a = 1.5a - 0.5a^2$$

are the two fixed points, $a = 1$ and $a = 0$.

If we substitute x for $a(n)$ and y for $a(n+1)$ in the above dynamical system, we then have the equation

$$y = 1.5x - 0.5x^2,$$

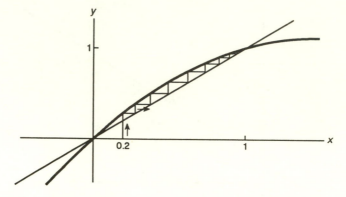

FIGURE I.II.   Cobweb graph using the curve $y = 1.5x - 0.5x^2$. Notice that $x = 1$ appears to be attracting or stable, and that $x = 0$ appears to be repelling or unstable.

which is the equation of the parabola shown in figure 1.11 along with the line $y = x$. Notice that the equilibrium values correspond to the points of intersection of the two curves, that is, the points $(0, 0)$ and $(1, 1)$.

In the same manner as for affine dynamical systems, we plot the point $(a(0), a(0))$ on the line $y = x$. To find $a(1)$ algebraically, we substitute $a(0)$ for x in the equation of the parabola. To find $a(1)$ graphically, we go vertically from the point $(a(0), a(0))$ to the graph of the parabola. This point on the parabola corresponds to $(x, y) = (a(0), a(1))$. In figure 1.11, this is done using $a(0) = 0.2$. Proceeding as before, the cobweb goes towards $(1, 1)$ indicating that the equilibrium value $a = 1$ is stable, while the equilibrium value $a = 0$ is unstable.

If you construct cobwebs using other values for $a(0)$, you will get that when $a(0)$ is reasonably close to 1, the cobweb is attracted to 1, while if $a(0)$ is near 0, the cobweb is repelled away from 0, although not necessarily to infinity.

When an equilibrium value a is stable, an interesting and often difficult problem is to determine the maximum interval (c, d) about a such that if $a(0)$ is in that interval then $\lim_{k \to \infty} a(k) = a$. Using graphical techniques, these intervals can sometimes be determined exactly; for example, by redrawing figure 1.11 using a different range for the axes it can be seen that if $a(0)$ is in the interval $(0, 3)$ then $a(k)$ goes to 1, but if $a(0)$ is outside that interval then $|a(k)|$ goes to infinity. See figure 1.12. Generally, it is extremely difficult to solve this problem.   ■

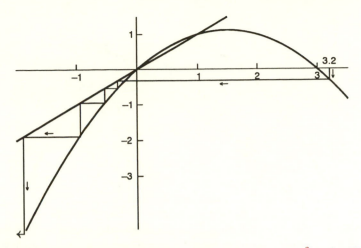

FIGURE 1.12.   Cobweb graph using the curve $y = 1.5x - 0.5x^2$ and $a(0) = 3.2$. Notice that $a(k)$ is going toward negative infinity.

EXAMPLE 1.15   Consider the dynamical system

$$a(n + 1) = 3.2a(n) - 0.8a^2(n).$$

After making the substitutions $a(n+1) = y$ and $a(n) = x$, we consider the parabola

$$y = 3.2x - 0.8x^2.$$

The equilibrium values, which are the solutions of the equation $a = 3.2a - 0.8a^2$ are $a = 0$ and $a = 2.75$. In figure 1.13 it is seen that if $a(0)$ is close to 2.75 ($a(0) = 2.7$), then the solution $a(k)$ is repelled away from 2.75, and in fact appears to become a 2-cycle. Notice that although $a = 2.75$ is unstable, the solution does not necessarily go towards infinity or another equilibrium value.

Likewise $a = 0$ is unstable. If you construct a cobweb starting with a small positive initial value $a(0)$ (say $a(0) = 0.1$), you will find that the solution $a(k)$ is repelled from $a = 0$, and eventually goes to the same 2-cycle as seen in figure 1.13. ∎

EXAMPLE 1.16   Consider the dynamical system

$$a(n + 1) = a^3(n) - a^2(n) + 1.$$

The two equilibrium values are $a = -1$ and $a = 1$. Notice in figure 1.14 that if $a(0)$ is slightly less than $a = 1$, the solution $a(k)$ goes to this equilibrium value, but if $a(0)$ is slightly greater that $a = 1$,

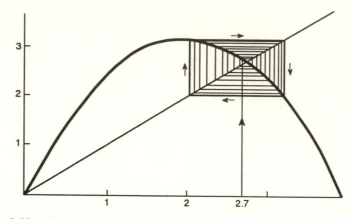

FIGURE I.13. Cobweb using the parabola $y = 3.2x - 0.8x^2$ and the starting value of $a(0) = 2.7$. Notice the cobweb oscillates away from the fixed point $(2.75, 2.75)$, and towards a 2-cycle.

the solution goes to infinity. Thus, $a = 1$ is attracting on the left and repelling on the right. Such a point is called **semistable**. (If the graph is extended, you will see that $a = -1$ is unstable.) ■

We now give the formal definition of stable and unstable for completeness.

DEFINITION I.17 *Suppose a first-order dynamical system has an equilibrium value a. This equilibrium value is said to be **stable** or **attracting** if there*

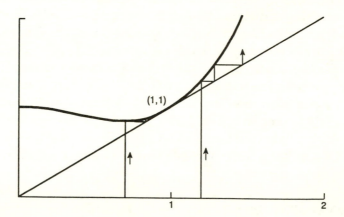

FIGURE I.14. Cobweb using the curve $y = x^3 - x^2 + 1$. It indicates that the equilibrium value $a = 1$ is semistable.

is a number ϵ, unique to each system, such that, whenever $a(0)$ satisfies

$$|a(0) - a| < \epsilon,$$

then the solution goes to a, that is,

$$\lim_{k \to \infty} a(k) = a.$$

*An equilibrium value is **unstable** or **repelling** if there is a number ϵ such that, whenever $a(0)$ satisfies*

$$0 < |a(0) - a| < \epsilon,$$

then, for some k-value

$$|a(k) - a| > \epsilon.$$

Note that $|a(k) - a|$ gives the distance the solution $a(k)$ is from the equilibrium value a. Intuitively, the definition states that an equilibrium value is stable if, whenever $a(0)$ is sufficiently close to a then $a(k)$ goes to a.

Similarly, an equilibrium value is **unstable** if, no matter how close $a(0)$ is to a, $a(k)$ eventually gets far, at least ϵ, from a. Usually, these $a(k)$ values stay away from a, going to infinity or to some other equilibrium value. But this does not preclude the possibility that the $a(k)$ values get close to a at a later point in time. We saw an example of this earlier in this section.

The stability of an equilibrium value for a nonlinear dynamical system will be studied in an intuitive manner.

We end this section with a discussion of our terminology. Suppose we are studying a dynamical system that models the real world, say the price of a product under the pressures of supply and demand. Then an equilibrium value for the dynamical system corresponds to an equilibrium price; that is, a price which, once reached, remains there for a long time (assuming no other outside influences). Thus, the term "equilibrium value" seems to fit better than the term "fixed point". Suppose we are considering a dynamical system without being concerned about its real world applications (such as we did in this section). Then if a cobweb is started at a point of intersection, the cobweb doesn't move; that is, we have a "fixed point".

Similarly, if an equilibrium value relates to the real world, say an equilibrium price, then it makes more sense to say the price is "stable" or "unstable". But if we are drawing cobwebs of the dynamical system, then the cobwebs make the intersection point appear to be "attracting" or "repelling" the cobweb. Thus, the best terminology to use seems to depend on the context. For this reason, we will refer to "equilibrium value" and "fixed point" interchangeably throughout this text, and similarly for the terms "stable" and "attracting", and "unstable" and "repelling".

PROBLEMS

1. Find the equilibrium values for the following dynamical systems, if any.
   a. $a(n+1) = 2a(n) + 6$
   b. $a(n+1) = -4a(n) + 7$
   c. $a(n+1) = 2a(n) + n$
   d. $a(n+1) = a^2(n) - 2$
   e. $a(n+1) = 4a^3(n)$

2. For what value of b is the equilibrium value of the dynamical system

$$a(n+1) = -0.3a(n) + b$$

   equal to 5?

3. For what value of b is the equilibrium value of the dynamical system

$$a(n+1) = 2a(n) + b$$

   equal to -8?

4. For what value of r is the equilibrium value of

$$a(n+1) = ra(n) - 2$$

   equal to 0.5?

5. For what value of r is the equilibrium value of

$$a(n+1) = ra(n) + 3$$

   equal to 4?

6. The equilibrium value for the dynamical system

$$a(n+1) = 2a(n) - 3$$

   is $a = 3$.
   a. Compute $a(1)$, $a(2)$, and $a(3)$ when $a(0) = 3.1$.
   b. Compute $a(1)$, $a(2)$, and $a(3)$ when $a(0) = 2.9$.

    c. Does this correspond to the conclusion of theorem 1.13, that a is an unstable equilibrium value?

7. Sketch a cobweb for the dynamical system

$$a(n+1) = -1.5a(n) + 5,$$

with $a(0) = 3$. Also find the equilibrium point and determine its stability.

8. For each part, sketch the cobweb, find the equilibrium value if there is one, and use cobweb analysis to determine the stability of the equilibrium value.

    a. $a(n+1) = -a(n) + 3$, $a(0) = 1$
    b. $a(n+1) = 1.5a(n) - 5$, $a(0) = 9$
    c. $a(n+1) = a(n) + 3$, $a(0) = 0$

9. Suppose for the dynamical system

$$a(n+1) = 2a(n) - 1,$$

you are given that $a(3) = 9$. Sketch a cobweb in reverse, starting at $(a(3), a(3)) = (9,9)$ to approximate $a(0)$.

10. Consider the dynamical system

$$a(n+1) = 2a(n) - 0.5a^2(n).$$

The two equilibrium values are $a = 0$ and $a = 2$ (you should check this).

    a. Compute $a(1)$, $a(2)$, and $a(3)$ when $a(0) = 0.1$ and again when $a(0) = -0.1$. Does this indicate that $a = 0$ is a stable or unstable equilibrium value?
    b. Compute $a(1)$, $a(2)$, and $a(3)$ when $a(0) = 1.8$ and again when $a(0) = 2.2$. What does this indicate about the stability of $a = 2$?
    c. Draw a cobweb for this dynamical system.

11. Consider the dynamical system

$$a(n+1) = 3a(n) - a^2(n) - 1.$$

Its one equilibrium value is $a = 1$.

    a. Compute $a(1)$, $a(2)$, and $a(3)$ using $a(0) = 1.1$. Compute $a(1)$, $a(2)$, and $a(3)$ using $a(0) = 0.9$. What do these computations indicate about the stability of $a = 1$?
    b. Draw a cobweb for this dynamical system.

12. Consider the dynamical system

$$a(n+1) = 1.4a(n) - 0.2a^2(n) + 3.$$

Find the equilibrium values for this system and determine their stability using cobweb analysis.

13. Consider the dynamical system

$$a(n+1) = 3a(n) - a^2(n) + 3.$$

   a. Find the equilibrium values for this dynamical system.
   b. Draw the graph of the curve $y = 3x - x^2 + 3$ and determine the stability of the equilibrium values using cobweb analysis.

14. Consider the dynamical system

$$a(n+1) = 2a(n) - 0.25a^2(n) - 0.75.$$

The two equilibrium values are $a = 1$, which is unstable, and $a = 3$, which is stable. Sketch a cobweb for this equation and use it to determine the maximum interval containing $a = 3$ such that if $a(0)$ is in that interval, then $\lim_{k \to \infty} a(k) = 3$.

15. For the dynamical system

$$a(n+1) = a(n) - a^3(n),$$

the only equilibrium value is $a = 0$. Draw a cobweb for this curve and use it to determine if $a = 0$ is stable, unstable, or semistable.

1.4 Dynamic economic applications

In this section we will describe several traditional economic models. The first two attempted to describe the entire development of society. The third is a more modest modern attempt to understand one particular part of the economy, the results of supply and demand. In understanding the implications of supply and demand, we can hopefully control some of the harmful results.

1.4.1 Classical dynamics

Classical dynamic economic models were developed during and after the time of Malthus. They tended to be grand in scale, but they were based on oversimplified and dubious assumptions. Their arguments were based on Malthus' theory of population growth, that is, people would tend to have as many children as they could afford. The increase in people would lead to an increase in production, leading to more profits and higher wages, leading again to more offspring. This cyclic behavior described an economy that progressed toward an equilibrium state which is less desirable than the present state of the economy. Let's analyze this process, and the corresponding assumptions in more detail.

Assume that there is a subsistence wage necessary for a person to live. Let that wage be s. Thus, if there are x people in the population, their combined wages must be sx for all of them to just subsist. This is given by the straight line $y = sx$ in figure 1.15. Note that the slope of this line is s, the subsistence wage for one person.

Now let's consider the total output produced by x people. For a small number of people, the total output exceeds the total subsistence needed for those people. This means that the curve $y = f(x)$, representing the total output of x people must be above the subsistence line $y = sx$ for x below some number, that is, for small x. If the population increases by one, the total output also increases since the previous people can produce the same amount, and the extra person can produce a little more. Thus, as the number of people increase, the total output increases, meaning that $f(x)$ is an increasing function.

Since there is a finite amount of land and resources, as the number of people increases, crowding starts to happen. Thus, each time a new person is added, the workplace becomes more crowded than previously, so the new person's output will be less than previous people's output. This means the curve $f(x)$ increases at a slower pace as x gets large, that is, $f(x)$ is concave down as seen in figure 1.15. Thus the curve and the line must intersect at some point a.

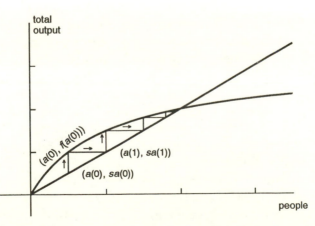

FIGURE 1.15. Horizontal axis represents people and vertical axis represents total output or total wages. The steps indicate the progression of society over time.

Now that the curves are given, let's consider the dynamic process. Suppose the population is at some level $a(0)$ at the present. The people are being paid subsistence wages, so that the total wages are at $sa(0)$. Thus, plotting people versus wages gives us the point $(a(0), sa(0))$ on the line $y = sx$. But the total output is $f(a(0))$ given by the point $(a(0), f(a(0)))$ on the curve. Since the output is greater than the wages, profit is being made and accumulated. Accumulation of profit leads to investment. Investment, in terms of building new factories, leads to a greater demand for labor and so wages go up. As long as profit is being made, this will happen, and profit will be made until total wages increase to $f(a(0))$, represented by the line from $(a(0), sa(0))$ to $(a(0), f(a(0)))$ in figure 1.15. Since the total output of society equals the total wages, further investment ceases since there are no profits.

Now wages are above subsistence, so people have a higher standard of living and hence have more children. This increases the size of the population. As long as people are living above subsistence, they continue to have more children. The increase in population is designated by the horizontal line from $(a(0), f(a(0)))$ to $(a(1), sa(1))$. (Note that $f(a(0)) = sa(1)$.) Once society reaches a population of $a(1)$, people are living at subsistence again so the population ceases to increase. But now the total output has increased and profits are again being made. Thus, the vertical and horizontal movements are repeated again.

As can be seen in figure 1.15, this process of increase in output followed by increase in population proceeds toward the point of intersection of the line and the curve. Once this point has been reached, the output of society equals the subsistence wages. This led to pessimistic conclusions about the development of society, that is, eventually society reaches a point at which (nearly) everyone is living at subsistence.

What about technological advancements? They can increase the total output of society, given by the higher curve given in figure 1.16. By following the same vertical and horizontal movements as before, the new point of intersection is eventually reached, again meaning that society is living at subsistence. The only improvement is that this point is reached at a later date. Thus, technology does not eliminate the problem, but only postpones it.

Since the predicted disaster has not happened, some of the assumptions must be incorrect. One possible error is that an increase

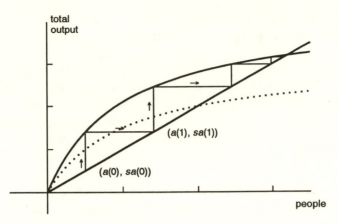

FIGURE 1.16. Horizontal axis represents people and vertical axis represents total output/wages. The higher curve represents output after a technological improvement. The steps indicate the progression of society over time.

in income leads to an increase in family size. It appears that many people like a better standard of living and so do not increase their family size.

1.4.2   Harrod's model

In the 1930's and 40's, Harrod (1939) developed a dynamic economic model that rivals the classic model. Let $t(n)$ be the total income (public and private) of society during year n. Denote by $i(n)$ the **additional investment** made in year n. Denote by $s(n)$ the additional savings made in year n.

The **saving assumption** is that savings in any year is proportional to income that year, that is, each person sets aside a certain fraction of their income. This translates into

$$s(n) = p_s t(n). \tag{1.11}$$

If income is going up, such as a result of increased sales, then **additional** investment will be made, such as building more factories to keep up with demand. Likewise, if income goes down through, say, decreased sales, then total investment will be **reduced** by closing or selling factories. Let $i(k)$ be the additional investment made in year k. Then $i(k)$ may be negative if investment is reduced in year k. The above leads to the **investment assumption** which is that additional investment in any year is proportional to the change in total income between that year and the previous year. This is also known as the **acceleration principle** and is given by

$$i(n) = p_i[t(n) - t(n-1)]. \tag{1.12}$$

It must be stated that $i(k)$ is the **desired** additional investment and not the actual investment. The actual investment depends on the money available for investing. Observe that money which is set aside as savings is the money that is available to invest. If desired investment is less than savings, and all of the savings is invested, then there is overproduction. This leads to less of a need for investment which causes even more overproduction. On the other hand, if desired investment is greater than savings, some investments cannot be made and factories go unbuilt and there is underproduction and shortages. This lack of possible sales keeps income down which in turn leads to less savings and further underinvestment. Thus, we see that in a healthy economy, desired additional investment should equal actual additional investment (savings), that is,

$$i(n) = s(n). \tag{1.13}$$

Substituting into equation (1.13) for $s(n)$ from equation (1.11) and for $i(n)$ from equation (1.12) gives

$$p_i[t(n) - t(n-1)] = p_s t(n)$$

or

$$t(n) = \frac{p_i}{p_i - p_s} t(n-1). \tag{1.14}$$

Since $t(n)$ must be positive, we must have that $p_i > p_s$. To use cobweb analysis on dynamical system (1.14), graph the line $y = rx$ where $r = p_i/[p_i - p_s]$. It is clear that $r > 1$ and the point of intersection of this line with $y = x$ is at the origin. So the cobweb increases toward infinity as is seen in figure 1.17.

In fact, the total income is growing exponentially, which is seen from the solution

$$t(k) = \left(\frac{p_i}{p_i - p_s} \right)^k t(0).$$

But nothing can grow exponentially, so the economy must break down at some point in time.

In conclusion, Harrod's model is no more optimistic then the classic dynamic model. Either the economy becomes unhealthy because $i(n) \neq s(n)$ or the economy grows exponentially until it breaks down.

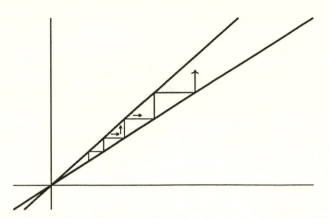

FIGURE 1.17. Cobweb analysis for total income $t(n)$ for Harrod's model of an economy. The input/output relationship is $y = [p_i/(p_i - p_s)]x$. The line $y = x$ is used to convert one output to the next input.

For more detail on the discussions of the classical dynamic model and Harrod's model, see Baumol (1961).

1.4.3   Supply and demand

We now consider a more conservative model. In particular, let's consider supply and demand as it relates to a product that takes one unit of time to produce. This model was developed to study the farming industry where the product (a crop) is produced once each year, and the farmer has to plan next year's crops using information about prices this year.

Suppose we own a farm and want to decide how much acreage to devote to corn. If the price of corn is high after this year's harvest, then we will plant a large amount of corn next year. With a large harvest of corn next year, the price of corn will have to drop in order to create enough demand to sell the entire harvest. With the price dropping next year, people will not plant much corn the year after. With a small harvest that year, the price will rise since not much demand is needed to sell the entire crop. The price is then high that year and this entire process repeats itself.

Notice that this is a recursive process and that the price oscillates between high and low values. But does it oscillate to some equilibrium price or with increasing magnitude (or does it exhibit some other type of behavior)? Let's develop a dynamical model of this supply and demand process and see if we can find the answer.

To develop our model, we need to consider three quantities, the supply $s(n)$ of our product, the demand $d(n)$ for our product, and the price $p(n)$ of one unit of our product, all in year n. Since there are three quantities or unknowns, we need to develop three equations relating them. This corresponds to three assumptions. One reasonable set of assumptions is the following.

1. The supply of the product in any year depends positively on the price of the product the previous year.
2. The demand for the product in any year depends negatively on the present price of the product.
3. Each year, the price of the product is adjusted so that the demand equals the supply.

Let's study the third assumption in more detail. Suppose we have a certain supply of a product this year and we wish to sell the entire supply. Our goal is to sell at a high price. The consumers' goal is to buy at a low price. The consumers offer to buy a large amount of the product at a low price. If we do not have that much of the product, we hold out for a higher price. The consumers then offer a higher price, but decrease the amount they are willing to buy. This process continues until the consumers reach a price at which they are willing to buy exactly what the producer has to sell. This is the price implied by the third assumption.

EXAMPLE 1.18   Let's put these assumptions together. Let $s(n)$ represent the supply of rutabagas, let $d(n)$ represent the demand (the amount the consumers buy) for rutabagas, and let $p(n)$ represent the price per bushel of rutabagas, all in year n.

The first assumption states that the supply in year $n + 1$ depends positively on the price in year n, that is, the higher the price in year n, the larger the supply in year $n + 1$. Let's assume that the relationship is given by the **supply equation**

$$s(n + 1) = 0.8p(n).$$

Thus, if the price is 6 dollars per bushel this year, the producers will grow 4.8 units of rutabagas next year, but if the price is 12 dollars per bushel this year, they will produce 9.6 units of rutabagas next year. The higher the price this year, the more rutabagas next year. This can be seen in that the slope of the line $s = 0.8p$ is positive.

The second assumption states that the demand this year depends negatively on the price this year, that is, the higher the price, the

lower the demand. Let's assume that this relationship is given by the
demand equation

$$d(n) = -1.2p(n) + 20.$$

In this case, if the price is 6 dollars per bushel, the consumers are
willing to buy 12.8 units of rutabagas, while if the price is 12 dollars
per bushel, the consumers are only willing to buy 5.6 units of rutabagas. Notice that the higher the price, the lower the demand. This is
reflected in the negative slope of the line $d = -1.2p + 20$.

We now use the third assumption to determine the price of rutabagas for next year. This assumption states that the price next year will
be adjusted so that the supply equals the demand, that is,

$$d(n+1) = s(n+1).$$

But since $d(n+1) = -1.2p(n+1) + 20$ and $s(n+1) = 0.8p(n)$,
substitution gives

$$-1.2p(n+1) + 20 = 0.8p(n).$$

Solving for $p(n+1)$ gives the first-order affine dynamical system

$$p(n+1) = -\frac{2}{3}p(n) + \frac{50}{3}. \tag{1.15}$$

From this first-order affine dynamical system, it is easy to see that if
the price this year is 7 dollars, then the price next year is $p(n+1) =
-(2/3) \times 7 + 50/3 = 12$ dollars.

In figure 1.18 is the cobweb analysis for dynamical system (1.15).
The point of intersection of the two lines is $(10, 10)$, so the equilibrium value (or equilibrium price) is $p = 10$. Since the absolute value
of the slope of the line corresponding to dynamical system (1.15) is
$|r| = 2/3 < 1$, the equilibrium price is stable as can be seen by the
cobweb cycling to the intersection point.   ■

Let us analyze this process more abstractly. From our first assumption, the supply next year $s(n+1)$ depends on the price this year.
Let us suppose that the relationship is "linear", so that the **supply
equation** will be

$$s(n+1) = r_s p(n) + c,$$

where r_s and c are two fixed constants. (While this seems to be a
bit restrictive, in real life this usually models the behavior reasonably
well.) Since the supply is positively related to price, we also assume

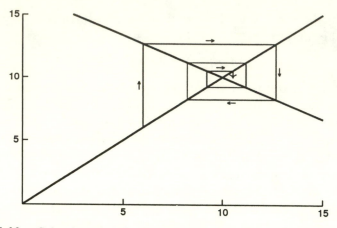

FIGURE 1.18. Cobweb analysis for dynamical system $p(n+1) = -(2/3)p(n)+50/3$. Price cycles to equilibrium $p = 10$.

that $r_s > 0$. The number r_s measures the **sensitivity of the producers to price**.

From the second assumption, demand in year $n + 1$ depends on the price in year $n + 1$. Again we assume this relationship is linear, so that the **demand equation** is

$$d(n + 1) = -r_d\, p(n + 1) + b,$$

where r_d and b are two fixed constants. Since demand is negatively related to price, we assume that $r_d > 0$ so that the coefficient $-r_d$ of $p(n + 1)$ is negative. The number r_d measures the **sensitivity of consumers to price**.

From the third assumption, we want the demand in year $n + 1$ to equal the supply in year $n + 1$, that is,

$$d(n + 1) = s(n + 1).$$

Remember that through bargaining, the producers and consumers will come to an agreement on price that is as advantageous as possible to both. From this equation, we can substitute in for both $s(n + 1)$ and $d(n + 1)$, just as we did in the previous example, to get

$$-r_d\, p(n + 1) + b = r_s\, p(n) + c.$$

Subtracting b from both sides, then dividing by $-r_d$ gives the first-order affine dynamical system

$$p(n + 1) = -\left(\frac{r_s}{r_d}\right) p(n) + \frac{b - c}{r_d}. \tag{1.16}$$

Cobweb analysis for dynamical system (1.16) consists of drawing the lines $y = x$ and $y = -(r_s/r_d)x + (b-c)/r_d$. The point of intersection is (p, p) where

$$p = \frac{b - c}{r_s + r_d},$$

is the equilibrium price. The absolute value of the slope of the line is $|r| = r_s/r_d$. We know that if

$$-1 < r = -\frac{r_s}{r_d} < 1,$$

then the equilibrium price is stable, that is, $p(k)$ oscillates to p as k goes to infinity. This means we have a stable price structure for this product.

Since $-r_s/r_d < 0 < 1$, the equilibrium value is stable if $-1 < -r_s/r_d$, that is, if $r_d > r_s$. Likewise, the equilibrium value is unstable if $r_d < r_s$, and the price will fluctuate with increasing amplitude until something drastic happens. This can be stated formally in what is known as the **cobweb theorem of economics**.

THEOREM 1.19 *Suppose we have a product whose price satisfies the three assumptions given earlier. Then the equilibrium price is* **stable** *and we have a stable market if*

$$r_s < r_d,$$

that is, if the producers are less sensitive to price than the consumers. Likewise, the price is **unstable** *and we have an unstable market if*

$$r_s > r_d,$$

that is, if the producers are more sensitive to price than the consumers.

When studying the market for a particular product, how can we use this model, that is, how can we find the constants, r_s, c, r_d, and b? First, we aren't interested in the actual equilibrium price, but only whether that price is stable or not. Therefore, we only need the constants r_s and r_d. In fact, to determine the stability, we don't even need r_s and r_d. What we need to know is how r_s and r_d compare. By observing consumers' and suppliers' habits at different prices, it can usually be determined who is the more sensitive.

Consider the oil industry. When the price of oil went up in the 1970s, the oil industry greatly expanded their search for, and consequently their production of, oil (with a few years' delay). Thus, their sensitivity, r_s, was large. But the consumers had somewhat fixed oil requirements, and consumption dropped only slightly as the price rose. So the consumers' sensitivity to price r_d was relatively small. Since $r_s > r_d$, the cobweb theorem of economics implies an unstable cyclic situation in the oil industry, which seems to be the case.

Suppose that for a certain product, say corn, it has been determined that $r_s > r_d$, and the market is unstable. What can be done? One solution is that the government can enter the picture as a consumer and make r_d larger. This can be done by price supports. Another solution is to pay farmers not to grow corn. Paying not to grow the crop is almost the same as buying the corn that would have been grown, and again increases r_d.

As we see from this discussion, mathematics can not only explain the reasons for certain problems, but also often help suggest solutions to the problem.

PROBLEMS

1. Consider the following modification of the savings assumption for Harrod's model of an economy. Suppose savings is a proportion of the total income plus a constant amount. Develop a dynamical system for total income $t(n)$, find the equilibrium income, and determine its stability using cobweb analysis.

2. Consider the following modification of the investment assumption (or acceleration principle) for Harrod's model. Suppose there is always some constant amount of desired additional investment needed for maintenance, and that the real desired additional investment is that constant amount plus some proportion of the change in income between this year and the previous year. Develop a dynamical system for $t(n)$, find the equilibrium value, and use cobweb analysis to determine the stability of the equilibrium value.

3. Suppose in Harrod's model, the desired additional investment is proportional to the change in income plus a certain portion of this year's income. Under what conditions will the income increase exponentially and under what conditions will the income decrease toward zero?

4. Determine the dynamical system for $p(n+1)$ in terms of $p(n)$, find the equilibrium value, and classify it as stable, unstable, or

unsure, if

$$s(n+1) = 1.2p(n) - 1 \quad \text{and} \quad d(n+1) = -0.8p(n+1) + 7.$$

5. Consider the supply and demand equations

$$s(n+1) = 0.8p(n) \quad \text{and} \quad d(n+1) = -0.6p(n+1) + 7.$$

Find the first-order affine dynamical system relating $p(n+1)$ to $p(n)$, determine the equilibrium value, and discuss its stability.

6. Consider the supply and demand equations

$$s(n+1) = p^2(n) \quad \text{and} \quad d(n+1) = -3p(n+1) + 4.$$

a. Assuming supply equals demand, find a first-order nonlinear dynamical system relating $p(n+1)$ to $p(n)$.
b. Find the two equilibrium values for this equation.
c. By constructing a cobweb close to the positive equilibrium value, determine the stability of this equilibrium value.

7. Consider the supply and demand equations

$$s(n+1) = 0.3p(n) - 4 \quad \text{and} \quad d(n+1) = -0.4p(n+1) + 10.$$

a. Repeat problem 5 for these equations.
b. Use these equations to develop a first-order affine dynamical system relating $s(n)$ to $s(n+1)$, find the equilibrium supply and determine its stability.
c. Use these equations to develop a first-order affine dynamical system relating $d(n)$ to $d(n+1)$, find the equilibrium demand, and determine its stability.

8. Suppose we have followed the price of a product for several years. We observed that when the price was 8 dollars per bushel in one year, the demand that year was 6000 bushels and the supply was 12 000 bushels the next year. We also observed that when the price was 5 dollars per bushel in one year, the demand that year was 9000 bushels and the supply was 10 000 bushels the next year.

a. From this information, determine the constants r_s and c in the supply equation $s(n+1) = r_s p(n) + c$.
b. From this information, determine the constants r_d and b in the demand equation $d(n+1) = -r_d p(n+1) + b$.
c. Determine the equilibrium price and its stability.

1.5   Applications of dynamics using spreadsheets

Spreadsheets are a wonderful tool for the study of discrete dynamical systems in that they were designed to perform recursion. A spreadsheet consists of an array of columns and rows, such as table 1.1. The headings of the columns are A, B, C, ..., and the headings of the rows are the integers 1, 2, ... A cell in the spreadsheet is given by pairing a letter with a number. For example, in table 1.1 the number 55 is in cell A1 and the word "label" is in cell B2. Each cell can contain either an arithmetic expression or a label. An arithmetic expression consists of 1) just a number, 2) a computation such as $45 + 15 * 8$, where the asterisk means multiplication, or 3) a computation using a cell name to represent the number in that cell, such as $3 * A1 + 4$.

When you type an algebraic expression into a cell, the expression is retained in memory but the result of the computation appears in the cell. For example, the expression $3 * A1 + 4$ was entered into cell A2, but the result, 169 is displayed in cell A2 as seen is table 1.1. If the number, 55, in cell A1 is changed, say to 10, then **the number in cell A2 will be changed to** 34 **simultaneously**. This is because cell A2 contains the expression $3 * A1 + 4$, and not the number 169.

Suppose we deposit 1000 dollars in a bank account at 10 per cent annual interest, compounded annually. Then the amount in the bank account each year is related to the amount in the account the previous year by the dynamical system

$$a(n+1) = 1.1a(n).$$

Let's put 1000 in cell C1 and the formula

$$1.1 * C1 \tag{1.17}$$

in cell C2. Then the second year's principal, 1100, appears in cell C2. You could then enter $1.1 * C2$ in cell C3, and so forth. But this would be too much work. One nice feature of spreadsheets is that

TABLE I.I. Sample spreadsheet consisting of numbered rows and lettered columns. Each cell can contain a label or an arithmetic expression.

	A	B	C
1	55		1000
2	169	label	1100
3			1210
4		117 390.8	1331

references to other cells are relative, that is, when expression (1.17) is entered into cell C2, the spreadsheet interprets it as **multiply the number in the previous cell by** 1.1 **and put that number in this cell**.

Spreadsheets have the ability to copy the contents of one cell into one or more other cells. Let's copy the contents of cell C2 into cell C3. The spreadsheet copies the formula **multiply the number in the previous cell by** 1.1 into cell C3. But the number in the previous cell is the number in cell C2. So

$$1.1 * C2$$

is what is actually copied into cell C3. Thus, the number 1210 appears in cell C3.

By copying the contents of C2 into cells C3 and C4, the spreadsheet interprets expression (1.17) as $1.1 * C2$ when copying into C3, and as $1.1 * C3$ when copying into C4, giving the values in table 1.1.

If we wanted to know the amount in our account after 50 years, we would

- Enter the initial amount, 1000, into cell C1.
- Enter the expression $1.1 * C1$ into cell C2.
- Copy cell C2 into cells C3 through C51.

Then the amount after 50 years would be displayed in cell C51. If the amount in cell C1 is changed, the amount in cells C2 through C51 are simultaneously changed.

One problem is that the amount we wish to know, C51, is not displayed on the screen and we need to move down the spreadsheet to see it. One solution to this problem is to store the expression +C51 in a cell that is on the screen, such as B4. Note that in table 1.1, 117390.8 is stored in B4. This is the amount in cell C51. The "+" is included so that the spreadsheet knows that this is an algebraic expression and not a label.

REMARK: Any expression starting with a letter is assumed to be a label. An algebraic expression must begin with a number or an operation such as $+$ or $-$.   □

To review, suppose we wish to study a dynamical system such as

$$a(n+1) = 2a(n) - 5. \tag{1.18}$$

You put $a(0)$ in some cell, say C1. The first relationship is $a(1) = 2a(0) - 5$. You store the right side of this equation in the next cell, that is, you put $2 * C1 - 5$ in C2. You then copy this relationship into as many more cells as you wish. The values $a(0)$, $a(1)$, ... will be displayed in cells C1, C2, ..., respectively. You can change $a(0)$ in cell C1 and the new values for $a(1)$, $a(2)$, ... will be displayed in cells C2, C3, ..., respectively.

Suppose you want to change the r value to 3, that is, you want to study the dynamical system

$$a(n+1) = 3a(n) - 5. \tag{1.19}$$

You could change the 2 in cell C2 to a 3 and then copy this new equation. There is a simpler approach. The trick to this approach is that when $ is used in referencing a cell, then that cell is always fixed. For example, suppose the expression $2 * C1 + \$B\$2$ is entered in cell C2. The spreadsheet reads that expression as **two times the number in the previous cell plus the number in cell B2**. If this is copied into cell C3, the spreadsheet will store $2 * C2 + \$B\$2$ in C3, that is, two times the number in the previous cell plus the number in B2.

REMARK: On the other hand, suppose $2 * C1 + B2$ is typed into cell C2. The spreadsheet interprets this as two times the amount in the previous cell plus the amount in the cell to the left. Thus, if this is copied into cell C3, the formula $2 * C2 + B3$ appears in cell C3.   □

The easiest way to study dynamical systems using spreadsheets is to put all constants in their own cells and reference the cells. For example, table 1.2 shows one way to study dynamical system (1.18). Let's consider the template on the left of the table. In cells A1 and A2 are the labels "r =" and "b =", respectively. In cell C1 is the value for $a(0)$. In cell C2 is the dynamical system $ra(0) + b$ given as $+\$B\$1 * C1 + \$B\$2$, that is, r is referenced by cell $\$B\$1$ with the $meaning to always use that particular cell. Likewise the $\$B\$2$ always uses the number in cell B2 for the b-value. But the C1 means to use the value in the previous cell as the $a(n)$ value. Thus, when this expression is copied,

- $+\$B\$1 * C2 + \$B\$2$ is copied into C3
- $+\$B\$1 * C3 + \$B\$2$ is copied into C4

and so forth. The results are given in the middle spreadsheet of table 1.2.

Now if we change the number in cell B1 to 3, this changes the dynamical system from $2a(n) - 5$ to $3a(n) - 5$. These results are given in the right spreadsheet of table 1.2. Thus, by changing the

TABLE 1.2. Spreadsheet studying the dynamical system $a(n+1) = ra(n) + b$. The left spreadsheet shows what is entered in each cell when $r = 2$, $b = -5$, and $a(0) = 3$, with the numerical results given in the middle spreadsheet. The right spreadsheet gives the numerical results when r is changed to 3.

| | | | | Numeric results when | | | | | |
| | Template | | | $r = 2$ | | | $r = 3$ | | |
	A	**B**	**C**	**A**	**B**	**C**	**A**	**B**	**C**
1	$r=$	2	3	$r=$	2	3	$r=$	3	3
2	$b=$	-5	$+\$B\$1 * C1 + \$B\$2$	$b=$	-5	1	$b=$	-5	4
3			copy			-3			7
4			$\vdots$			-11			16

r-value in cell B1, the b-value in cell B2, and/or the $a(0)$-value in cell C1, we can study a different dynamical system. In the rest of this section, we will study examples that show how powerful this tool is.

EXAMPLE 1.20 Suppose we decide to borrow 150 000 dollars from the bank in order to buy a home. The bank will loan us the money at 10 per cent annual interest. The loan is to be amortized or paid back by making 240 equal monthly payments for the next 20 years. The question is, what should we pay each month so that the amount we owe the bank after 20 years is zero?

Let $a(n)$ be the amount we owe the bank after making n monthly payments. How can we compute $a(n+1)$? Just after our nth payment, we owe the bank $a(n)$. One month later, we owe the bank this same balance plus one month's interest, that is, we owe

$$a(n) + 0.1a(n)/12 = (1 + 0.1/12)a(n).$$

We then make a payment of x dollars reducing the amount we owe. This final amount,

$$a(n+1) = (1 + 0.1/12)a(n) - x, \qquad (1.20)$$

is the amount we owe after $n + 1$ months. We also know that $a(0) = 150\,000$ since that is what we have borrowed, and that $a(240) = 0$ since after 240 payments we have paid back the loan. The one unknown is the monthly payment x.

Later in this text we will learn how to solve for x algebraically. Spreadsheets give an easy method for solving for x with a minimum of algebra. The trick is to setup a spreadsheet for dynamical system (1.20) using a guess for x. Then look at $a(240)$ on the spreadsheet. If it is not zero, we modify our guess for x. Since one month's interest on 150 000 at 10 per cent interest is $150\,000 \times 0.1/12 = 1250$, our guess at the monthly payment must be greater than 1250, say $x = 1500$. We then set up a template similar to that in table 1.3.

In this table, 150 000 was entered in cell A1, and the dynamical system

$$(1 + 0.1/12) * A1 - \$C\$1$$

was entered in cell A2. Note that cell A1 contains $a(0)$ and cell A2 contains $a(1)$. Thus, cell A241 contains $a(240)$. So +A241 is entered in cell C3 so that we can easily see the final amount owed the bank given our guess at the monthly payment. It is seen in table 1.3 that when the monthly payments are 1500 dollars then the amount owed after 20 years is $-39\,842.20$ dollars, meaning we overpaid. Thus, our monthly payment should be less than 1500. We then make a second guess of 1400 dollars, by changing the 1500 in cell C1 to 1400. Almost instantaneously we get 36 094.67 in cell C3 meaning we owe the bank this amount at the end of 20 years. The monthly payment must therefore be greater than 1400. Then enter 1450 in cell C1 which gives -1873.76 in cell C3. The payments must be less than 1450. Continuing, we got that $x = 1440$ gives a balance of 5719.92, that $x = 1447$ gives C3 = 404.34, that $x = 1448$ gives C3 = -355.03, that $x = 1447.50$ gives C3 = 24.65, that $x = 1447.53$ gives C3 = 1.87, and that $x = 1447.54$ gives C3 = -5.72.

Thus, the equal monthly payments must be between 1447.53 dollars and 1447.54 dollars. Since this is impossible, we pay the larger amount, 1447.54 dollars each month for 239 months. Since a final payment of this amount would overpay our balance by C3 = -5.72, the last payment is reduced by this amount, that is, the final payment is 1441.82 dollars.   ∎

TABLE I.3. Spreadsheet studying amortization of a 150 000 loan at 10 per cent annual interest paid back in 240 payments.

		guess 1		guess 2	guess 3
	A	**B**	**C**	**C**	**C**
1	150 000	payment =	1500	1400	1450
2	149 750				
3	149 497.90	balance =	$-39\,842.20$	36 094.67	-1873.76

EXAMPLE 1.21   When buying a house, a good approach is to first determine how much you can afford to pay each month for a mortgage, and then use this to determine how much you can afford to borrow. For example, suppose the bank will lend money at 10 per cent interest for 20 years, and we can afford monthly payments of 1500 dollars. To the nearest 1000 dollars, how much can we afford to borrow. For this problem, we use the same template as in table 1.3 with 1500 in cell C1 and a guess at the initial amount of the loan in cell A1, say 170 000. In this case, the balance after 20 years is 106 719 in cell C3 meaning we underpaid, so we cannot borrow this much. See table 1.4. Note in table 1.4 that when the guess is 150 000, the balance after 20 years is $-39\,842$ meaning we overpaid. If you continue, you will find that if $a(0) = 155\,000$ then $a(240) = -3201.84$, if $a(0) = 157\,000$ then $a(240) = 11\,454.30$, and if $a(0) = 156\,000$ then $a(240) = 4126.23$. Thus, we can borrow between 155 000 and 156 000 dollars.   ■

EXAMPLE 1.22   Suppose you win 1 000 000 dollars in a lottery in the sense that you will receive 50 000 dollars a year for the next 20 years. How much did you **actually** win?

Let's rephrase this problem. Suppose the lottery commission deposits W dollars into a bank account paying $100I$ per cent interest, compounded annually. Your entire winnings will be paid out of this account, after which the account will be depleted. You have actually won W dollars.

Let $a(n)$ represent the amount in this bank account after you have received your nth payment of 50 000 dollars. Thus, $a(1) = W - 50\,000$ since they immediately give you your first payment. Since the next payment comes in one year and the money, $a(1)$, collects interest for

TABLE 1.4.   Spreadsheet studying amortization of a loan at 10 per cent annual interest paid back in 240 payments with monthly payments of 1500 dollars.

| | guess 1 | | | guess 2 | | |
	A	B	C	A	B	C
1	170 000	pay. =	1500	150 000	pay. =	1500
2	169 916			149 750		
3	169 832	bal. =	106 719	149 498	bal. =	−39 842

a year, $a(2) = (1 + I) a(1) - 50\,000$. Similarly,

$$a(n + 1) = (1 + I) a(n) - 50\,000. \tag{1.21}$$

Also, $a(20) = 0$, since after your 20th payment you receive no more money. The amount W is the present value of your lottery win which is how much you have **actually won**.

Suppose the money is put into an account paying an effective annual interest rate of 9 per cent. Let's set up a spreadsheet using equation (1.21), with $I = 0.09$. The results are given in table 1.5 with the first two guesses at the winnings.

In table 1.5, the guess at your winnings in entered in cell C1, $a(1)$ is entered in cell A1 as $+C1 - 50\,000$, dynamical system (1.21) is entered in cell A2 as $1.09 * A1 - 50\,000$, and is copied through cell A20. Finally, A20 is entered in cell C3. After several attempts, you should find that $W = 497\,505.74$.

Note that the interest rate you can get on your money affects the amount of your winnings. If you could only get 8 per cent interest, then your winnings would be $530\,179.96$ dollars. Try to find this result using a variation on table 1.5. ∎

PROBLEMS

Use spreadsheets to answer the following questions.

1. Suppose you have 100 000 dollars in a savings account that pays 8 per cent interest, compounded quarterly. Suppose each quarter you withdraw 3000 dollars from your account.
   a. How many quarters will it take for you to deplete your account to only 50 000 dollars?
   b. How many quarters will it take for you to deplete your entire savings account?

TABLE 1.5. Spreadsheet studying present value of 1 000 000 paid in 20 yearly installments of 50 000 dollars and invested at 9 per cent interest.

		guess 1		guess 2
	A	**B**	**C**	**C**
1	450 000	won =	500 000	480 000
2	440 500			
3	430 145	bal. =	12 824	−90 008

2. Suppose you start a savings account which pays 8 per cent interest a year compounded monthly. You initially deposit 1000 dollars and decide to add an additional 100 dollars each month thereafter. How much is in your account after 5 years?

3. How many years would it take to double your money if you invested at
    a. 5 per cent interest compounded annually?
    b. 9 per cent interest compounded annually?

4. Suppose you wish to buy a house, but can only afford 800 dollars a month.
    a. If you can get a loan at 9 per cent (compounded monthly) for 30 years, how large can your loan be?
    b. How large can your loan be if you amortize your loan in 20 years?

5. Suppose you have a savings account that earns 8 per cent interest, compounded monthly. Over the next 10 years, you withdraw 500 dollars a month from this savings account, at which point, all the account is totally depleted. How much money was in the account initially?

6. Suppose you have a 5000 dollar, 20 year bond, with a coupon value of 11 per cent, that is, it pays you 550 dollars each year for the next 20 years, at which time you get your 5000 dollars back. What is the present value of this bond if your alternate investment is a savings account paying 7 per cent interest compounded annually?

7. At retirement, your savings account has 300 000 dollars which is collecting 6 per cent interest, compounded monthly. How much money can you withdraw each month so that your savings will last 20 years?

8. Suppose you borrow 6000 dollars at 10 per cent annual interest rate, compounded monthly, in order to buy a car. If the loan is to be paid back in 5 years, what is your monthly payment?

9. Suppose you borrow 2000 dollars at 8 per cent interest, compounded quarterly, to take a vacation. If the loan is to be paid back in eight equal quarterly payments, how much should these payments be?

10. How much is the monthly payment on a 100 000 dollar loan at 12 per cent interest (compounded monthly) which must be amortized in

   a. 10 years?
   b. 20 years?
   c. 30 years?

11. Consider the following retirement problem.
    a. Suppose you presently need 2000 dollars a month to live comfortably, and that inflation is 0.5 per cent per month. In 40 years, how much will you need each month to live comfortably?
    b. Using your answer to part (a), how much savings should you have (in 40 years) so that you can live comfortably for the next 30 years? Your savings are collecting 12 per cent interest compounded monthly.
    c. How much money should you put in your savings account (collecting 12 per cent interest compounded monthly) each month in order to have the savings computed in part (b)? Remember you are saving for 40 years.

12. Suppose you buy a painting for 200 dollars. Five years from now, someone offers you 300 dollars for your painting. What is the effective yearly interest on this investment?

13. You buy stocks for 1000 dollars. After 3 years, you sell them for 1500 dollars. What is your effective yearly yield? Note that it's actually higher since you have also (probably) received dividends on the stocks.

14. What effective yearly interest rate would you need on an investment in order to double your money in 8 years?

CHAPTER 2

First-order linear dynamical systems

2.1 Solutions to linear dynamical systems with applications

When modeling the real world, we first consider only the most basic assumptions. This often leads to a linear (possibly nonhomogeneous) dynamical system. The good news is that in general we can find solutions to these linear dynamical systems. These solutions can lead us to a better understanding of the situation we are modeling. We then try to make our model more realistic by adding secondary assumptions. These assumptions often lead to nonlinear dynamical systems. Unfortunately, we can rarely find solutions to nonlinear dynamical systems. We will see in chapter 6 that while we cannot find solutions, we can determine the long term behavior of many nonlinear systems using the techniques we develop in studying linear systems. We must comment that there are still a great many types of nonlinear systems for which the behavior is still not understood. To understand the world better, mathematicians need to develop means for understanding these equations better.

A major part of this book will be devoted to finding solutions to linear systems. This is not because linear systems are more important or more common than nonlinear systems, but because we know more about linear systems. Similarly, a great many of the applications in this text will lead to linear dynamical systems. This does not mean that most applications are linear, but that the first step to most applications is linear.

The most general form of a first-order linear dynamical system is

$$a(n+1) = ra(n), \qquad \text{for} \quad n = 0, 1, \ldots, \qquad (2.1)$$

where r is a constant. One example is the model of compound interest

$$a(n+1) = (1+I)a(n), \qquad \text{for} \quad n = 0, 1, \ldots,$$

where I is the yearly interest rate and $a(n)$ is the amount in the account at the beginning of the nth year.

A **solution** to a dynamical system is a function $a(k)$ defined for all integers $k \geq 0$ that satisfies the dynamical system. One solution to dynamical system (2.1) is

$$a(k) = 3r^k, \qquad \text{for} \quad k = 0, 1, 2, \ldots.$$

Note that this means $a(0) = 3$, $a(1) = 3r$, $a(2) = 3r^2$, $a(n) = 3r^n$, and $a(n+1) = 3r^{n+1}$. Substituting into dynamical system (2.1) gives equality,

$$3r^{n+1} = r(3r^n).$$

Thus the above is a solution to dynamical system (2.1).

Likewise $a(k) = (-2)r^k$ is a solution. In fact the discrete function

$$a(k) = cr^k \qquad (2.2)$$

is a solution where c is any constant. The function (2.2) is called **the general solution** to the dynamical system (2.1). This means that every solution to dynamical system (2.1) is of the form of function (2.2). If we are given an initial value such as $a(0) = 5$, then by letting $k = 0$ in equation (2.2) we get $a(0) = cr^0 = c$. Since $c = a(0) = 5$, we know $c = 5$ and that the **particular solution** to the first-order linear dynamical system (2.1) with initial value $a(0) = 5$ is

$$a(k) = 5r^k.$$

REMARK: The dynamical system $a(n+1) = ra(n)$ corresponds to the infinite set of equations, $a(1) = ra(0)$, $a(2) = ra(1)$, and so forth. If $a(0)$ is given, say $a(0) = 5$ as above, then we can use the first equation to find the unique value for $a(1)$, $a(1) = 5r$. We can then use the value for $a(1)$ and the second equation to find the unique value for $a(2)$. Continuing in this fashion, we see that there is a unique value for every $a(k)$. Thus, once $a(0)$ is given, there is only one particular solution that satisfies the equations and the initial value. This is why we say **the** particular solution.   □

DEFINITION 2.1   *The **general solution** to a first-order dynamical system*

$$a(n+1) = f(a(n)), \qquad for \quad n = 0, \ldots$$

is a function $a(k)$, with domain $k = 0, \ldots$, which

- *satisfies the dynamical system when substituted in for $a(n+1)$ and $a(n)$, and*
- *involves a constant c which can be determined once an initial value is given.*

*The **particular solution** to a first-order dynamical system*

$$a(n+1) = f(a(n)), \qquad for \quad n = 0, \ldots$$

is a function $a(k)$, with domain $k = 0, \ldots$, which

- *satisfies the dynamical system when substituted in for $a(n+1)$ and $a(n)$, and*
- *satisfies a given initial condition $a(0) = a_0$.*

For the first-order dynamical system (2.1), if you are given that $a(0) = a_0$, then simple computations and substitutions give that $a(1) = ra_0$, $a(2) = ra(1) = r^2 a_0$, ..., $a(k) = ra(k-1) = r(r^{k-1}a_0) = r^k a_0$. This leads us to believe that the particular solution to the dynamical system (2.1) is

$$a(k) = a_0 r^k.$$

Simple substitution of $a(k)$ into dynamical system (2.1) shows that it does satisfy the system and is therefore the particular solution. Furthermore, if we define

$$a(k) = cr^k,$$

then $a(k)$ satisfies dynamical system (2.1) and we can solve the equation $a(0) = cr^0 = a_0$ for c, that is, $c = a_0$. We have thus proved the following theorem.

THEOREM 2.2   *The general solution to the first-order linear dynamical system*

$$a(n+1) = ra(n), \qquad for \quad n = 0, \ldots$$

is

$$a(k) = cr^k, \qquad for \quad k = 0,\ldots.$$

The particular solution to this dynamical system given that $a(0) = a_0$ is

$$a(k) = a_0 r^k, \qquad for \quad k = 0,\ldots.$$

If $a(1) = a_1$ were given instead of $a(0)$, we could find the particular solution by solving

$$a(1) = cr^1 = a_1$$

for the unknown c. In this case, $c = a_1 r^{-1}$ and the particular solution is

$$a(k) = a_1 r^{-1} r^k = a_1 r^{k-1}.$$

EXAMPLE 2.3   What is the general solution to

$$a(n+1) = -3a(n), \qquad for \quad n = 0, 1, \ldots?$$

Since $r = -3$, the general solution is

$$a(k) = c(-3)^k.$$

What is the particular solution to the above dynamical system with $a_0 = -3$?   By the above we see that $c = a(0) = -3$ and thus the particular solution is

$$a(k) = (-3)(-3)^k = (-3)^{k+1}.$$

Substituting $a(n + 1) = (-3)^{n+2}$ and $a(n) = (-3)^{n+1}$ into the dynamical system gives the equality

$$(-3)^{n+2} = (-3)(-3)^{n+1},$$

verifying this theorem.

Suppose we want to find the particular solution given that $a(3) = 6$. Solving $c(-3)^3 = a(3) = 6$ gives that $c = -6/27 = -2/9$. Thus, the particular solution is

$$a(k) = (-2/9)(-3)^k \qquad \blacksquare$$

We already know from theorem 1.13 that if $|r| < 1$ then $a(k)$ goes to zero, if $|r| > 1$ then $|a(k)|$ goes to infinity, and if $r = -1$ then $a(k)$ forms a 2-cycle. Let's look at two figures that tell us how $a(k)$ goes to zero. Similar figures can be constructed to show how $|a(k)|$ goes to infinity when $|r| > 1$.

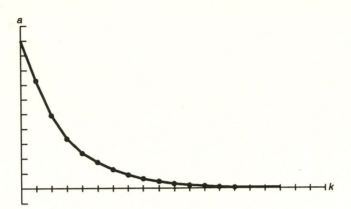

FIGURE 2.1. The points $(k, a(k))$ for the dynamical system $a(n+1) = 0.7a(n)$ with $a(0) = 10$.

When $0 < r < 1$, the solution $a(k)$ goes to zero exponentially. If $a_0 > 0$ it decreases to zero as in figure 2.1 in which $r = 0.7$ and $a_0 = 10$, and if $a_0 < 0$ it increases to zero. When $-1 < r < 0$, the solution oscillates to zero. This can be seen in figure 2.2, where $r = -0.8$ and $a_0 = 10$. When $r > 1$, the solution goes exponentially to either positive or negative infinity, and if $r < -1$ then the solution oscillates to infinity with increasing amplitude.

DEFINITION 2.4 *A solution $a(k)$ is **periodic** if*

$$a(k + m) = a(k)$$

*for some fixed integer m and all k. The smallest integer, m, for which this holds is called the **period** of the solution.*

For the dynamical system $a(n + 1) = -a(n)$ with $a_0 = -6$, the particular solution is $a(k) = -6(-1)^k$ and thus

$$a(k + 2) = -6(-1)^{k+2} = -6(-1)^k(-1)^2 = -6(-1)^k = a(k).$$

So this solution is periodic with period 2. In particular $a(0) = a(2) = a(4) = \cdots = -6$ and $a(1) = a(3) = a(5) = \cdots = 6$. Notice that $a(k + 4) = a(k)$, $a(k + 6) = a(k)$, etc. But since 2 is the smallest number that works, this solution has period 2.

There are dynamical systems with very large periods. Cyclic behavior also occurs in nature in the periodic attacks of locusts and in planetary motion such as the return of Halley's comet.

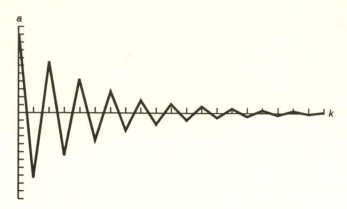

FIGURE 2.2. The points $(k, a(k))$ for the dynamical system $a(n+1) = -0.8a(n)$ with $a(0) = 10$.

It is interesting to note the many different types of behavior exhibited by the solutions to the simple dynamical system

$$a(n+1) = ra(n)$$

for different values of r and a_0. Note that the equation

$$a(n+1) = (1+I)a(n),$$

which gives the amount in a savings account, is first-order linear with $r = 1 + I$, $I > 0$. Thus the money in a savings account increases exponentially (to infinity). We could be very rich if we could wait long enough.

2.1.1 Radioactive decay

Let's use our results to investigate the behavior of radioactive materials. Let's fix an interval of time, T. The interval T could be one second, one minute, or even one year. Take any one atom of this radioactive material and watch it for any time interval of length T. There is a fixed probability p, such that this atom will decay during time interval T.

Let's assume that T is one year and $p = 0.1$, that is, during any one year there is a 10 per cent chance that one particular atom will decay. So if we have many atoms, about 10 per cent of them should decay during a year, under the assumption that each atom acts independently of all the others.

For any fixed time period T, let p equal the probability that any atom decays during that time period and let $a(n)$ be the amount of material we have after n such time periods. Then the amount of material left after $n + 1$ time periods is proportional to the amount left after n time periods, that is,

$$a(n+1) = (1 - p)a(n) = ra(n),$$

where $0 < r = 1 - p < 1$. It is relatively easy to get a good approximation for r. For example, suppose that at time $n = 0$ you have $a(0) = a_0 = 10$ grams of the material, but that after 1 year you have 9.9 grams. Then

$$a(1) = ra(0) \qquad \text{or} \qquad 9.9 = 10r,$$

so $r = 0.99$. It must also be noted that this is not the method normally used for computing p for radioactive materials, but it is not far off.

The dynamical system for the decay of this radioactive substance is then

$$a(n + 1) = 0.99a(n).$$

What is the half-life of this substance, that is, in how many years, k, will $a(k) = 0.5a_0$? Since the solution is $a(k) = (0.99)^k a_0$, we need to solve for k in the equation

$$0.5a_0 = (0.99)^k a_0.$$

Dividing both sides by a_0 and taking logarithms (any base) gives

$$k = \frac{\log 0.5}{\log 0.99} = 68.97 \text{ years.}$$

This is a much better approach than waiting for 69 years to find 5 grams remaining.

In most applications, numbers are approximated. Thus, when someone computes that after one year, 9.9 grams of the radioactive material is still left, this number is accurate to only 1 decimal place, that is, the actual amount of radioactive material left is between 9.85 and 9.95 grams. If, for example, 9.85 grams are left, then our dynamical system would be

$$a(n + 1) = 0.985a(n),$$

with solution $a(k) = (0.985)^k a_0$. In this case, we would find the half-life by solving $0.5a_0 = (0.985)^k a_0$, which gives a half-life of

$$k = 45.8 \text{ years.}$$

If, on the other hand, 9.95 grams of the radioactive material were left, then we would get a half-life of 138.3 years. Thus, we would say that the half-life was **between 45 and 139 years**. To get a more accurate estimate for the half-life, we would need a more accurate computation of the number of grams of the radioactive material left after one year.

It must be stated that r is the effective yearly change in the radioactive substance, but that actually it is decaying continuously. Corresponding continuous models can be developed using techniques of calculus. This leads to what are called differential equations.

An important application of radioactive decay is to **carbon dating**. A certain proportion of the carbon in the atmosphere is radioactive carbon-14. The carbon-14 in the atmosphere is the result of the bombardment of earth by rays from outer space. It is assumed that this bombardment has remained constant over time, so that the proportion of carbon-14 in the atmosphere has remained constant over time. The proportion of the carbon in plants and animals that is carbon-14 is the same as that in the atmosphere, due to respiration. Thus, we know what per cent of the carbon in a plant or animal was carbon-14 when the plant or animal died.

When an animal or plant dies, the carbon-14 in the tissue starts decaying and is not replaced with new carbon-14. The half-life of carbon-14 is approximately 5700 years, so that $a(5700) = 0.5a(0)$. Solving for r in the solution

$$0.5a(0) = a(5700) = r^{5700}a(0)$$

gives

$$r = (0.5)^{\frac{1}{5700}}.$$

Suppose we find a bone of some animal in which only 83 per cent of the original amount of carbon-14 is present. How old is this bone? If the animal died in year 0, then the amount left after k years is, from our solution,

$$a(k) = (0.5)^{\frac{k}{5700}} a(0).$$

If today is year k, then we know that $a(k) = 0.83a(0)$. Substitution into the above equation, canceling the $a(0)$'s, and taking logarithms gives that

$$\frac{\log 0.83}{\log 0.5} = \frac{k}{5700}$$

or that $k = 1532$. Thus this animal lived about 1500 years ago.

Again, there is a problem of accuracy. If, by saying 83 per cent of the original amount of carbon-14 is present, we actually mean between 82.5 and 83.5 per cent is left, then we must compute a range for the age of the bone. In this case, 82.5 and 83.5 per cent give ages of 1582 and 1482 years, respectively. Thus, the most that could be said is that the bone is between 1482 and 1582 years old.

2.1.2 Glottochronology

The term, glottochronology, is a combination of Greek words meaning, essentially, language dating. Mathematicians and linguists worked together to develop a theory for studying the changes in languages.

We all know that, over time, certain words disappear from usage and new words appear. Suppose that, at a certain point in time, we compile at a list of L words (say $L = 250$). At a later point in time, we study that same list of words and determine what per cent of the original list of words are still in use.

Let one unit of time be 1 year. Thus, time n will be n years. Let $a(n)$ represent the fraction of the original list of words still in use n years later. The basic assumption is that the fraction $a(n+1)$ of the original list of words in use in any year $n+1$ is proportional to the fraction of the original list of words in use the previous year n, that is,

$$a(n+1) = ra(n),$$

where r is a positive constant less than one. At time 0, all of the original list of words are in use, so $a(0) = 1$. Therefore, at time k, $a(k) = r^k(1) = r^k$ is the fraction of the original list of words still in use.

Since languages change slowly, r should be close to 1 and would probably be hard to estimate on a year by year comparison. By comparing a written language today with the same language a millennium ago, glottochronologists can estimate r^{1000}. This number r also depends on the particular language. Glottochronologists have found that the number r^{1000} is usually close to 0.805. So for languages with no written history, that is, for languages in which we cannot estimate r, we will assume that

$$r^{1000} = 0.805 \qquad \text{or} \qquad r = (0.805)^{0.001} = 0.999\,783.$$

The fraction of the original list of words that are still in use k years later is then $r^k = (0.805)^{0.001k}$.

Before proceeding with our discussion of languages, we need to review some aspects of probability. We will do this through an example.

EXAMPLE 2.5   Suppose two people have each memorized a certain fraction of the words on a list of nonsense words. For example, suppose Frank knows 70 per cent of list L and Sue knows 80 per cent of list L, where L contains 100 words. Given any random sublist of words from list L, we would expect Frank to know 70 per cent of them and Sue to know 80 per cent of them.

Frank knows 70 of the original 100 words. We would expect Sue to know 80 per cent of Frank's 70 words, that is, 56 of Frank's words. Thus, Sue and Frank know 56 words in common, that is, the fraction of the 100 words that Frank and Sue both know is $(0.80)(0.70) = 0.56$ or 56 per cent.   ■

MULTIPLICATION PRINCIPLE: Suppose person A knows p per cent of a list of L words and person B knows q per cent of the same list of L words, where p and q are given as decimals. Given no additional information, we would expect A and B to both know pq per cent of the words.   □

This principle can be stated more generally and is quite useful in many diverse situations, such as in the study of genetics in section 2.3. Thus we state the more general version and refer the reader to section 3.1 where we study the multiplication principle in more detail.

MULTIPLICATION PRINCIPLE: Suppose A occurs p per cent of the time and B occurs q per cent of the time, where p and q are given as decimals. Given no additional information, we would expect A and B to both occur pq per cent of the time. (This assumes that A occurring does not affect the chances of B occurring.)   □

Suppose at time 0, a group of people separate themselves from their culture. A group of American Indians leaves the tribe and forms its own tribe, or a group sails to a deserted island and starts its own culture. We then have two cultures, A and B. At time 0, they have the same language, so that for a given list of L words, $a(0) = b(0) = 1$ is the fraction of the words they both know (and have in common).

By our previous result, if we contact each of these cultures k years later, culture A will know $a(k) = (0.805)^{0.001k}$ fraction of the original list and culture B will also know $b(k) = (0.805)^{0.001k}$ fraction of the original list. Thus the fraction of the words that both cultures have in common is, by the multiplication principle,

$$a(k)b(k) = (0.805)^{0.001k}(0.805)^{0.001k} = (0.805)^{0.002k}.$$

What the glottochronologist does now is to construct a list of words. From that list of words, the two cultures are studied and it is determined what fraction of this list of words is known by both cultures. Thus in the equation above, $a(k)b(k)$ is known, but k, the number of years since the two cultures separated, is unknown. Solving for k gives

$$k = \frac{500 \log (a(k)b(k))}{\log 0.805}.$$

EXAMPLE 2.6 Suppose that the natives of two islands have similar languages. From a list of 300 words, 180 words are understood by both groups, that is, $a(k)b(k) = 180/300 = 0.6$. Then

$$k = \frac{500 \log 0.6}{\log 0.805} = 1177.5.$$

We then conclude that the natives of these two islands came from a common ancestry, approximately 1200 years ago. ■

Suppose a collection of tribes with a similar language is considered. First, group the tribes into geographical regions. Then date the time separation k for pairs of tribes in each geographical region. It can be argued that the region with the pair of tribes with the largest time separation is the homeland of the tribes. The reason for this conclusion is as follows. Suppose one tribe separates into three tribes. One tribe might move away while the other two remain in the same general region. The tribe that moved away may split again in its geographical location, but the largest time separation will always be the two that remained in the original area.

Glottochronology was popular for some time, but there are now some doubts about its accuracy. This could be from, among other things, problems with measurement and assumptions that are not realistic. For example, how do you determine if a word is the same for two cultures? If the spelling of a word or the pronunciation of a word changes "slightly", we will still count it as being on the list. If the **meaning** of a word changes "significantly", we will delete it from the list. Thus, there is some subjectivity in determining the fraction of common words, which could drastically change the results. Thus, similarly to radioactive decay, we could at best give a range for the fraction of words that have changed and use this to find a range for the length of time at which the two cultures separated.

Another problem is that some words are more likely to change than others. But in the multiplication principle, it is tacitly **assumed** that all words are equally likely to change. This can also throw the results off.

The moral of this is that you need to be careful not to make more claims about your model than are justified. Most of the models in this text are only first approximations at modeling a situation.

PROBLEMS

1. For the following dynamical systems, find $a(1)$ through $a(5)$, give the particular solution, describe the long term behavior (that is, increasing to infinity, oscillating to zero, etc.), and draw a graph of the points $(k, a(k))$.
   a. $a(n+1) = -1.1a(n)$, $a(0) = 2$
   b. $a(n+1) = 1.1a(n)$, $a(0) = -1$
   c. $a(n+1) = -0.2a(n)$, $a(0) = 1000$
   d. $a(n+1) = 0.2a(n)$, $a(0) = -1000$

2. For the following dynamical systems, give the general solution.
   a. $a(n+1) = 3a(n)$
   b. $a(n+1) = -0.5a(n)$
   c. $a(n+1) = -1a(n)$

3. For the dynamical systems in problem 2, give the particular solution for $a(0) = 4$.

4. For the dynamical systems in problem 2, give the particular solution for $a(2) = 10$.

5. Suppose that you presently have 15 grams of a radioactive material. In a year, you have 14.7 grams remaining.
   a. What is the half-life of this material?
   b. Given that 14.7 is accurate to only 1 decimal place, find the possible range for the half-life of this material.

6. Suppose the half-life of a radioactive material is 100 years. If you presently have 1 gram of the material, how much will you have in 1 year?

7. Suppose that a piece of wood is discovered at an archaeological dig which has 30 per cent of its original amount of carbon-14.
   a. Approximate the age of this wood.
   b. Suppose that between 29.5 and 30.5 per cent of the original amount of carbon-14 is left. Find an age range for this piece of wood.

8. Radium decreases at the rate of 0.0428 per cent per year. What is its half-life?

9. Radioactive beryllium, with a half-life of 4.6 million years, is used to date fossils in deep-sea sediment.
   a. If a fossil is found with 99 per cent of its original amount of radioactive beryllium, then how old is this fossil?
   b. If the fossil has between 98.9 and 99.1 per cent of its original amount of radioactive beryllium, then find an age range for the fossil.

10. Two groups of people have a common language. From a list of 250 words, the two groups have 220 in common. How long ago did these two groups split from one?

11. Consider the model of glottochronology. Assume a language is given today.
    a. How long will it take for 1/4 of the words to change?
    b. How long will it take for 10 per cent of the words to change?

12. Suppose that person A knows 60 per cent of a list of 1000 words, person B knows 70 per cent of that list, and person C knows 30 per cent of that list.
    a. How many words do you expect all three people know in common?
    b. What per cent of the words is known by both A and B but not by C?

2.2 Solutions to an affine dynamical system

In this section we are going to consider the dynamical system

$$a(n+1) = ra(n) + b$$

in the simple case in which $r = 1$, that is, the dynamical system

$$a(n+1) = a(n) + b.$$

Suppose we wish to know how much gas is left in our car after we have driven somewhere. Let $a(n)$ represent the amount of gas left after driving n miles. To study this situation, we need additional information. In particular, we need to know how much gas we started with, $a(0)$, and we need to know how many miles we get per gallon of gas. Suppose we start with 15 gallons of gas and our car gets 25 miles per gallon. Since we get 25 miles per gallon, we use 0.04 gallons when we drive one mile. This can be written as the discrete dynamical system

$$a(n+1) = a(n) - 0.04 \quad \text{for} \quad n = 0, 1, \ldots, \text{ and } a(0) = 15. \quad (2.3)$$

Note that dynamical system (2.3) does not give the amount of gas as a function of the number of miles driven. In this case, we see that

$$a(1) = a(0) - 0.04 = 15 - 0.04.$$

By substitution, we get that

$$a(2) = a(1) - 0.04 = 15 - 2(0.04), \quad a(3) = 15 - 3(0.04), \quad \ldots.$$

It is now easy to see that

$$a(k) = 15 - 0.04k \quad \text{for} \quad n = 0, 1, \ldots. \quad (2.4)$$

Equation (2.4) is the **particular solution** to dynamical system (2.3).

We now study first-order affine dynamical systems of the type

$$a(n+1) = a(n) + b.$$

Suppose that $a(0) = a_0$. Then we have $a(1) = a_0 + b$ and, by substitution, $a(2) = a(1) + b = a_0 + 2b$, $a(3) = a_0 + 3b$, and so forth. Thus, the particular solution appears to be

$$a(k) = a_0 + bk.$$

From this, it is clear that the general solution should be

$$a(k) = c + bk.$$

Substitution of $a(n) = c + bn$ and $a(n+1) = c + b[n+1]$ into this affine dynamical system gives

$$c + b[n+1] = [c + bn] + b.$$

We have equality, so this is the correct form of the general solution.

Given that $a(0) = a_0$, substitution of $k = 0$ into the general solution gives

$$a(0) = c + b \times 0 = a_0,$$

so $a_0 = c$ and the particular solution is

$$a(k) = a_0 + bk.$$

EXAMPLE 2.7   Find the particular solution to

$$a(n+1) = a(n) - 2$$

with $a_0 = 9$. Since $b = -2$, the solution is

$$a(k) = 9 - 2k.$$

Notice that there is no equilibrium value. The graph of this solution is given in figure 2.3 where the points $(k, a(k))$ are plotted.   ■

Note that $a(n+1) = a(n) + b$ corresponds to the equation

$$a(n+1) - a(n) = b,$$

that is, differences of consecutive numbers are constant. When we have constant differences or constant change, then the solution is

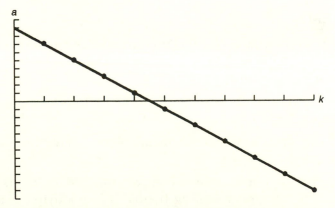

FIGURE 2.3.   The points $(k, a(k))$ for the dynamical system $a(n+1) = a(n) - 2$ with $a(0) = 9$. Observe the linear relationship.

given by $a(k) = c + bk$, that is, the points $(k, a(k))$ lie on a straight line. Thus, constant change results in a linear relationship between amount $a(k)$ and time k.

EXAMPLE 2.8 Suppose the cost of renting a car is 15 dollars plus 4 cents per mile. Let $a(n)$ be the cost of renting a car if we drive it n miles. Then the discrete dynamical system that models this problem is

$$a(n+1) = a(n) + 0.04 \qquad \text{with} \quad a(0) = 15,$$

and its solution is

$$a(k) = 15 + 0.04k \qquad \text{for} \quad k = 0, 1, \ldots. \qquad \blacksquare$$

REMARK: While the models describing the amount of gas left in a car and the cost for driving a rental car are similar, there is an important difference. Let $f(x)$ be the amount of gas left in a car after driving x miles, where x does not need to be an integer. Since a car uses gas continuously, we can generalize solution (2.4) to

$$f(x) = 15 - 0.04x.$$

But for renting a car, the cost goes up in jumps so that the cost $g(x)$ for driving x miles is

$$g(x) = 15 + 0.04\lceil x \rceil$$

where $\lceil x \rceil$ is the smallest integer greater than or equal to x. $\square$

PROBLEMS

1. For the following dynamical systems, give the general solution.
   a. $a(n+1) = a(n) - 5$
   b. $a(n+1) = a(n) + 3$

2. For the dynamical systems in problem 1, give the particular solution for $a(0) = 4$.

3. For the dynamical systems in problem 1, give the particular solution for $a(2) = 0$.

4. A basketball player signs a contract for 30 000 dollars a game plus a 50 000 dollar signing bonus. Write a dynamical system describing this contract and use its solution to find the amount the player earns for a 90 game season.

5. An encyclopedia salesman earns a 20 per cent commission on each set sold plus 150 dollars per week. A set of encyclopedias sell for 500 dollars. Develop a dynamical system to describe the salary of this salesman and use the solution to determine how many sets must be sold to earn over 1000 dollars for the week.

6. A video rental outlet charges 3 dollars for the first movie rented and 2 dollars for each additional movie rented. Develop a dynamical system describing the cost of renting n movies and give the particular solution.

7. As air warms, it becomes less dense. As a result, sound travels faster. The speed of sound increases by 0.7 meters per second for each degree Celsius the temperature rises. Develop a dynamical system modeling this situation and find the particular solution given that sound travels at 330 meters per second at 0 degrees Celsius.

2.3   An introduction to genetics

Many traits of individuals of a species are determined by **genes** inherited from each of the parents of the individual. Let's consider a gene A that determines a certain trait. The simplest case is when gene A has two forms or **alleles**, allele A and allele a. Suppose the individual inherits one allele from each parent and the trait is determined by the particular allelic pair (A, A), (A, a), (a, A), or (a, a).

Individuals with allelic pairs (A, A) or (a, a) are called **homozygotes**. Individuals with allelic pairs (A, a) or (a, A) are called **heterozygotes**. In this section we will consider traits in which the order of the alleles on the chromosome is unimportant in determining the trait. Thus, we will always list the allelic pair for heterozygotes as (A, a), meaning that they have exactly one allele of each type. These three types are called the **genotypes** of the individuals. In the rest of this section, "gene" will mean "gene A".

Let us initially assume that mating among our species is random and there are no outside forces, such as mutation, acting on the species. We also assume that the fitness of the three genotypes is the same, that is, none of the genotypes has an environmental advantage.

Suppose that the size of the initial population, which we will call generation 0, is T. Since each individual has two genes, there are $2T$ genes in our initial population. Of these genes, assume that the proportion that are alleles A is $p(0)$, and that the rest are alleles *a*. Note that the proportion that are *a* must be $1 - p(0)$ which, for simplicity, we will call $q(0)$. Therefore

$$p(0) + q(0) = 1.$$

For simplicity, we will also assume that the proportion of A-alleles is the same for males as it is for females.

The total number of alleles of the form A is the total number of genes $2T$ times the proportion $p(0)$ of alleles of the form A, that is, there are $2p(0)T$ A-alleles. Likewise, the total number of a-alleles is $2q(0)T$.

An individual in the next generation, that is, generation 1, receives two genes, one from each parent. Since mating is random, it is equally likely that the individual will be the offspring of any two parents, and so the alleles that this individual receives are essentially drawn at random from the total pool of genes.

To model this selection process, consider two bowls, one containing all of the females' genes and the other containing all of the males' genes. If the females have a total of W genes, then $p(0)W$ of them are A-alleles and $q(0)W$ are a-alleles. If the males have a total of M genes, then $p(0)M$ of them are A and $q(0)M$ are a. The number of ways we can draw a gene from the females' bowl is W. For each of these ways, there are M ways to draw a gene from the males' bowl. Thus, there are, by the multiplication principle, WM equally likely ways to draw an individual's two genes, one for each sex.

The number of ways in which an individual can be an A-homozygote is the number of ways $p(0)W$ in which an A-allele can be drawn from the females' bowl times the number of ways $p(0)M$ an A-allele can be drawn from the males' bowl, that is, $p^2(0)WM$. Thus the proportion of A-homozygotes in generation 1, which we will call u, is the number of ways of getting two A-alleles divided by the total number of ways of selecting two genes, that is,

$$u = \frac{p^2(0)MW}{MW} = p^2(0).$$

Similarly, the proportion, w, of a-homozygotes is

$$w = q^2(0).$$

The number of ways in which an individual can get an A-allele from a female and an a-allele from a male is $p(0)Wq(0)M$. The number of ways in which an individual can get an a-allele from a female and an A-allele from a male is $q(0)Wp(0)M$. The number of ways of getting an A-allele from one bowl and an a-allele from the other is the sum of these two totals, that is, $2p(0)\,q(0)\,MW$. So the proportion of heterozygotes is

$$v = \frac{2p(0)\,q(0)\,MW}{MW} = 2p(0)\,q(0).$$

EXAMPLE 2.9   Suppose that gene A determines the color of a particular kind of flower and that it comes in two forms, allele A which gives these flowers the color red, and allele a which gives these flowers the color white. Each flower has a pair of color genes. Suppose that initially, the proportion of A-alleles is $p(0) = 0.3$, and the proportion of a-alleles is $q(0) = 0.7$. Then the proportion of flowers of each genotype in the next generation is

$$u = (0.3)^2 = 0.09, \qquad v = 2(0.3)\,(0.7) = 0.42, \qquad w = (0.7)^2 = 0.49.$$

Notice that $u + v + w = 1$. This is always true, since each individual must be one of these three types.

As an aside, if A is a dominant allele and a is a recessive allele, that is, if the heterozygotes are also red, then there are two **phenotypes**, red and white. In this case, the proportion of red flowers is the proportion of (dominant) A-homozygotes plus the proportion of heterozygotes, that is, 0.51 of the flowers. Then 49 per cent of the flowers will be white.

In this example, suppose that the total number of flowers in generation 1 is T. Since each individual has two genes, the total number of genes in generation 1 is $2T$. The total number of plants of each genotype is

$$U = 0.09T, \qquad V = 0.42T, \qquad W = 0.49T,$$

respectively. Since each A-homozygote has two A-alleles, and each heterozygote has one A-allele, the total number of A-alleles in generation 1 is

$$2U + V = 0.18T + 0.42T = 0.6T.$$

To get $p(1)$, the proportion of A-alleles in generation 1, we divide the total number of A-alleles by the total number of genes to get

$$p(1) = \frac{0.6T}{2T} = 0.3.$$

Thus, $q(1) = 1 - p(1) = 0.7$.

Notice that the proportion of A- and a-alleles in generation 1 is the same as it was in generation 0. By induction, we see that the proportion of A- and a-alleles remains fixed at $p(n) = 0.3$ and $q(n) = 0.7$ for every generation n. Therefore, the proportions of A-homozygotes, heterozygotes, and a-homozygotes remain fixed at 0.09, 0.42, and 0.49, respectively, for every generation. Again, if A is dominant, then the per cent of red and white flowers remains fixed at 51 per cent and 49 per cent, respectively, from the first generation on. ∎

The above example is a demonstration of the Hardy–Weinberg law.

THEOREM 2.10   The Hardy–Weinberg law   *Suppose a parent population has random mating with an initial proportion, $p(0) = p$ of A-alleles, and $q(0) = q$ of a-alleles but the proportion of the three genotypes is unknown. If there are no outside influences such as mutation or selection due to fitness, then the proportions of A and a alleles are in equilibrium, that is, for all k,*

$$p(k) = p \qquad and \qquad q(k) = q.$$

The proportions of A-homozygotes, heterozygotes, and a-homozygotes will be

$$u = p^2, \qquad v = 2pq, \qquad w = q^2,$$

respectively, starting with generation 1. (It is unclear what u, v, and w are for generation 0.)

PROOF: Suppose $p(n)$ and $q(n) = 1 - p(n)$ are known. Let's write a first-order dynamical system for $p(n+1)$ in terms of $p(n)$. To do this, we first compute u, v, and w, the proportions for generation $n + 1$ of A-homozygotes, heterozygotes, and a-homozygotes, respectively.

Let T be the total number of individuals in generation $n + 1$. To be an A-homozygote, an individual must draw (inherit) two A-alleles from parents in generation n. Since it is unknown who these parents are, the individual is equally likely to inherit any one of the W females' genes and any one of the M males' genes for a total of WM ways of selecting two genes. To be an A-homozygote, one of the $p(n)W$ females' A-alleles and one of the $p(n)M$ males' A-alleles must be inherited. Thus, as in the above example, the proportion of A-homozygotes is

$$u = \frac{p(n)\,Wp(n)\,M}{WM} = p^2(n).$$

The total number of A-homozygotes in generation $n + 1$ will be the proportion of A-homozygotes, u, times the total number T of individuals in generation $n+1$. Since each of these individuals has two A-alleles, the total number of A-alleles contributed by A-homozygotes in generation $n + 1$ is

$$2Tu = 2Tp^2(n).$$

Likewise, the number of ways of getting a heterozygote in generation $n + 1$ is the number of ways of getting an A-allele from a female and an a-allele from a male, or an a-allele from a female and an A-allele from a male, that is, $2p(n)q(n)\,WM$. Thus, the proportion of heterozygotes is

$$v = \frac{2p(n)\,q(n)\,WM}{WM} = 2p(n)q(n).$$

Thus, there are $2p(n)q(n)\,T$ heterozygotes in generation $n + 1$, and since each of these individuals contributes one A-allele, the total number of A-alleles contributed by the heterozygotes is $2Tp(n)q(n)$.

The a-homozygotes contribute no A-alleles, so the total number of A-alleles in generation $n + 1$ is the total of the A-alleles computed above, that is,

$$2Tp^2(n) + 2Tp(n)q(n) = 2Tp(n)[p(n) + q(n)] = 2Tp(n)$$

since $p(n) + q(n) = 1$. The total number of genes in generation $n+1$ is $2T$, thus the proportion $p(n + 1)$ of A-alleles in generation $n + 1$ is the quotient of these two numbers, that is,

$$p(n + 1) = \frac{2Tp(n)}{2T} = p(n).$$

The particular solution is easily seen to be

$$p(k) = p(0)1^k = p(0).$$

Thus, $q(k) = 1 - p(k) = 1 - p(0) = q(0)$.

Knowing $p(n) = p(0)$ and $q(n) = q(0)$, substitution into the above formulas gives $u = p^2(0)$, $v = 2p(0)q(0)$, and $w = q^2(0)$, from the first generation on, which completes the proof. ∎

The above theorem implies that even when A is a dominant gene and a is a recessive gene, if there is no selective advantage to any of the genotypes, then the recessive trait will not tend to die out, but will remain fixed at $q^2(0)$ of the population. This goes counter to the intuition of many people. In fact, Hardy and Weinberg discovered this theorem independently in an effort to show that recessive traits need not die out, and in fact that the recessive trait may even be the most common trait.

Next we consider the effect of selection. We now assume that A is a dominant allele and a is a recessive allele. In this case, genotype (A, A) is called a dominant homozygote, while genotype (a, a) is called a recessive homozygote.

Suppose there **is** an outside factor affecting the recessive trait. In fact, let's consider a genetic defect that causes recessive homozygotes to be sterile or to die before reproductive age. Allele a is said to be a **lethal** allele and allele A is said to have a **selective advantage**.

EXAMPLE 2.11   Suppose we have a lethal recessive trait and that $p(0) = q(0) = 1/2$, that is, half the genes are A-alleles and half are a-alleles. By the same computations used to prove the Hardy–Weinberg law, we get that the proportions of the three genotypes in generation 1 are

$$u = p^2(0) = 0.25, \qquad v = 2p(0)q(0) = 0.5, \qquad w = q^2(0) = 0.25.$$

Suppose T individuals are born to this generation. Then the number of each genotype is

$$U = 0.25T, \qquad V = 0.5T, \qquad W = 0.25T.$$

For convenience, let $T = 4$. Then $U = 1$, $V = 2$, and $W = 1$. The recessive homozygote has a lethal trait and dies before reaching adulthood. Thus, we have 1 dominant homozygote and 2 heterozygotes as adults. There are a total of 6 genes among these 3 adults, and only 2 of them are a-alleles, one from each of the two heterozygotes. Thus,

$$q(1) = \frac{1}{3} \qquad \text{and so} \quad p(1) = 1 - q(1) = \frac{2}{3}.$$

We now have that the proportions of the three genotypes in generation 2 are

$$u = p^2(1) = \frac{4}{9}, \qquad v = 2p(1)q(1) = \frac{4}{9}, \qquad w = q^2(1) = \frac{1}{9}.$$

For convenience, let the number of individuals born to this generation be $T = 9$. Then $U = 4$, $V = 4$, and $W = 1$. The recessive homozygote has a lethal trait and dies before reaching adulthood. Thus, we have 4 dominant homozygotes and 4 heterozygotes as adults. There are a total of 16 genes among these 8 adults, and only 4 of them are a-alleles, one from each of the four heterozygotes. Thus,

$$q(2) = \frac{4}{16} = \frac{1}{4} \quad \text{and so} \quad p(2) = 1 - q(2) = \frac{3}{4}.$$

Note that $q(0) = 1/2$, $q(1) = 1/3$, and $q(2) = 1/4$. It appears the denominator of the $q(n)$'s is increasing by 1, that is, if we define $Q(n) = 1/q(n)$, then $Q(0) = 2$, $Q(1) = 3$, $Q(2) = 4$, and

$$Q(n+1) = Q(n) + 1.$$

Thus, $Q(k) = Q(0) + 1 \times k = 2 + k$, and the fraction of a-alleles in generation k should be

$$q(k) = \frac{1}{Q(k)} = \frac{1}{k+2}.$$

In this case, the number of a-alleles is slowly going to zero and so the fraction of recessive homozygotes born to generation $k + 1$,

$$w = q^2(k) = \frac{1}{(k+2)^2},$$

is also going to zero.   ■

Let's investigate lethal traits in more detail. Again, let $p(n)$ and $q(n)$ be the proportion of A- and a-alleles in the reproductive population of generation n. As above, the proportions of the three genotypes in generation $n + 1$ are

$$u = p^2(n), \qquad v = 2p(n)q(n), \qquad w = q^2(n).$$

We let T be the total number of individuals born into generation $n + 1$.

What we would like to do is to compute the proportion of recessive a-alleles in the adult (reproductive) population of generation $n + 1$. To do this, we must compute the total number of **adults** in generation $n + 1$. Our assumption is that all of the dominant homozygotes and heterozygotes reach reproductive age or adulthood, while **none of the recessive homozygotes reaches reproductive age**. Thus, the total number of individuals that reach adulthood in generation $n + 1$ is

$$T(u + v) = T[p^2(n) + 2p(n)q(n)]$$
$$= Tp(n)[p(n) + 2q(n)] = Tp(n)[1 + q(n)],$$

since $p(n) = 1 - q(n)$. Since each of these individuals has two genes, the total number of genes in the adult population of generation $n+1$ is

$$2Tp(n)[1 + q(n)].$$

Since none of the dominant homozygotes have an a-allele, each heterozygote has one a-allele, and none of the recessive homozygotes reaches adulthood, the total number of a-alleles in the adult population of generation $n + 1$ equals the number of heterozygotes in that generation, that is,

$$Tv = T2p(n)q(n).$$

Since the proportion $q(n + 1)$ of a-alleles in the adult population of generation $n + 1$ is the ratio of the number of a-alleles to the total number of genes, we have

$$q(n + 1) = \frac{2Tp(n)q(n)}{2Tp(n)[1 + q(n)]} = \frac{q(n)}{1 + q(n)}. \tag{2.5}$$

This is a first-order nonlinear dynamical system.

In example 2.11, it was seen that the denominator of $q(n)$ satisfied a simple dynamical system, in particular, each denominator was one more than the previous one. Maybe it will help to look at the denominator in this case also. To do this, let's make the substitution

$$q(k) = \frac{1}{Q(k)} \tag{2.6}$$

Rewriting dynamical system (2.5) by substituting $n + 1$ and n, respectively for k in equation (2.6) gives

$$\frac{1}{Q(n + 1)} = \frac{1/Q(n)}{1 + 1/Q(n)}.$$

Simplifying the right hand side of this equation gives

$$\frac{1}{Q(n + 1)} = \frac{1}{1 + Q(n)}.$$

If reciprocals are equal, then the numbers are equal. Thus we have the first-order affine dynamical system (where $r = 1$)

$$Q(n + 1) = Q(n) + 1.$$

The particular solution is then

$$Q(k) = k + Q(0).$$

Thus, substituting $1/q(k)$ back for $Q(k)$ gives that

$$q(k) = \frac{1}{k + Q(0)} = \frac{1}{k + 1/q(0)} = \frac{q(0)}{kq(0) + 1}$$

is the particular solution to dynamical system (2.5). Let's summarize.

THEOREM 2.12 Principle of selection *Suppose that individuals with a lethal recessive trait determined by allele a do not reproduce, but that dominant homozygotes and heterozygotes reproduce normally. Then the proportion of a-alleles in the population in generation k is given by*

$$q(k) = \frac{q_0}{kq_0 + 1},$$

where q_0 is the initial proportion of a-alleles.

Notice that $q(k)$ goes to zero as k goes to infinity. But it goes to zero very slowly. Suppose there is a lethal trait, or, as was advocated in the early 1900s, negative eugenics (in which all people with a certain undesirable trait are sterilized) is practiced. Since the trait is deleterious, we will assume that, initially, the proportion of these a-alleles in the population is small, say $q(0) = 0.02$. Notice that, after one generation, the proportion is $q(1) = 0.02/(0.02 + 1) = 1/51$, which is not much of a change.

How long will it take to cut the initial proportion in half, that is, for what value of k is $q(k) = 0.01$? To answer this question, we need to solve

$$0.01 = q(k) = \frac{q(0)}{q(0)k + 1} = \frac{0.02}{0.02k + 1}$$

for k. This equation simplifies to

$$0.02k + 1 = \frac{0.02}{0.01} = 2, \qquad \text{or} \qquad k = 50.$$

Thus, it takes 50 generations (which would be approximately 1250 years for humans) for the proportion of a-alleles to be cut in half, that is, to be cut to

$$q(50) = \frac{0.02}{50(0.02) + 1} = 0.01.$$

Similarly, it will take 100 more generations to cut this proportion in half again, that is, to 0.005. (We note that these are only predictions, and that reality will vary somewhat from these numbers.)

A Supreme Court decision in the first part of the twentieth century allowed a mentally retarded woman to be sterilized so that she wouldn't produce retarded children. Assuming retardation to be a recessive genetic trait (a poor assumption), we see that sterilizing all such people would not significantly reduce this trait in society. This slow reduction in the proportion of deleterious alleles explains why negative eugenics would be ineffective for humans.

Next we will consider the effects of mutation with no selective advantage. We do not assume that either allele is dominant, but we do assume that all genotypes have the same fitness.

In mutation, a certain proportion of the genes spontaneously change from A-alleles to a-alleles or vice versa. Let $p'(n)$ and $q'(n)$ be the proportions of A- and a-alleles, respectively, in generation n just before mutation. Let $p(n)$ and $q(n)$ be the proportions of A- and a-alleles, respectively, in generation n just after mutation.

To simplify matters, we will consider the following series of events. We know that the proportions of A- and a-alleles in generation n just prior to reproduction are $p(n)$ and $q(n)$, respectively. We then draw genes for generation $n+1$ as before. Thus, the proportions of A- and a-alleles in the children of generation $n+1$ but before mutation has taken place are, by the Hardy–Weinberg law, $p'(n+1) = p(n)$ and $q'(n+1) = q(n)$.

Then we assume that a certain proportion μ of the A-alleles mutate to a-alleles. (Alternatively, we could have assumed the genes mutated first, then the children were born.) If a proportion μ of the A-alleles mutate to a-alleles, then the proportion of A-alleles in generation $n+1$ after mutation is

$$p(n+1) = p'(n+1) - \mu p'(n+1) = p(n) - \mu p(n) = (1-\mu)p(n).$$

We know that the solution to this linear dynamical system is

$$p(k) = (1-\mu)^k p(0).$$

Thus we get the following theorem.

THEOREM 2.13 Principle of mutation   *Suppose that the proportion of alleles that mutate from A to a is μ. In the absence of other influences, the proportion of A-alleles present in generation k is*

$$p(k) = (1-\mu)^k p(0),$$

and thus goes to zero.

Usually the mutation rate is small, say $\mu \approx 10^{-5}$. This means that $p(k)$ goes to zero slowly. In fact, with μ as above, after 10 000 generations,

$$p(10\,000) = 0.9048p(0).$$

Thus, you can see that even after long periods of time, the proportion of A-alleles has decreased by only a small amount. This and the previous example of selection help explain why evolution sometimes takes so long. If we observe certain species for a short (say several hundred years) period of time, the characteristics of that species will not appear to be changing. In these cases it may take thousands of years for the effects of evolution to have an observable affect.

The estimates for the mutation rate tend to be accurate to only one significant digit and are useful for giving a magnitude of time and not an exact number of years or an exact $p(k)$ value.

For some species the effects of evolution may be much quicker, as in the rapid evolution of resistant bacteria. One explanation for this is the short time span between generations. There are other reasons, but these are not relevant to this discussion.

PROBLEMS

1. Suppose that a certain bug has either red or black eyes, determined by a pair of genes. Suppose that the dominant allele A gives red eyes while the recessive allele a gives black eyes.
   a. Suppose that, initially, $p(0) = 0.4$ and $q(0) = 0.6$. Use the Hardy–Weinberg law to find the proportion of dominant homozygotes, heterozygotes, and recessive homozygotes in generation n, where $n > 0$.
   b. Suppose we observe that 16 per cent of the bugs have black eyes while 84 per cent have red eyes. What proportion, $q(n)$, of the genes are a-alleles and what proportion of the individuals are dominant homozygotes?
   c. Suppose that, initially, $p(0) = 0.2$ and $q(0) = 0.8$. Suppose that 10 000 bugs are born to this generation of bugs. How many dominant homozygotes, heterozygotes, and recessive homozygotes do you expect in this generation? How many A-alleles do you then expect in generation 1? From this, what proportion of the genes in this generation are A-alleles?

2. Assume that alleles A and a are additive in the sense that A- and a-homozygotes appear to be identical, but the heterozygotes are different from the homozygotes. Suppose the proportion of heterozygotes is 0.32 while the proportion of the homozygotes (combined) is 0.68. Determine the proportion of A- and a-alleles in the population.

3. Suppose that we have a trait in which the recessive homozygotes do not reach adulthood so that the **principle of selection** applies. Suppose that, initially, $q(0) = 0.06$, that is 6 per cent of the genes are of type a.
   a. How many generations will it take to cut the proportion to one-fourth of the initial amount?
   b. How many generations will it take to cut the proportion of a-alleles to 1 per cent?
   c. How many generations will it take to cut the proportion of recessive homozygotes to 0.09 per cent of the population?

4. Suppose that the mutation rate from A-alleles to a-alleles is $\mu = 0.0001$. Also suppose that $p(0) = 0.9$. How many generations will it take so that $p(k) = 0.5$?

5. Suppose μ of the genes mutate from A- alleles to a-alleles and that ν of the genes mutate from a-alleles to A-alleles. Develop a first-order affine dynamical system for $p(n)$, the proportion of A-alleles in generation n, and find the equilibrium value for this dynamical system.

6. Consider the dynamical system

$$p(n+1) = \frac{p(n) + a^2}{1 + p(n)}.$$

Use the substitution

$$p(k) = a + \frac{1}{P(k)},$$

which is similar to the substitution (2.6) used to study selection. Then rewrite this dynamical system as a first-order affine dynamical system in terms of $P(n)$ and solve. Substitute back to find the general solution for $p(k)$.

2.4   Solution to affine dynamical systems with applications

We now turn our attention to the analysis of first-order affine dynamical systems, that is, equations of the form

$$a(n+1) = ra(n) + b,$$

where $r \neq 1$ and $b \neq 0$.

EXAMPLE 2.14   Let's consider a savings account that receives $100I$ per cent yearly interest rate, compounded m times a year. To make this account more realistic, let's assume that at the beginning of each compounding period, we make an additional deposit of b dollars. Let $a(n)$ represent the amount we have in our account after n compounding periods, and immediately following our deposit of b dollars.

For example, suppose we make an initial deposit of $a_0 = 1000$ dollars, and we collect 8 per cent interest, compounded quarterly. At the end of each compounding period, we make an additional deposit of 100 dollars. Thus, the amount we have after the $n+1$th quarter $a(n+1)$ is our principal from the nth quarter $a(n)$, the quarterly interest on our principal $0.02a(n)$, plus the 100 dollar deposit we make. Notice that the deposit is made at the end of the quarter, so it does not receive any interest yet. In summary,

$$a(n+1) = 1.02a(n) + 100,$$

with $a_0 = 1000$. We wish to find the particular solution to this dynamical system. (You could use this equation and spreadsheets or a calculator to compute what you have at the beginning of, say, the second year, that is, $a(4)$.)   ∎

Our goal now is to find the general and the particular solution to first-order affine dynamical systems, that is, to dynamical systems of the form

$$a(n+1) = ra(n) + b, \tag{2.7}$$

where r and b are two constants. First observe that the equilibrium value for dynamical system (2.7) is

$$a = \frac{b}{1-r}.$$

The trick is to measure how far $a(n)$ is from equilibrium. If $e(k) = a(k) - a$, then $e(k)$ measures how far $a(k)$ is from the equilibrium value after the kth iteration. We then use this to make the substitution

$$a(k) = e(k) + a = e(k) + \frac{b}{1-r}$$

into dynamical system (2.7). Substituting $a(n+1) = e(n+1) + b/(1-r)$ and $a(n) = e(n) + b/(1-r)$ into dynamical system (2.7) gives

$$e(n+1) + \frac{b}{1-r} = r\left(e(n) + \frac{b}{1-r}\right) + b,$$

or, after simplifying,

$$e(n+1) = re(n).$$

We know that the general solution to this linear dynamical system is $e(k) = cr^k$, so substituting back, we get that

$$a(k) = e(k) + \frac{b}{1-r} = cr^k + \frac{b}{1-r},$$

is the general solution to dynamical system (2.7) when $r \neq 1$.

What is the particular solution when $a(0) = a_0$ is given, that is, what is c in terms of r, b, and a_0? Substitution of $k = 0$ into the general solution gives

$$a_0 = a(0) = cr^0 + \frac{b}{1-r} = c + \frac{b}{1-r}.$$

Equating the two forms of $a(0)$ and solving for the unknown c gives

$$c = a_0 - \frac{b}{1-r}.$$

So, the particular solution to the first-order affine dynamical system (2.7), when $r \neq 1$, is

$$a(k) = r^k\left(a_0 - \frac{b}{1-r}\right) + \frac{b}{1-r}.$$

EXAMPLE 2.15   Let's consider the dynamical system

$$a(n+1) = 2a(n) + 5$$

with initial value $a_0 = 3$. We compute $b/(1-r) = 5/(1-2) = -5$. Thus the general solution to this dynamical system is

$$a(k) = c2^k - 5.$$

Since $a_0 - b/(1-r) = 3 - (-5) = 8$, the particular solution is

$$a(k) = 8(2^k) - 5.$$

We can use this solution to find $a(k)$ at any future point in time, say $a(5) = 8 \times 2^5 - 5 = 251$. Note that this is quicker than iterating the dynamical system five times.

Observe that if $a(3) = 4$ is given, then we can still use the general solution to find the particular solution. In this case, we have $4 = a(3) = c2^3 - 5$ or $8c - 5 = 4$, so that $c = 9/8$. The particular solution is

$$a(k) = (9/8) \times 2^k - 5 = 9 \times 2^{k-3} - 5. \qquad \blacksquare$$

We summarize our results as a theorem.

THEOREM 2.16   *Consider the first-order affine dynamical system*

$$a(n+1) = ra(n) + b.$$

For simplicity, let $a = b/(1-r)$. Then the general solution to this dynamical system is

$$a(k) = c\,r^k + a, \qquad if \quad r \neq 1,$$

and

$$a(k) = b\,k + c, \qquad if \quad r = 1.$$

If $a(0) = a_0$ is given, then the particular solution is

$$a(k) = (a_0 - a)\,r^k + a, \qquad if \quad r \neq 1,$$

and

$$a(k) = b\,k + a_0, \qquad if \quad r = 1.$$

Notice that when $r \neq 1$ and $a_0 = a$, the equilibrium value, then the particular solution to the first-order affine dynamical system is the constant solution

$$a(k) = \frac{b}{1-r}.$$

This should come as no surprise since the initial value equals the equilibrium value.

EXAMPLE 2.17   Often, one is not interested in the solution per se, but in the long term behavior of the solution, that is,

$$\lim_{k\to\infty} a(k).$$

For example, consider the finite geometric series

$$a(k) = 1 + (-0.5) + (-0.5)^2 + \cdots + (-0.5)^k.$$

One problem is to find the sum of the corresponding infinite series; that is,

$$\lim_{k\to\infty} a(k).$$

By substitution of $n+1$ for k,

$$a(n+1) = 1 + (-0.5) + \cdots + (-0.5)^{n+1}.$$

Factoring the right side of this equation gives

$$a(n+1) = 1 + (-0.5)[1 + \cdots + (-0.5)^n].$$

Noting that the term in brackets equals $a(n)$, we now have that the finite geometric series satisfies the dynamical system

$$a(n+1) = -0.5a(n) + 1. \tag{2.8}$$

The equilibrium value is $1/(1+0.5) = 2/3$. Thus, by theorem 2.16, the general solution is

$$a(k) = c(-0.5)^k + \frac{2}{3}.$$

Since $a(0) = 1$, then $c = 1 - 2/3 = 1/3$ and the particular solution is

$$a(k) = 1 + (-0.5) + \cdots + (-0.5)^k = \frac{2 + (-0.5)^k}{3}.$$

Since $(-0.5)^k$ goes to zero as k goes to infinity,

$$\lim_{k\to\infty} a(k) = 1 + (-0.5) + (-0.5)^2 + \cdots = \frac{2}{3}. \qquad \blacksquare$$

The techniques of example 2.17 can be applied to sum any finite or infinite geometric series. Let

$$a(k) = 1 + r + r^2 + \cdots + r^k. \tag{2.9}$$

Let's find a simple formula for this finite sum and also a formula for what the infinite sum

$$1 + r + r^2 + \cdots$$

adds to. Note that by equation (2.9),

$$a(n+1) = 1 + r + \cdots + r^{n+1}$$

$$= 1 + r[1 + r + \cdots + r^n] = 1 + ra(n).$$

The equilibrium value for this dynamical system is $a = 1/(1-r)$, so the general solution is

$$a(k) = cr^k + \frac{1}{1-r}.$$

Since $a(0) = 1$, then $c = 1 - 1/(1-r) = -r/(1-r)$ and the particular solution is, after simplification

$$a(k) = \frac{1 - r^{k+1}}{1 - r}.$$

If $|r| < 1$, then r^k goes to zero as k goes to infinity, and so

$$\lim_{k \to \infty} a(k) = 1 + r + r^2 + \cdots = \frac{1}{1-r}.$$

2.4.1   A model of an arms race

We will now develop a dynamical model of a two-nation arms race. A variation of this model was originally developed by Lewis Fry Richardson in the 1930s. Let's assume that we have two countries, A and B. Let $a(n)$ and $b(n)$ represent the amount (in some common monetary unit) spent on armaments (defense?) by the corresponding countries in year n. Also assume that each country has a fixed amount of distrust of the other country, causing it to retain arms. We will now develop equations that relate the amount each country spends on arms in one year in terms of what they **both** spent the previous year.

First look at the **increase** in expenditures by country A, that is, $a(n+1) - a(n)$. We wish to have an equation that takes into account that if B spends a lot on defense in one year, then A spends more on defense the next year. Also, since large expenditures will deplete a country's treasury, the equation should indicate that large expenditures by A one year will cause smaller expenditures the next year. A simple equation that satisfies the assumptions above is the dynamical system

$$a(n+1) - a(n) = -ra(n) + sb(n) + c_1,$$

where r, s, and c_1 are given constants, with r and s both being positive. The constant s measures country A's distrust of country B in that it reacts to the way B arms itself. The constant r is a measure of country A's own economy.

Similar assumptions for country B lead to the dynamical system

$$b(n+1) - b(n) = -Rb(n) + Sa(n) + c_2,$$

with S measuring B's distrust of A and R measuring country B's economy.

We now have a dynamical system of two equations, which we will study in more detail later in this text. For the present, we will make simplifying assumptions. Assume that the two countries have an equal amount of distrust of each other, that is, $s = S$. Also assume that the two countries' economies are about the same, that is, $r = R$. We now have

$$a(n+1) - a(n) = -ra(n) + sb(n) + c_1$$

and

$$b(n+1) - b(n) = -rb(n) + sa(n) + c_2.$$

Add the two equations, giving

$$[a(n+1) + b(n+1)] - [a(n) + b(n)] = (-r+s)[a(n) + b(n)] + c,$$

where $c = c_1 + c_2$.

The idea is to look at the total expenditures of the two countries, that is, $t(n) = a(n) + b(n)$. Making this substitution in the previous equation gives $t(n+1) - t(n) = (s-r)t(n) + c$ or

$$t(n+1) = (1 - r + s)t(n) + c.$$

The equilibrium value for this first-order affine dynamical system is $c/(r-s)$, and the particular solution is

$$t(k) = (1 - r + s)^k \left(t_0 - \frac{c}{r-s} \right) + \frac{c}{r-s}, \qquad (2.10)$$

where $t_0 = t(0)$. We know that if $-1 < 1 - r + s < 1$, that is, if $-2 < -r + s < 0$, then the equilibrium value $c/(r-s)$ is stable.

For $-r + s < -2$, we would need for $r > 2$. But r is the fraction by which a country cuts its own expenditures on arms. For $r > 2$ a country would cut its expenditures by more than it is spending, which is impossible. Therefore, the left side of the above equations, $-2 < -r + s$ is clearly true.

Looking at the right side of this inequality, we see that if $s < r$, then $t(k)$ approaches $c/(r-s)$, and we have a **stable arms race**. In other words, if a country's restraint r caused by its economy is greater then its distrust s of the other country, there will eventually be a constant expenditure on arms each year.

But if the distrust is too great, then $s > r$ and $(1 - r + s)^k$ goes to positive infinity. In the solution (2.10), if $t_0 - c/(r - s) < 0$, then $t(k)$ goes to negative infinity (or zero, which comes first); but if $t_0 - c/(r - s) > 0$, then $t(k)$ goes to positive infinity. The conclusion in this case is that if the initial total expenditure t_0 of the two countries is small enough (less than $c/(r - s)$), and the countries have a large amount of mutual distrust ($s > r$), then the arms race will die out. But if t_0 is large, then the arms race will escalate. Since no two countries can sustain exponentially increasing expenditures on arms, the alternative is war or negotiations.

EXAMPLE 2.18 Before World War I, there were two alliances, France and Russia being one, and Germany and Austria-Hungary being the other. Thus, it can be argued that the total expenditures on arms by these two alliances satisfies a dynamical system of the form

$$t(n + 1) = (1 - r + s) t(n) + c.$$

In Richardson (1960), we find that the estimated total expenditures of these two alliances were (in millions of pounds sterling): 199 in 1909, 205 in 1910, and 215 in year 1911. Letting 1909 be year 0, we have $t(0) = 199$, $t(1) = 205$, and $t(2) = 215$. From the dynamical system, we have

$$t(1) = (1 - r + s) t(0) + c \qquad \text{and} \qquad t(2) = (1 - r + s) t(1) + c.$$

Letting $1 - r + s = x$, and substituting for $t(0)$, $t(1)$, and $t(2)$ gives the two equations

$$205 = 199x + c \qquad \text{and} \qquad 215 = 205x + c.$$

Solving these two equations gives $x = 5/3$ and $c = -380/3$. Thus, the dynamical system becomes

$$t(n + 1) = \frac{5}{3} t(n) - \frac{380}{3}$$

and the solution is

$$t(k) = (\frac{5}{3})^k (t_0 - 190) + 190.$$

The values we computed for x and c are only approximate. There are techniques for finding better estimates of x and c by using more years, and in some sense, averaging all the x and c values we find. But this is a different topic.

Note that $1 - r + s = 5/3$, so that $s - r = 2/3$. Although we do not know r or s, we do know that $s > r$. From the solution $t(k)$, we see that if $t_0 < 190$, then the countries would have disarmed. But, since $t_0 = 199$, $t(k)$ grew exponentially leading to an accelerating arms race, which led to **war**. This, Richardson claimed, could be one of the causes of World War I.   ■

While this is an over-simplification of the causes of war, it could help us understand the reasons behind escalations in arms races and might possibly help in ending them. Because of satellite surveillance, it is now easier to determine if a country is living up to the terms of an agreement. In this sense, countries can trust each other more, decreasing the value of s to a point where it is less than r. This could help end an arms race.

2.4.2   More on genetics

We have seen that selection by itself causes a trait to die out. Similarly, mutation in one direction also causes a trait to die out. Why then are there traits in which the proportions of A- and a-alleles are both relatively large?

Let's consider two possible explanations. First, selection may act in one direction while mutation acts in the other direction. This will be discussed later in this section. A second possible explanation is that mutation occurs in both directions, that is, μ of the A-alleles mutate to a-alleles and ν of the a-alleles mutate to A-alleles. As in section 2.3, $p(n)$ and $q(n)$ will represent the proportions of A- and a-alleles, respectively, in the adults of generation n. By the Hardy–Weinberg law, the proportion of A- and a-alleles among the children of generation $n + 1$ will also be $p(n)$ and $q(n)$, respectively. Now mutation from a to A will increase the proportion of A-alleles by $\nu q(n)$, while mutation from A to a will decrease the proportion of A-alleles by $\mu p(n)$. This gives

$$p(n+1) = p(n) - \mu p(n) + \nu q(n) = (1 - \mu)p(n) + \nu q(n).$$

Since $q(n) + p(n) = 1$, we have

$$p(n+1) = (1 - \mu)p(n) + \nu[1 - p(n)] = (1 - \mu - \nu)p(n) + \nu.$$

The general solution to this dynamical system, with $r = (1 - \mu - \nu)$ and $b = \nu$, is

$$p(k) = c(1 - \mu - \nu)^k + \frac{\nu}{\mu + \nu}$$

after simplification. Since μ and ν are small, $|1 - \mu - \nu| < 1$ so that

$$\lim_{k \to \infty} p(k) = \frac{\nu}{\mu + \nu}.$$

To rephrase, the proportion of A-alleles in the population should be close to the equilibrium value, $p = \nu/(\mu + \nu)$. Therefore the proportion of a-alleles should be close to $q = 1 - p = \mu/(\mu + \nu)$.

EXAMPLE 2.19   Suppose that A is a dominant allele. Then the proportion of recessive homozygotes equals q^2. In this case, it is possible to estimate the proportion of recessive homozygotes, and suppose this equals 0.04. Then the proportion of a-alleles is $q = \sqrt{0.04} = 0.2$ and the proportion of A-alleles is then $p = 0.8$.

Suppose that neither trait has a selective advantage, and that A-alleles mutate to a-alleles and vice versa. Then

$$\frac{p}{q} = \frac{0.8}{0.2} = 4.$$

But

$$\frac{p}{q} = \frac{\nu/(\mu + \nu)}{\mu/(\mu + \nu)} = \frac{\nu}{\mu}.$$

Thus, while we cannot compute ν and μ, we do know the ratio $\nu/\mu = 4$, that is, alleles mutate from a to A at four times the rate that they mutate from A to a. ∎

Now we will study the combined effects of mutation and selection. **Galactosaemia is a lethal genetic disease** caused by a lethal recessive allele, a. It also happens that normal, dominant A-alleles mutate to a-alleles at the rate μ. We will model this disease.

Let $p(n)$ and $q(n)$ be the proportions of A- and a-alleles, respectively, in adults of generation n after mutation. Let u, v, and w be the proportions of dominant homozygotes, heterozygotes, and recessive homozygotes, respectively, among the children of generation $n + 1$. As before

$$u = p^2(n), \qquad v = 2p(n)q(n), \qquad w = q^2(n).$$

Let T be the total number of children born to generation $n + 1$. Since all of the recessive homozygotes die before reaching adulthood, the total number of adult dominant homozygotes, heterozygotes, and recessive homozygotes are

$$U = Tp^2(n), \qquad V = T2p(n)q(n), \qquad W = 0.$$

Thus, the proportion of a-alleles in the adults before mutation is

$$\frac{V + 2W}{2U + 2V + 2W} = \frac{T2p(n)q(n)}{2Tp^2(n) + 4Tp(n)q(n)}$$

$$= \frac{q(n)}{p(n) + 2q(n)} = \frac{q(n)}{1 + q(n)}$$

as before. Since the proportion of A- and a-alleles must add to one, the proportion of A-alleles among the adults before mutation must be

$$\frac{1}{1 + q(n)}.$$

Now μ of the A-alleles mutate to a-alleles, giving

$$q(n + 1) = \frac{q(n)}{1 + q(n)} + \mu \frac{1}{1 + q(n)},$$

or

$$q(n + 1) = \frac{q(n) + \mu}{1 + q(n)}. \qquad (2.11)$$

We use a substitution in this nonlinear dynamical system that is similar to substitution (2.6), used in section 2.3 when studying selection. First, find an equilibrium value for this dynamical system, that is, find a number q such that $q(k) = q$ satisfies the equation. The number q can be found by substituting q for $q(n+1)$ and $q(n)$ giving

$$q = \frac{q + \mu}{1 + q}, \qquad \text{or} \qquad q^2 = \mu.$$

We will use the root $q = -\sqrt{\mu}$ although either root will work.
    Next, make the substitution

$$q(n) = q + \frac{1}{a(n)} = -\sqrt{\mu} + \frac{1}{a(n)} = \frac{-\sqrt{\mu}a(n) + 1}{a(n)}$$

into dynamical system (2.11). Collecting all terms except $1/a(n+1)$ on the right, finding a common denominator, and simplifying gives the first-order affine dynamical system

$$a(n+1) = \left(\frac{1 - \sqrt{\mu}}{1 + \sqrt{\mu}} \right) a(n) + \frac{1}{1 + \sqrt{\mu}}.$$

With

$$r = \left(\frac{1 - \sqrt{\mu}}{1 + \sqrt{\mu}} \right) \qquad \text{and} \qquad b = \frac{1}{1 + \sqrt{\mu}},$$

we get the general solution

$$a(k) = c \left(\frac{1 - \sqrt{\mu}}{1 + \sqrt{\mu}} \right)^k + \frac{1}{2\sqrt{\mu}},$$

after simplification. Again, c depends on $q(0) = q_0$.

Since

$$\left| \frac{1 - \sqrt{\mu}}{1 + \sqrt{\mu}} \right| < 1,$$

it follows that

$$\lim_{k \to \infty} a(k) = \frac{1}{2\sqrt{\mu}}.$$

Thus,

$$\lim_{k \to \infty} q(k) = -\sqrt{\mu} + \lim_{k \to \infty} \frac{1}{a(k)} = -\sqrt{\mu} + 2\sqrt{\mu} = \sqrt{\mu},$$

and so

$$q = \lim_{k \to \infty} q(k) = \sqrt{\mu}$$

is the stable equilibrium proportion of a-alleles when balancing selection and mutation.

The above analysis tells us that the proportion of recessive homozygotes should be approximately $w = q^2 = \mu$. It has been observed that the frequency of galactosaemia in the general population is $w = 5.6 \times 10^{-5}$, thus, this should also be (approximately) the mutation rate.

PROBLEMS

1. For the following dynamical systems, find $a(1)$ through $a(5)$, give the particular solution, describe the long term behavior (that is, increasing to infinity, oscillating to zero, etc.), and draw a graph of the points $(k, a(k))$.

   a. $a(n+1) = 0.8a(n) - 1, \quad a(0) = 3$
   b. $a(n+1) = -a(n) + 3, \quad a(0) = 5$
   c. $a(n+1) = -1.5a(n) + 5, \quad a(0) = 3$

2. For the following dynamical systems, give the general solution and the equilibrium value, if any.
   a. $a(n+1) = a(n) + 3$
   b. $a(n+1) = 1.5a(n) - 5$
   c. $a(n+1) = -0.9a(n) + 3.8$

3. For the dynamical systems in problem 2, give the particular solution for $a(0) = 4$.

4. For the dynamical systems in problem 2, give the particular solution for $a(2) = 0$.

5. For what value of r is the equilibrium value of

$$a(n+1) = ra(n) + 2r$$

equal to 2?

6. Consider the dynamical system $a(n+1) = 3a(n) - 4$.
   a. Given that $a(0) = 3$, for what value of k does $a(k) = 245$?
   b. Given that $a(6) = -1456$, find $a(0)$.

7. Find the general solution to the dynamical system

$$3a(n+1) + 1 = 2a(n) - 4.$$

8. Find the general solution to the dynamical system

$$2a(n+1) - 3a(n) = 8.$$

9. Suppose an arms race between two countries is given by the dynamical system

$$t(n+1) = 1.2t(n) - 40,$$

where $t(n)$ is the total amount spent by the two countries in year n. The arms race will die out if the initial total arms expenditures is below what number?

10. Suppose the arms race between two countries is described by the equations

$$a(n+1) - a(n) = -2a(n) + 1.5b(n) + 30$$
$$b(n+1) - b(n) = 1.5a(n) - 2b(n) + 70.$$

What are the long range projections for the total expenditures by the two countries in this arms race?

11. Suppose the total expenditures by two countries are 75, 80, and 84 billion dollars in three consecutive years. Derive the dynamical system for $t(n)$ and make your predictions about this arms race.

12. Using methods similar to those in this section, find the general solution to the following dynamical systems, and find $\lim_{k \to \infty} a(k)$.
    a. $a(n+1) = 3a(n)/(2 - a(n))$
    b. $a(n+1) = (4 + 5a(n))/(-1 - 2a(n))$
    c. $a(n+1) = (-12 + 5a(n))/(-2 + a(n))$

13. Suppose that the mutation rate from A- to a-alleles is 10^{-4} and that the mutation rate from a to A-alleles is 4×10^{-5}. After a long period of time, what per cent of the genes are A, what per cent are a, and what per cent of the individuals exhibit the dominant trait?

14. Suppose that the mutation rate from A-alleles to a-alleles is 0.01, but that **all** recessive homozygotes die before reaching adulthood. Find a dynamical system for $p(n+1)$ in terms of $p(n)$, find the equilibrium value, and determine its stability by solving the dynamical system.

15. Suppose that there is a partially lethal trait in that only half of the recessive homozygotes die before reaching adulthood, while the dominant homozygotes and heterozygotes are normal. (The fitness of a genotype is the relative proportion of that genotype which survives to adulthood. In this problem, the fitness of the dominant homozygotes is 1, the fitness of the heterozygotes is 1, and the fitness of the recessive homozygotes is 0.5.)
    a. Assuming there is no mutation, develop a nonlinear dynamical system for $q(n)$, the proportion of a-alleles in the adult population. Find the equilibrium values for this dynamical system. You will not be able to solve this dynamical system.
    b. Assuming in addition that A-alleles mutate to a-alleles at the rate μ, again develop a nonlinear dynamical system for $q(n)$. Find the equilibrium values for this dynamical system.

2.5 Applications to finance

Consider this general situation. Suppose we have an account that contains $a(n)$ dollars after n compounding periods. Suppose that the account is collecting $100I$ per cent annual interest, compounded m times per year. In addition, assume a constant amount b is added to the account at the end of each compounding period, or taken from the account if $b < 0$. Let a_0 be the initial amount in the account, that is, $a(0) = a_0$. The dynamical system

$$a(n+1) = \left(1 + \frac{I}{m}\right) a(n) + b \tag{2.12}$$

describes the relationship between the amount in the account at the end of $n + 1$ compounding periods and the amount after n compounding periods. As we will see in this section, this simple model applies to a wide range of financial applications.

First, we recall that the solution to the affine dynamical system $a(n+1) = ra(n) + b$ (when $r \neq 1$) is

$$a(k) = r^k \left(a_0 - \frac{b}{1-r}\right) + \frac{b}{1-r}.$$

This is exactly dynamical system (2.12) with $r = 1 + I/m$. Thus, the solution to dynamical system (2.12) is, after substitution and simplification,

$$a(k) = \left(1 + \frac{I}{m}\right)^k \left(a_0 + \frac{mb}{I}\right) - \frac{mb}{I}. \tag{2.13}$$

When $b = 0$, this formula simplifies to

$$a(k) = \left(1 + \frac{I}{m}\right)^k a_0.$$

The idea in the following is to study a situation similar to the one described above. Note that there are six quantities in dynamical system (2.12), that is, k, $a(k)$, a_0, I, m, and b. In most applications, five of these quantities are known. Once we have identified the situation as being modeled by dynamical system (2.12), we can immediately write down the solution (2.13) using the known quantities wherever possible. There will then be one unknown left in equation (2.13), for which we solve. That number will be the unknown quantity.

To start with, we will consider situations in which $a(k)$ is the unknown, and no work is necessary to solve for it.

EXAMPLE 2.20 Suppose we initially deposit $a_0 = 100$ dollars, plus an additional deposit of $b = 100$ dollars every year thereafter, into a savings account that pays 10 per cent interest, compounded annually.

Then $a(n + 1) = 1.1a(n) + 100$, and $mb/I = 100/0.1 = 1000$, so

$$a(k) = (1.1)^k(1100) - 1000.$$

In particular $a(40) = 48\,785.18$. A total of $48\,785.18$ dollars in our account after 40 years is quite amazing when we realize that our total deposits have been only 4100 dollars. Our interest totals $44\,685.17$ dollars. ∎

Suppose you have an account in which the advertised annual interest rate is $100I$ per cent, compounded m times a year. The **effective interest rate** is the per cent increase in your principal if left in the account for 1 year.

We model this with an account in which we make no deposit except for the initial deposit of a_0 dollars. This is just dynamical system (2.12) with $b = 0$. The solution is $a(k) = (1 + I/m)^k a_0$, and the amount we have at the end of 1 year is

$$a(m) = \left(1 + \frac{I}{m}\right)^m a_0.$$

We have made $a(m) - a(0)$ dollars in interest, so the effective annual interest rate, denoted here by I' (where the interest rate is given as a decimal), is the total interest at the end of 1 year divided by the initial deposit, that is,

$$I' = \frac{a(m) - a(0)}{a_0} = \frac{(1 + I/m)^m a_0 - a_0}{a_0} = \left(1 + \frac{I}{m}\right)^m - 1.$$

EXAMPLE 2.21 Suppose we have an account that advertises an annual interest rate of 10 per cent, compounded monthly. The effective annual interest rate is then

$$I' = \left(1 + \frac{0.1}{12}\right)^{12} - 1 = 0.1047.$$

or 10.47 per cent interest. ∎

Note that the more frequent the compounding period, the higher the effective yield, that is, the more interest we earn.

Suppose we are given that in k years we will receive $a(k)$ dollars. Suppose we can also invest money at I per cent interest today. The amount, a_0 which we would need to invest today in order to receive $a(k)$ dollars in k years is called the **present value** of $a(k)$ dollars.

EXAMPLE 2.22   Suppose a number of years ago we purchased a savings bond which will pay us 200 dollars, 5 years from today. Suppose also that we have the opportunity to make an investment that will pay an effective yearly interest rate of 8 per cent. What is our savings bond presently worth to us, that is, what present amount of money, invested at 8 per cent interest, would yield 200 dollars in 5 years? Let $a(n)$ be the value of this bond in year n. Then $a(5) = 200$ since in year 5, we receive 200 dollars for this bond. A fair price today for this bond is an amount $a(0)$, such that if $a(0)$ dollars is put in an account paying 8 per cent interest today, then in 5 years the account will have 200 dollars. Note that the present value of the bond is dependent on the present interest rate we can receive. In year n, this account has $a(n)$ dollars in it. In year $n + 1$, the account collects interest so it has $a(n+1) = 1.08a(n)$ dollars in it. The solution to this dynamical system is

$$a(k) = 1.08^k a(0).$$

Substitution of $k = 5$ and $a(5) = 200$ gives

$$200 = 1.08^5 a(0) \qquad \text{or} \qquad a(0) = \frac{200}{1.08^5} = 136.12.$$

Thus, if someone would pay us more than 136.12 dollars for our savings bond, we should sell it, since we could then invest those proceeds at 8 per cent interest and have **more than** 200 dollars in 5 years.   ∎

Often people will make a long term investment at a certain interest rate. After a while, interest rates change. People then determine, not the actual value of their original investment, but the present value (or present worth) of this investment given the new interest rate. This is one of the causes of changes in the bond market.

EXAMPLE 2.23   Suppose several years ago you paid 1000 dollars for a bond which pays you 100 dollars a year as a dividend, and that when the bond matures 10 years from now, you get your 1000 dollars back. The **coupon value** of this bond is said to be 10 per cent, since you receive 10 per cent interest on your 1000 dollars each year.

If you sell the bond to someone else today, you can invest the proceeds at an effective yearly interest of 9 per cent. How much is this bond worth to you now, that is, what is its **present value**?

The trick is to rephrase this question in more familiar terms. Suppose you sold your bond today for a_0 dollars and invested that money in a bank account paying 9 per cent interest per year, compounded annually. Assume that at the end of each year, for the next 10 years, you withdraw 100 dollars from this account, and that at the end of 10 years, there is 1000 dollars in this account. What is a_0?

The dynamical system that models the last paragraph and which helps answer our original question is

$$a(n+1) = (1.09)a(n) - 100,$$

with a(10)=1000. The solution to this dynamical system is

$$a(k) = 1.09^k(a_0 - 10\,000/9) + 10\,000/9.$$

Using $k = 10$ and $a(k) = 1000$ gives

$$1000 = 1.09^{10}(a_0 - 1111.11) + 1111.11.$$

Subtracting 1111.11 from both sides and using $1.09^{10} = 2.367$ gives

$$-111.11 = 2.367(a_0 - 1111.11).$$

Dividing both sides by 2.367, then adding 1111.11 gives $a_0 = 1064.17$. ∎

A similar application of present value is to **annuities**. In an annuity, you receive a fixed payment over some period of time. The question is, what are all these payments worth today, that is, what is the annuity's present value? One particular example of this is the true value of winning 1 000 000 dollars in a lottery, which was first studied in section 1.5.

EXAMPLE 2.24 Suppose you win 1 000 000 dollars in a lottery in the sense that you will receive 50 000 dollars a year for the next 20 years. How much did you **actually** win?

Let's rephrase this problem. Suppose the lottery commission deposits a_0 dollars into a bank account paying 8 per cent interest, compounded annually. Let $a(n)$ represent the amount in this bank account after you have received your nth payment of 50 000 dollars and for simplicity, assume you receive your first payment after 1 year. Thus, $a(1) = (1.08)a(0) - 50\,000$, and

$$a(n+1) = (1.08)a(n) - 50\,000.$$

Also, $a(20) = 0$. The amount, a_0, is the present value of your lottery win which is how much you **actually won**. Since $b/(1 - r) = 625\,000$, the solution to this dynamical system is

$$a(k) = 1.08^k (a_0 - 625\,000) + 625\,000.$$

Using that $a(20) = 0$ gives

$$0 = 1.08^{20} (a_0 - 625\,000) + 625\,000.$$

Subtract 625 000 from both sides, divide by 1.08^{20}, then add 625 000 to both sides to get that

$$a_0 = 490\,907.37.$$

Normally, you would be paid your first 50 000 dollars today, and would receive 19, not 20, additional payments. In this case, your winnings will be $a_0 + 50\,000$, where

$$a_0 = -625\,000(1.08)^{-19} + 625\,000 = 480\,179.96.$$

So your million dollars paid in 20 yearly installments of 50 000 dollars has a present value of 530 179.96 dollars if the present interest rate is 8 per cent. The moral of this is that the present value of most lottery winnings is far less than the advertised winnings.   ■

In some applications of dynamical system (2.12), the unknown quantity is the deposit (or withdrawal) b.

EXAMPLE 2.25   Suppose that when we retire, we have $a_0 = 700\,000$ dollars in our savings account and it is collecting 10 per cent interest, compounded annually. We want our savings to last for 25 years. The question is now, how much can we withdraw from our savings account each year so that our savings last for 25 years, that is, $a(25) = 0$? We are assuming that we continue earning 10 per cent interest for the entire 25 years, which is probably an unrealistic assumption.

The dynamical system that models this problem is

$$a(n+1) = (1.1)a(n) + b,$$

where b is the amount we withdraw, and is therefore a negative number. The equilibrium value is $b/(1 - r) = -10b$, so the solution to this dynamical system is

$$a(k) = 1.1^k (a_0 + 10b) - 10b.$$

Substitution gives

$$0 = 10.8347(700\,000 + 10b) - 10b,$$

where $1.1^{25} = 10.8347$. Solving for b gives

$$10.8347 \times 10b - 10b = -10.8347 \times 700\,000$$

or $b = -77\,117.65$. This means we can withdraw nearly 80 000 dollars from our account each year. ∎

Suppose we are given a_0, $a(k)$, and k, and we wish to determine I. For simplicity, let's assume $m = 1$ and $b = 0$, and that we wish to determine the effective yearly interest rate.

EXAMPLE 2.26 Suppose you have a savings bond that pays you 1000 dollars at its maturity, in 10 years. Someone will give you 500 dollars for your bond today. Thus your bond is worth $a_0 = 500$ today and is worth $a(10) = 1000$ in 10 years. Let the effective yearly interest rate on this offer be I per cent as a decimal. Then the growth of the investment satisfies the dynamical system

$$a(n+1) = (1+I)a(n).$$

The equilibrium value is 0, so the solution is

$$a(k) = (1+I)^k a_0 \qquad \text{or} \qquad 1000 = (1+I)^{10}500.$$

Dividing by 500 and taking tenth roots of both sides gives

$$1 + I = 2^{\frac{1}{10}} = 1.072.$$

The effective interest on this bond is then $I = 0.072$ or 7.2 per cent. If you can find an investment that pays more than 7.2 per cent, take the person up on their offer. ∎

EXAMPLE 2.27 Suppose we wish to obtain a 100 000 dollar mortgage to buy a house. We can get a 9 per cent, 30 year loan. Many mortgage companies require the borrower to pay **points**. For example, suppose we must pay 2 points on our loan. What this means is that we must pay 2 per cent of the loan at settlement, that is, 2000 dollars. Usually people borrow the points, too, so we borrow 102 000 dollars. Note that we only see the 100 000 dollars to buy our home. The monthly payments on 102 000 dollars are found to be 820.72 dollars.

But since we've only seen 100 000 dollars of the borrowed money, as far as we are concerned we are making monthly payments of 820.72 dollars on a 100 000 dollar loan. What is the **effective interest** on this loan, that is, what interest rate I would cause us to make monthly payments of 820.72 dollars for 360 months on a loan of 100 000 dollars? Let $a(n)$ be the amount we owe the bank after we make our nth monthly payment. Then this situation is modeled by the dynamical system

$$a(n+1) = \left(1 + \frac{I}{12}\right) a(n) - 820.72.$$

The equilibrium value is $b/(1 - r) = 820.72 \times 12/I = 9848.64/I$. Since $a(0) = 100\,000$, as far as we are concerned, and $a(360) = 0$, then the solution is

$$0 = a(360) = \left(1 + \frac{I}{12}\right)^{360} (100\,000 - 9848.64/I) + 9848.64/I.$$

We need to solve this for I. To make things easier, multiply both sides by I, giving the 361th degree polynomial

$$0 = \left(1 + \frac{I}{12}\right)^{360} (100\,000I - 9848.64) + 9848.64.$$

There may be many solutions, so when we find one we have to make sure it is the correct one.

There are no methods for algebraically finding roots of polynomials of degree greater than 4. Thus, roots must be approximated. One method is to "guess" at I: try I values slightly more than 0.09 until you have a reasonably correct answer, in other words an I value that makes the above expression close to zero. If you try this, you will find that the solution, that is, the **effective interest** on this loan, is slightly more than 9.2 per cent. There are many times in mathematics in which solutions cannot be found exactly and methods have to be developed to approximate solutions.   ∎

Finally, let's consider the case in which we know $a(k)$, a_0, and I, and we wish to find k.

EXAMPLE 2.28 Suppose you invest money at 7 per cent effective annual interest. How many years will it take for your investment to double, that is, for what value of k will $a(k) = 2a(0)$? The dynamical system is $a(n+1) = 1.07a(n)$ and the solution is

$$a(k) = 1.07^k a(0) \qquad \text{or} \qquad 2a(0) = 1.07^k a(0).$$

Dividing both sides by $a(0)$ and taking logarithms gives

$$\log 2 = k \log 1.07 \quad \text{or} \quad k = \frac{\log 2}{\log 1.07} = 10.24.$$

Thus it takes 11 years to more than double your money at a 7 per cent interest rate.

Notice that when $b = 0$ you do not need to know a_0 and $a(k)$. Instead you only need to know $a(k)/a_0$. For example, how long will it take to triple your money, that is, for what value of k is $a(k)/a_0 = 3$?

$$k = \frac{\log 3}{\log 1.07} = 16.24. \qquad \blacksquare$$

PROBLEMS

1. Set up the dynamical system and compute the amount in your savings account after
   a. 1 year with an initial deposit of 1000 dollars at 8 per cent interest, compounded quarterly.
   b. 5 years with an initial deposit of 200 dollars at 5 per cent interest, compounded semi-annually.
2. Suppose you start a savings account which pays 8 per cent interest a year compounded monthly. You initially deposit 1000 dollars and decide to add an additional 100 dollars each month thereafter. How much is in your account after 5 years?
3. Suppose you wish to buy a house, but can only afford 800 dollars a month.
   a. If you can get a loan at 9 per cent (compounded monthly) for 30 years, how large can your loan be?
   b. How large can your loan be if you amortize your loan in 20 years?
4. Suppose you have a 5000 dollar, 20 year bond, with a coupon value of 11 per cent, that is, it pays you 550 dollars each year for the next 20 years, at which time you get your 5000 dollars back. What is the present value of this bond if your alternate investment is a savings account paying 7 per cent interest compounded annually?
5. At retirement, your savings account has 300 000 dollars which is collecting 6 per cent interest, compounded monthly. How much money can you withdraw each month so that your savings will last 20 years?

6. Suppose you borrow 6000 dollars at 10 per cent per year, compounded monthly, in order to buy a car. If the loan is to be paid back in 5 years, what is your monthly payment?

7. Suppose you borrow 2000 dollars at 8 per cent interest, compounded quarterly, to take a vacation. If the loan is to be paid back in eight equal quarterly payments, how much should these payments be?

8. How much is the monthly payment on a 100 000 dollar loan at 12 per cent interest (compounded monthly) which must be amortized in
   a. 10 years?
   b. 20 years?
   c. 30 years?

9. Suppose you buy a painting for 200 dollars. Five years from now, someone offers you 300 dollars for your painting. What is the effective yearly interest on this investment?

10. What effective yearly interest rate would you need on an investment in order to double your money in 8 years?

11. How many years would it take to double your money if you invested at
    a. 5 per cent effective yearly interest rate?
    b. 9 per cent effective yearly interest rate?

12. Suppose you have 100 000 dollars in a savings account that pays 8 per cent interest, compounded quarterly. Suppose each quarter you withdraw 3000 dollars from your account.
    a. How many quarters will it take for you to deplete your account to only 50 000 dollars?
    b. How many quarters will it take for you to deplete your entire savings account?

CHAPTER 3

Probability and dynamical systems

3.1 The multiplication and addition principles

Many of the applications in the later chapters require the use of some basic concepts of probability, particularly the **multiplication principle** and **conditional probability**. In this chapter we introduce those concepts from probability that are needed for these later applications. Readers that are familiar with some of these concepts may skip parts of this chapter.

In this chapter, we are going to learn some methods for counting the number of ways of performing a given task, and use these techniques to compute the probability of certain events occurring. This knowledge will then be combined with dynamical systems to study applications. Our approach is fairly standard.

Suppose we have some task to perform such as picking several books from a bookshelf or choosing a group of people at a meeting to form a committee. Our aim in this section is to count the number of ways this task can be performed. For example, if a bookshelf contains five books and we wish to pick one of them, there are five possible outcomes. If we have three people, Tom, Dick, and Harry, at our meeting, there are three ways in which we can choose a committee of two of them, specifically, Tom and Dick, Tom and Harry, or Dick and Harry.

In performing our task, we will always assume that **no two objects are identical**. Thus, if we wish to pick one marble from a bag containing three yellow marbles and four blue marbles, we say that there are seven ways to pick the marble. If we wish to pick one yellow marble, there are three ways we may do this. Often we will pretend the objects are numbered, as in yellow marble one, yellow marble two, yellow marble three, blue marble one, ..., blue marble four.

A task will usually be labeled. For example, E is the set of ways of picking a marble from a bag containing 2 yellow marbles, 3 blue marbles, and 4 green marbles. Therefore, E is the set

$$E = \{\text{yellow marble one}, \ldots, \text{green marble four.}\}$$

The object is to count how many ways task E can be performed or, considering E as a set, count the number of "objects" E contains. Since there are nine ways of picking a marble, we say that E can be performed in nine ways, or $\#E = 9$ for short.

One important technique that helps us do this is to break the set E into several smaller sets. In this example, we could let E_1 be the ways of picking a yellow marble, E_2 be the ways of picking a blue marble, and E_3 be the ways of picking a green marble. Notice that each way of picking a marble that is in set E is also in **exactly one** of the sets E_1, E_2, or E_3. We say that

$$E = E_1 \cup E_2 \cup E_3,$$

and that E_1, E_2, and E_3 are **mutually exclusive**, that is, there is no one marble that is in two of these sets. In this case,

$$\#E = \#E_1 + \#E_2 + \#E_3 = 2 + 3 + 4 = 9.$$

THEOREM 3.1   Addition principle   *If a set E can be written as the sum of sets $E_1, \ldots, E_n$, in that*

$$E = E_1 \cup \cdots \cup E_n,$$

and the sets $E_1, \ldots, E_n$ are mutually exclusive, then

$$\#E = \#E_1 + \cdots + \#E_n.$$

Suppose we have a task which involves several stages. For example, we have two bags containing marbles. Bag 1 contains two yellow marbles and one blue marble, while bag 2 contains two red marbles and 2 green marbles. Let E be the task of picking one marble from bag 1, and then one marble from bag 2. What is #E? There are twelve possible results, namely

$$(y_1, r_1) \quad (y_1, r_2) \quad (y_1, g_1) \quad (y_1, g_2)$$
$$(y_2, r_1) \quad (y_2, r_2) \quad (y_2, g_1) \quad (y_2, g_2)$$
$$(b_1, r_1) \quad (b_1, r_2) \quad (b_1, g_1) \quad (b_1, g_2)$$

Let E_1 be the ways of picking a marble from bag 1, and let E_2 be the ways of picking a marble from bag 2. Notice that to perform task E, we need to perform task E_1 and then we must perform task E_2. There are three ways to perform task E_1. For each of those three results, there are four ways to perform task E_2. Thus the total number of ways to perform task E is $4 + 4 + 4 = 3 \times 4$.

THEOREM 3.2 Multiplication principle *Suppose it requires n steps or stages to perform task E. Suppose task E_j consists of the ways the jth step can be performed. Then*

$$\#E = \#E_1 \times \cdots \times \#E_n,$$

provided $\#E_j$ never depends upon how any previous step was performed.

The reason for this can be seen visually in the tree diagram in figure 3.1. The first set of branches in this tree represents the possible results of the first step of the task, that is, selecting a marble from bag 1. From each of these branches is a set of four branches representing the result of the second step of the task, that is, selecting a marble from bag 2. The end of each branch represents one possible result

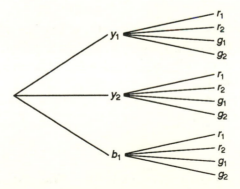

FIGURE 3.1. Tree diagram for drawing one marble from each of two bags. The first bag contains 2 yellow and 1 blue marble while the second bag contains 2 red and 2 green marbles.

of the task of picking two marbles. It is clear the number of branches is $3 \times 4 = \#E_1 \times \#E_2$.

Many counting problems require the use of both the addition principle and the multiplication principle.

EXAMPLE 3.3   Consider two bags of marbles. Bag 1 contains two yellow marbles and three blue marbles, while bag 2 contains four yellow marbles and three blue marbles. We select one marble from bag 1 and then one marble from bag 2. Let E be task of selecting two marbles of the same color, that is, we want both marbles selected to be yellow or both marbles selected to be blue. What is $\#E$?

REMARK: The two marbles can be selected in ways in which they are of different colors, but we are not interested in how many ways this can be done at this time.   □

The first step to solving this problem is to use the addition principle, that is, to write E as the union of two mutually exclusive sets. Let E_y be the task of selecting one marble from each bag so that both marbles are yellow, and let E_b be the task of selecting one marble from each bag so that both marbles are blue.

We first use the multiplication principle to find $\#E_y$. To do this, we write task E_y as a two step task. Step one is to pick a yellow marble from bag 1, denoted E_{y1}. Step two is to pick a yellow marble from bag 2, denoted E_{y2}. Notice that to perform task E_y, we must first perform E_{y1}, and then perform E_{y2}. Therefore,

$$\#E_y = \#E_{y1} \times \#E_{y2} = 2 \times 4 = 8.$$

Note that E_y consists of 8 pairs of marbles. The first pair of marbles is yellow marble 1 from bag 1 and yellow marble 1 from bag 2; ...; and the eighth pair of marbles is yellow marble 2 from bag 1 and yellow marble 4 from bag 2.

We now use the multiplication principle again to find $\#E_b$. Similarly to the above,

$$\#E_b = \#E_{b1} \times \#E_{b2} = 3 \times 3 = 9.$$

Note that E_b consists of 9 pairs of marbles. The first pair of marbles is blue marble 1 from bag 1 and blue marble 1 from bag 2; ...; and the ninth pair of marbles is blue marble 3 from bag 1 and blue marble 3 from bag 2.

Notice that to get a pair of marbles of the same color, we must get a pair of yellow marbles or a pair of blue marbles, that is,

$$E = E_y \cup E_b.$$

Also, E_y and E_b are mutually exclusive in that there is no **pair** of marbles that is in both E_y and E_b. So

$$\#E = \#E_y + \#E_b.$$

Thus, the number of ways of drawing one marble from each of these two bags so that the marbles are the same color is

$$\#E = \#E_y + \#E_b = 8 + 9 = 17.$$

You should try listing these 17 ways, starting with (y_1, y_1), meaning yellow marble one from bag one and yellow marble one from bag 2. Also separate these 17 pairs into E_y and E_b. ∎

EXAMPLE 3.4 Suppose we are members of a club which has 12 male members and 15 female members. a) How many ways can we pick a male president and a female vice-president? b) How many ways can we select a committee of two people, one of each sex? c) How many ways can we select a president and a vice-president of opposite sexes? d) How many ways can we select a president and a vice-president of the same sex? e) How can we select a committee of two people of the same sex?

a) Let A be the ways of selecting a male president and a female vice-president. This is a 2-stage process. The first stage is to select a male president, and there are 12 ways to do this. The second stage is to select a female vice-president, and there are 15 ways to do this. Therefore $\#A = 12 \times 15 = 180$.

b) Let B be the ways of selecting a committee of one male and one female. This is a 2-stage process also. The first stage is to select which male will be on the committee. There are 12 ways to do this. Then we select which female will be on the committee, and there are 15 ways to do this. Thus the number of ways to select the committee is $\#B = 12 \times 15 = 180$. Why is this the same as the answer to part a)?

c) Let C be the ways of selecting a president and a vice-president of opposite sexes. Let C_1 be the ways of selecting a male president and a female vice-president. In part a), we found that $\#C_1 = 180$. Let C_2 be the ways of selecting a female president and a male vice-president. By writing C_2 as a 2-stage process, we see that $\#C_2 = 15 \times 12 = 180$.

Note that

$$C = C_1 \cup C_2$$

since every choice of two officers of opposite sex must be made in one of these two ways. Also C_1 and C_2 are mutually exclusive since no choice of officers is in both C_1 and C_2. (C_1 contains male 1 as president and female 1 as vice-president while C_2 contains female 1 as president and male 1 as vice-president. While the people filling the offices are the same, the offices have been switched, so these are considered two different ways of selecting officers.)

Putting this together, we get that

$$\#C = \#C_1 + \#C_2 = 360.$$

d) Let D be the ways of selecting a president and a vice-president of the same sex. Let D_m be the ways of selecting two males for the offices and let D_f be the ways of selecting two females for the offices. Similarly to part c),

$$\#D = \#D_m + \#D_f.$$

We can consider D_m as a 2-stage process. The first stage is to select a male for president and there are 12 ways to do this. Once this has been accomplished, the second stage is to pick one of the **remaining** males for vice-president. Since one of the men has already been picked for president, there are only 11 possible choices for vice-president. Thus, $\#D_m = 12 \times 11 = 132$. In a similar manner, we see that $\#D_f = 15 \times 14 = 210$. Therefore,

$$\#D = \#D_m + \#D_f = 132 + 210 = 342.$$

e) Let E be the ways of selecting a committee of two people of the same sex. This is similar to d) except that in d), each committee member had a specific office. If we listed all 342 pairs of officers in part d), we would have: male 1 president and male 2 vice-president; male 2 president and male 1 vice-president; ...; female 14 president and female 15 vice-president; and female 15 president and female 14 vice-president. Thus every pair of two people of the same sex has been listed **twice**, once with one of them as president and once with the other one as president. Thus, the number of committees of two people of the same sex should be half that size, that is,

$$\#E = 342/2 = 171. \qquad \blacksquare$$

EXAMPLE 3.5   Suppose we have a bag containing three yellow marbles and four blue marbles. We draw a marble, record its color, and then replace it in the bag. We repeat this process two more times. a) How many ways can this be done? b) How many ways can this be done so that all three marbles are yellow? c) How many ways can this be done so that 1 of the marbles is blue and 2 are yellow?

To answer part a), observe that this is a 3-stage process. At each stage, there are seven possible results or marbles that can be drawn. Thus, there are $7 \times 7 \times 7 = 7^3 = 343$ ways to do this.

b) How many ways can we pick three yellow marbles? Again, this is a 3-stage process. There are three ways in which we can get a yellow marble on the first draw, three ways to get a yellow marble on the second draw (since the yellow marble on the first draw was replaced), and three ways to get a yellow marble on the third draw. Therefore, the answer is $3^3 = 27$.

c) How many ways can we get exactly one blue marble, and consequently two yellow marbles? Let E be the ways of getting exactly one blue marble in three draws. The trick to working this problem is to describe E in detail. Let E_1 be the ways of getting the one blue marble on the first draw, and yellow marbles on the second and third draws. Let E_2 be the ways of getting the one blue marble on the second draw, and yellow marbles on the first and third draws. Let E_3 be the ways of getting the blue marble on the third draw, and yellows on the first and second draws.

To compute $\#E_1$, observe that E_1 is a 3-stage process. The first stage is to draw a blue marble, the second stage is to draw a yellow marble, and the third stage is to draw another yellow marble. There are 4, 3, and 3 (since the yellow marble gotten on the second draw was returned to the bag) ways to do these stages, respectively. Therefore, $\#E_1 = 4 \times 3 \times 3 = 36$.

To compute $\#E_2$, observe that E_2 is also a 3-stage process. The first stage is to draw a yellow marble, the second stage is to draw a blue marble, and the third stage is to draw another yellow marble. There are 3, 4, and 3 ways to do these stages, respectively. Therefore, $\#E_2 = 3 \times 4 \times 3 = 36$.

It should now be clear that $\#E_3 = 3 \times 3 \times 4 = 36$.

Every way of accomplishing E must be in one of E_1, E_2, or E_3, so

$$E = E_1 \cup E_2 \cup E_3.$$

Note that E_1 contains triplets of the form (b_1, y_1, y_1) meaning blue marble 1 on the first draw, yellow marble 1 on the second draw, and yellow marble 1 again on the third draw. Also, E_2 and E_3 consist of triplets of the form (y_1, b_1, y_1) and (y_1, y_1, b_1), respectively, having similar meanings. **There is no triplet** that is in two of E_1, E_2, and E_3. They may have the same marbles in the triplet, but their order is different. Therefore E_1, E_2, and E_3 are mutually exclusive, and we can use the addition principle. Thus

$$\#E = \#E_1 + \#E_2 + \#E_3.$$

The number in each of these sets is the product of two 3's and one 4, the order of the product depending only on the position in which the blue marble was drawn. Thus.

$$\#E = \#E_1 + \#E_2 + \#E_3 = 3\#E_1 = 3 \times 36 = 108. \qquad \blacksquare$$

The last example is very important. It consists of using the addition principle to write a set E as the union of n mutually exclusive sets, $E_1, E_2, \ldots, E_n$, where

$$\#E_1 = \#E_2 = \cdots = \#E_n.$$

Thus

$$\#E = \#E_1 + \#E_2 + \cdots + \#E_n = n\#E_1.$$

So instead of computing $\#E$, we only need to compute $\#E_1$, and the number n, both of which are (usually) easy to do.

EXAMPLE 3.6 Consider again the bag with three yellow marbles and four blue marbles. Draw a marble and set it down on a table. Draw a second marble and set it to the right of the first marble. Draw a third marble and set it to the right of the first two marbles. How many ways can this be done so there are 2 yellow marbles and 1 blue marble?

Let E, E_1, E_2, and E_3 be as in example 3.5 except we are not replacing the marbles after each draw. We see that E is again the union of the other three mutually exclusive sets.

To compute $\#E_1$, consider E_1 as a 3-stage process. The first stage is to draw a blue marble, and there are 4 ways to do this. The second stage is to draw a yellow marble, and there are 3 ways to do this. The third stage is to draw another yellow marble. Since one yellow marble has already been drawn, there are only two left and there are only 2 ways to accomplish the third stage. Thus, $\#E_1 = 4 \times 3 \times 2 = 24$.

Alternatively, we could consider the process of drawing the 3 marbles as a 2-stage process. The first stage is to draw a blue marble which can be done 4 ways. The second stage is to draw 2 yellow marbles which, by the multiplication principle, can be done in $3 \times 2 = 6$ ways. By the multiplication principle, the answer is then $4 \times 6 = 24$. This approach is more sophisticated, but, as we will see, much more powerful.

By computing $\#E_2$ and $\#E_3$ in a like manner, it will become clear that

$$\#E_1 = \#E_2 = \#E_3,$$

and therefore $\#E = 3\#E_1 = 3 \times 24 = 72$. ∎

Notice that the problem being considered in example 3.5 was of the following form. A bag contains y yellow marbles and b blue marbles. One marble is drawn from this bag, its color is recorded, and the marble is returned to the bag. This process is then repeated for a total of n draws. How many ways can this be done so that exactly m yellow marbles, and consequently $n - m$ blue marbles, have been drawn? This problem is called drawing **with replacement and in order**. We say "with replacement" since we are replacing the marble after each draw. We say "in order" since we are recording the order in which the marbles are being drawn.

Let's now answer the simplest question of the kind we are considering; that is, how many ways can we draw all yellow marbles. We have a bag with y yellow marbles and an undetermined number of other marbles. Let $a(n)$ represent the number of ways that n yellow marbles can be drawn from this bag, in order and with replacement. We can consider the process of drawing n yellow marbles out of a bag as an n-stage process and use the multiplication principle n times. A simpler approach, which we first used in the preceding example, is to consider this process as a 2-stage process. The first stage is to draw one yellow marble from the bag. The second stage is to draw $n - 1$ more yellow marbles from the bag, with replacement and in order. By the multiplication principle, $a(n)$ is the product of the number of ways of doing each stage.

Since there are y yellow marbles in the bag, the first stage can be performed in y ways. By our notation, $a(n-1)$ is the number of ways of drawing $n - 1$ yellow marbles from the bag (in order and with replacement), and is the number of ways of performing the second stage. Thus $a(n)$ satisfies the first-order linear dynamical system

$$a(n) = ya(n-1).$$

We know that the general solution to this equation is

$$a(k) = cy^k.$$

Since $a(1) = y$ (there are y ways to draw 1 yellow marble), we have that

$$cy^1 = a(1) = y \qquad \text{or} \qquad c = 1.$$

Thus, the particular solution is

$$a(k) = y^k.$$

This proves the following theorem.

THEOREM 3.7   *Suppose a bag contains y marbles. We draw k marbles from the bag, in order and with replacement. Let $a(k)$ represent the number of ways this can be done. Then*

$$a(k) = y^k.$$

The proof of theorem 3.7 may belabor the obvious, but the recursive process of going back one step, from $a(n)$ to $a(n-1)$, is an important one that needs to be understood in simple cases so that it is easier to apply in more complicated cases.

EXAMPLE 3.8   Suppose a bag contains 5 yellow marbles and 3 blue marbles. We draw 10 marbles from the bag, with replacement and in order. a) How many ways can this be done? b) How many ways can this be done so that all 10 marbles are yellow? c) How many ways can this be done so that all 10 marbles are blue? d) How many ways can we draw 9 yellow marbles and 1 blue marble?

a) Since we are not concerned with color in this part, we can pretend the marbles are all the same color. We then wish to draw 10 marbles, in order and with replacement, from a bag containing 8 marbles. By theorem 3.7, this can be done in $8^{10} = 1\,073\,741\,824$ ways.

b) Since we are only considering the yellow marbles, we can ignore the fact that there are blue marbles in the bag. By theorem 3.7 there are 5^{10} ways to draw 10 yellow marbles from the bag, with replacement.

c) Since we are only considering the 3 blue marbles, we can ignore the fact that there are yellow marbles in the bag. By theorem 3.7 there are 3^{10} ways to draw 10 blue marbles from the bag, with replacement.

d) Let E be the ways of getting exactly 9 yellow marbles and 1 blue marble. Let E_1 be the ways of getting the blue marble on the first draw and yellow marbles on draws 2 through 10. Similarly, let E_j, $j = 2, \ldots, 10$, be the ways of getting the blue marble on the jth draw and yellow marbles on the other 9 draws. Clearly

$$E = E_1 \cup \cdots \cup E_{10}.$$

Each event in E_j is a list of 10 marbles with the jth one blue and the others yellow. Clearly no list of 10 marbles with one blue is in more than one E_j, so $E_1, \ldots, E_{10}$ are mutually exclusive. Thus

$$\#E = \#E_1 + \cdots + \#E_{10}.$$

Consider E_1 as a 2-stage task. The first stage is to get a blue marble on the first draw which can be done 3 ways. The second stage is to get 9 yellow marbles on the next 9 draws (in order and with replacement). By theorem 3.7, this can be done 5^9 ways. By the multiplication principle, $\#E_1 = 3 \times 5^9$.

By considering E_2 as a 2-stage task in which the first task is to get a blue marble on draw 2 and the second stage is to get 9 yellow marbles on the other 9 draws, you should be able to see that $\#E_1 = \#E_2$. Similarly, $\#E_1 = \#E_2 = \cdots = \#E_{10}$. Thus,

$$\#E = \#E_1 + \cdots + \#E_{10} = 10\#E_1 = 10 \times 3 \times 5^9. \qquad \blacksquare$$

The hardest part of these problems is to compute the number of equal subsets of task E, 10 in example 3.8 part d).

Let's now study a similar task. Suppose a bag contains n marbles. One marble is drawn from this bag, its color is recorded, and the marble is **discarded**, that is, it is **not returned** to the bag. This process is then repeated for a total of n draws, that is, until all the marbles have been drawn. How many ways can this be done? This problem is called drawing **without replacement but with order.**

If the bag contains 1 marble, then there is 1 way to draw it. If the bag contains 2 marbles, then we have a 2-stage process. The first stage is to draw the first marble which can be done 2 ways. The second stage is to draw the remaining marble which can be done 1 way. Thus, the 2 marbles can be drawn in 2×1 ways.

Suppose the bag contains 3 marbles. We can consider the drawing of these marbles as a 3-stage process, draw the first marble (3 ways), the second marble (2 ways), and the third marble (1 way). Thus there are $3 \times 2 \times 1$ ways to draw the marbles.

Alternatively, we could consider the process of drawing the 3 marbles as a 2-stage process. The first stage is to draw the first marble which can be done 3 ways. The second stage is to draw the remaining 2 marbles. We had previously computed that the number of ways of drawing 2 marbles from a bag containing 2 marbles was 2×1. By the multiplication principle, the answer is $3 \times 2 \times 1$.

Suppose the bag contains n marbles. Let $a(n)$ be the number of ways the n marbles can be drawn, in order but without replacement. This task can be considered a 2-stage process. The first stage is to draw one marble and this can be done n ways. The second stage is to draw the remaining $n-1$ marbles, in order but without replacement. This can be done $a(n-1)$ ways. By the multiplication principle,

$$a(n) = na(n-1). \tag{3.1}$$

This is called a first-order **nonautonomous** dynamical system. Refer to section 1.2 for the definition of nonautonomous.

We have already computed that $a(1) = 1$, $a(2) = 2 \times 1$, and $a(3) = 3 \times 2 \times 1$. A good guess at the particular solution is

$$a(k) = k \times (k-1) \times \cdots 3 \times 2 \times 1. \tag{3.2}$$

For simplicity, we will use the factorial notation

$$k! = k \times (k-1) \times \cdots 3 \times 2 \times 1.$$

Thus, $3! = 3 \times 2 \times 1 = 6$, and solution (3.2) can be given concisely as

$$a(k) = k!.$$

Substituting n for k and $n-1$ for k in solution (3.2) gives

$$n! = n \times (n-1)!$$

so the dynamical system is satisfied when using equation (3.2). It is therefore a solution. It also satisfies the initial condition that $a(1) = 1$, so it is the particular solution.

THEOREM 3.9   *Suppose a bag contains k marbles. We draw all k marbles from the bag, in order but without replacement. Let $a(k)$ represent the number of ways this can be done. Then*

$$a(k) = k! = k \times (k-1) \times \cdots \times 2 \times 1.$$

Let's consider a variation on theorem 3.9. Suppose a bag contains y marbles and we draw k of them, $k < y$, in order but without replacement. How many ways can this be done?

EXAMPLE 3.10   Let $a(n)$ be the number of ways that n marbles can be drawn from a bag containing 20 marbles, in order but without replacement. Then $a(1) = 20$, since there are 20 marbles to choose from. We use the multiplication principle to compute $a(2)$, that is, we can choose the first marble in any one of $a(1) = 20$ ways and then choose the second marble in any one of 19 ways. Thus, $a(2) = 19a(1) = 20 \times 19$.

To compute $a(3)$, we first choose 2 marbles in any one of the $a(2)$ ways as a first stage. There are 18 marbles left, so we can choose the third marble in any one of 18 ways. Thus, $a(3) = 18a(2) = 18 \times 19 \times 20$. Observe that

$$18 \times 19 \times 20 = \frac{1 \times 2 \times \cdots \times 17 \times 18 \times 19 \times 20}{1 \times 2 \times \cdots \times 17} = \frac{20!}{17!}.$$

Thus,

$$a(3) = \frac{20!}{17!} = \frac{20!}{(20-3)!}.$$

To compute $a(4)$, as the first stage we draw 3 marbles in any one of the $a(3)$ ways, and as the second stage we draw the fourth marble in any one of 17 ways. Thus,

$$a(4) = 17a(3) = \frac{17 \times 20!}{1 \times 2 \times \cdots \times 16 \times 17} = \frac{20!}{16!} = \frac{20!}{(20-4)!}.$$

Suppose we want to compute $a(n+1)$, where $1 < n+1 \leq 20$. This is a 2-stage process. The first stage is to draw n marbles which can be done in any one of $a(n)$ ways. The second stage is to draw the $n+1$th marble. Since n of the 20 marbles have already been drawn, there are only $20 - n$ ways to do this. Thus, we have the **nonautonomous** dynamical system

$$a(n+1) = (20-n)a(n), \qquad (3.3)$$

with initial condition $a(1) = 20$. The solution appears to be

$$a(k) = \frac{20!}{(20-k)!}. \qquad (3.4)$$

Let's verify this solution. Substituting 1 for k in solution (3.4) gives $a(1) = 20!/19! = 20$ which satisfies the initial condition. Substituting n and $n+1$ in for k gives $a(n) = 20!/(20-n)!$ and $a(n+1) = 20!/(19-n)!$, respectively. Substituting this into dynamical system (3.3) gives

$$\frac{20!}{(19-n)!} = \frac{(20-n) \times 20!}{(20-n)!}.$$

Canceling $20 - n$ in the numerator and denominator on the right side gives equality. Thus, solution (3.4) satisfies dynamical system (3.3). ∎

Example 3.10 can be easily generalized. Suppose a bag contains m marbles. The marbles are to be drawn one at a time, in order but without replacement. Let $a(n)$ be the number of ways that n marbles can be drawn. Clearly, $a(1) = m$ since that is the number of marbles in the bag. Suppose we want to compute the number of ways of drawing $n+1$ marbles. This can be considered a 2-stage process in which the first stage is to draw n marbles which can be done in $a(n)$ ways, and the second stage is to draw the $n+1$th marble which can be done in $m - n$ ways. Thus, we have the nonautonomous dynamical system

$$a(n + 1) = (m - n)a(n). \tag{3.5}$$

The solution $a(k) = m!/(m - k)!$ can be found in the same way as in example 3.10.

THEOREM 3.11   *Suppose a bag contains m marbles. The number of ways to choose k of them, $k \leq m$, in order but without replacement is*

$$a(k) = \frac{m!}{(m - k)!}.$$

We define $0! = 1$ by convention. Thus, when $k = m$, theorem 3.11 agrees with theorem 3.9.

PROBLEMS

1. How many ways are there to flip a coin, and then draw a marble from a bag containing 6 marbles?

2. A bag contains 3 red, 4 blue, and 5 yellow marbles.
   a. How many ways are there to draw either a red or a blue marble?
   b. How many ways are there to draw a red, and then draw a blue marble?
   c. How many ways are there to draw 2 marbles, such that they are of different colors? The order in which they are drawn is unimportant.

3. A buffet line contains 5 vegetables, 4 meats, and 4 desserts.

    a. How many ways are there to select 1 of each?

    b. How many ways are there to select 2 items that are not of the same food type?

    c. How many ways are there to select 2 different items of the same food type, in any order?

4. Three men and 4 women have applied for jobs at a certain company.

    a. How many ways can one person be hired for assistant manager and another be hired in sales?

    b. How many ways can a man be hired for assistant manager and a woman be hired in sales?

    c. How many ways can a woman be hired for assistant manager and another woman be hired in sales?

5. There are 15 children on a baseball team.

    a. How many ways can a pitcher and a catcher be chosen from the 15 children?

    b. How many ways can the nine positions be filled from the 15 players? It is important which child plays at each position.

6. There are 6 men and 5 women in a club.

    a. How many ways are there to choose men to fill the offices of president and vice president?

    b. How many ways are there to choose a woman for president and a man for vice president?

    c. How many ways are there to choose 1 person of each sex to fill the offices?

    d. How many ways are there to choose 2 different people of the same sex to fill the offices?

7. How many ways can a coin be flipped 10 times?

8. Roll a die 4 times in a row. How many ways can this be done? How many ways can it be done so that all four rolls result in even numbers?

9. There are 15 children in a class.

    a. How many ways can a teacher call on students, in order and with replacement, to answer 15 questions?

    b. How many ways can a teacher call on the students if each student is asked exactly one of the 15 questions?

    c. How many ways can a teacher call on five different students, in order, if each student is asked exactly one question?

10. A teacher has a class of 10 students. In alphabetical order, the teacher decides to pass or fail each student.
    a. How many ways can this be done?
    b. How many ways can exactly 9 of the 10 students pass?

11. How many different ways can all the letters in the word "marbles" be arranged?

12. How many different 4-letter arrangements can be made from the letters in the word "marbles" if no letter can be used twice.

13. A television executive must schedule half hour shows for the six time slots on Monday night.
    a. How many ways can this be done if each slot must be filled with either a situation comedy or a drama? It is unimportant which comedy or drama is in a particular spot. The only thing that is important is whether it is a comedy or a drama.
    b. How many ways can this be done if exactly 4 of the shows must be situation comedies?
    c. How many ways can this be done if at least 4 of the shows must be situation comedies?
    d. How many ways can this be done if at least 1 of the shows must be a situation comedy?

3.2 Introduction to probability

Suppose we perform a task, such as rolling a die, flipping a coin, or drawing a card from a deck of cards. Each possible outcome of our task is called an event. Examples would include, getting a six on a die, getting heads on the flip of a coin, or getting the ace of spades from a deck of cards. Our object is to compute the probability of a particular event happening. In the above cases, intuition tells us that the probability of getting a six in the roll of a fair die is 1/6, the probability of getting heads in the flip of a fair coin is 1/2, and the probability of getting the ace of spades from a deck of 52 cards is 1/52.

The way these computations are done is to count the number of possible results of each task, 6, 2, and 52 respectively. It is assumed that each result is equally likely, so that the probability of the event in question is one out of the total number of results. This idea is the key to understanding basic probability.

Suppose we perform a task, say we flip a coin twice. The object is to list the possible outcomes. For this task, we might list the results as exactly 0 heads, exactly 1 head, and exactly 2 heads. You might feel that the probability of each of these outcomes is 1/3. In doing a computer simulation of this process, we flipped a pair of coins 1000 times. Our results were, 0 heads occurred 258 times, 1 head occurred 498 times and 2 heads occurred 244 times. This does not seem to correspond to our predicted results of about 333 occurrences for each of these events. The reason this answer is wrong is that these events are not equally likely.

Let us again try to list the possible results of two flips of a coin. There are four, not three, possible outcomes or events. These are (head, head), (head, tail), (tail, head), and (tail, tail). Experience suggests that these events **are** equally likely. From this observation, the probability of getting exactly 0 heads should be 1/4, of getting exactly 1 head should be 2/4, and of getting exactly 2 heads should be 1/4. From this computation, when flipping a pair of coins 1000 times, we expect 0 heads to occur 250 times, 1 head to occur 500 times, and 2 heads to occur 250 times. This seems to correspond much more closely to the results of our computer simulation.

From the above, we see that the object is to list results which are equally likely. When we list all possible results of a task, we call each of these results a **simple event**. Often these simple events are all equally likely. Thus there are 4 simple events in the task of flipping a coin twice.

A **compound** event E is a set of simple events. The compound event of getting exactly one head in two flips of a coin consists of the simple events: (head, tail) and (tail, head).

EXAMPLE 3.12 Consider the task of flipping a coin 3 times. By the multiplication principle, there are $2^3 = 8$ simple events for this task. This is because this task is a 3-stage process, with 2 possible results at each stage, heads or tails. Equivalently, we could consider this as drawing 3 marbles from a bag containing 1 red (head) marble and 1 blue (tail) marble, with replacement and in order.

Let E be the event that less than 2 heads occurred in the 3 flips. Let H denote heads and T denote tails. Then

$$E = \{(T, T, T), (T, T, H), (T, H, T), (H, T, T)\}.$$

The probability that E will occur, denoted $P(E)$, is $4/8 = 1/2$. ■

EXAMPLE 3.13 Consider the task of rolling a pair of dice. By considering this as a 2-stage process, that is, roll the first die and then roll the second die, the multiplication principle implies that there are $6^2 = 36$ simple events. See if you can list them.

Let E be the event that the total of the two dice is 6. Then

$$E = \{(1,5), \quad (2,4), \quad (3,3), \quad (4,2), \quad \text{and} \quad (5,1)\},$$

where the first number is the result on the first die and the second number is the result on the second die. We would expect that $P(E) = 5/36$.   ∎

COUNTING PRINCIPLE: Consider a task in which there are a total of $\#T$ possible (equally likely) simple events. (T stands for total, not tails.) Suppose that an event E is composed of $\#E$ simple events. Then the probability of event E occurring when the experiment is performed is: The number of simple events that compose E divided by the total number of simple events that compose the task, that is,

$$P(E) = \frac{\#E}{\#T}.$$ □

EXAMPLE 3.14 In example 3.3, we had two bags, the first containing 2 yellow and 3 blue marbles, the second containing 4 yellow and 3 blue marbles. One marble was drawn from each bag. We then computed the number of ways this task could be accomplished so that both marbles were the same color. Let's now compute the probability that both marbles are the same color.

The total number of simple events is the total number of ways one marble can be drawn from each bag. Since there are 5 marbles in the first bag and 7 in the second bag, by the multiplication principle, there are

$$\#T = 5 \times 7 = 35$$

equally likely simple events.

As before, let E be the event that both marbles are the same color. Then E is the union of the 2 mutually exclusive sets E_y and E_b, where E_y is the event that both marbles are yellow and E_b is the event that both marbles are blue. Recall that by the multiplication principle, $\#E_y = 2 \times 4 = 8$ and $\#E_b = 3 \times 3 = 9$, and by the addition principle, $\#E = \#E_y + \#E_b = 17$.

The probability that both marbles are yellow is

$$P(E_y) = \frac{\#E_y}{\#T} = 8/35,$$

the probability that both marbles are blue is

$$P(E_b) = \frac{\#E_b}{\#T} = 9/35,$$

and the probability that both marbles are the same color is

$$P(E) = \frac{\#E}{\#T} = \frac{\#E_y + \#E_b}{\#T} = \frac{\#E_y}{\#T} + \frac{\#E_b}{\#T} = P(E_y) + P(E_b) = 17/35.$$

Note that the addition principle for counting results in

$$P(E) = P(E_y) + P(E_b).$$

What is the probability that the two marbles are of different colors? Let this be event E'. Since every simple event must be in either E or E', but not both, we have that

$$\#E + \#E' = \#T.$$

Thus

$$P(E') = \frac{\#E'}{\#T} = \frac{\#T - \#E}{\#T} = 1 - \frac{\#E}{\#T} = 1 - P(E) = 18/35. \quad \blacksquare$$

For many events, it is easier to compute the probability of the event **not happening** then it is to compute the probability that the event will happen. This was the case for the event that both marbles were of different colors in example 3.14.

DEFINITION 3.15 *Event E is a set of simple events. The **complement of** event E is the set of all simple events that are **not** in event E, and is denoted by E'.*

Since every simple event must be in event E or E', but not both, then $\#E + \#E' = \#T$. Thus, $P(E) + P(E') = \#E/\#T + \#E'/\#T = 1$, or

$$P(E') = 1 - P(E).$$

This, and the results of example 3.14 lead to the following theorem.

THEOREM 3.16   The addition principle for probability   *Suppose events E_1 and E_2 are **mutually exclusive**, that is, there is no simple event that is in both* *of these events. Let event E be composed of all simple events that are in either E_1 or E_2, that is, $E = E_1 \cup E_2$. Then*

$$P(E) = P(E_1) + P(E_2).$$

Let event E' be the complement of event E. Then

$$P(E') = 1 - P(E).$$

EXAMPLE 3.17   In 6 flips of a fair coin, what is the probability that heads occurs at least once? Let E' be the event that heads occurs at least once, that is, heads occurs exactly 1 time, 2 times, ..., or 6 times. We could consider 6 cases and use the addition principle to compute the number of ways E' can occur. A simpler method is to observe that E' is the complement of E, the event in which heads occurs 0 times. Flipping a coin 6 times is a 6-stage process with two possible outcomes at each stage. Thus, there are $2^6 = 64$ simple events. Heads occurring 0 times means that tails occurs on every flip, and there is only one way for this to occur. Thus, by theorem 3.16,

$$P(E') = 1 - P(E) = 1 - 1/64 = 63/64. \qquad \blacksquare$$

EXAMPLE 3.18   Suppose a company makes calculators, and that 1 per cent of their calculators are defective. A box is packed with 10 calculators. a) What is the probability that there are no defective calculators in the box? b) What is the probability that there is 1 defective calculator in the box? c) What is the probability that there are more than 1 defective calculators in the box?

Let's rephrase this problem as a marble drawing problem. Let a bag contain 1 red and 99 blue marbles. We draw 10 marbles in order and with replacement. Each red marble drawn corresponds to a defective calculator. The above questions can be rephrased as a) what's the probability of getting no red marbles, b) of getting exactly one red marble, and c) of getting more than one red marble?

a) Let E_0 be the event of getting no red marbles in 10 draws, drawing with replacement and in order. By theorem 3.7, we can draw 10 marbles in order and with replacement from a bag containing 100 marbles in $\#T = 100^{10}$ ways. Getting no red marbles means getting all blue marbles. Again, by theorem 3.7, we can draw 10 blue marbles from a bag containing 99 blue marbles in $\#E_0 = 99^{10}$ ways. Thus,

$$P(E_0) = \frac{99^{10}}{100^{10}} = (0.99)^{10} = 0.9044.$$

b) Let E_r be the event of getting exactly one red marble. Event E_r can be broken into 10 cases: getting all blue marbles except for red on the first draw, getting all blue marbles except for red on the second draw, ..., getting all blue marbles except for red on the tenth draw. Label these events as $E_1, E_2, \ldots, E_{10}$, respectively.

To compute the number of simple events in event E_j, $j = 1, \ldots,$ 10, consider the 10-stage process of drawing the 9 blue marbles and the one red marble. Each blue marble can be drawn in anyone of 99 ways, but the red marble can be drawn in only one way. Thus, $E_j = 99^9 \times 1$, and

$$P(E_j) = \frac{99^9}{100^{10}} \qquad \text{for} \quad j = 1, \ldots, 10$$

By the addition principle for probability, theorem 3.16,

$$P(E_r) = P(E_1) + \cdots + P(E_{10}) = 10P(E_1) = \frac{10 \times 99^9}{100^{10}} = 0.0914.$$

c) Let event E' be the event of getting more than one red. Event E' is the complement of event E, the event of getting at most one red marble. But $E = E_r \cup E_0$ where E_r and E_0 are mutually exclusive. Thus $P(E) = P(E_0) + P(E_r) = 0.9958$, and

$$P(E') = 1 - P(E) = 0.0042.$$

Note that it is **almost certain** that we will not get more than 1 defective. ∎

EXAMPLE 3.19 Consider a box that contains 100 calculators, of which 1 is defective and 99 are not defective. We know this, but the quality control workers do not know this. For quality control, 10 calculators are drawn from the box, in order, but without replacement. If these 10 calculators are all not defective, the box will be passed, and sent to a consumer. a) What is the probability that the box will be passed, that is, that 0 defective calculators are drawn? b) What is the probability that the box will fail, that is, at least 1 defective is drawn? c) What is the probability of getting exactly one defective?

a) This is identical to drawing 10 marbles, in order but without replacement from a bag containing 1 red marble and 99 blue marbles. Let E_0 be the event that exactly 0 red marbles are drawn. The number of ways of drawing 10 marbles in order from a bag of 100 is, by theorem 3.11

$$\#T = \frac{100!}{(100 - 10)!} = 100 \times 99 \times \cdots \times 91.$$

The number of ways of drawing 10 blue marbles from a bag containing 99 blue marbles is, again by theorem 3.11,

$$\#E_r = \frac{99!}{(99 - 10)!} = 99 \times \cdots \times 90.$$

Thus,

$$P(E_0) = \frac{99 \times \cdots \times 90}{100 \times \cdots \times 91} = 90/100 = 0.9.$$

b) The probability of getting at least one defective, E_0' is

$$P(E_0') = 1 - P(E_0) = 0.1.$$

c) It is not difficult to compute the probability of getting exactly one red from a bag containing 99 blue and one red when drawing in order but without replacement. Note that since there is only one red marble, we can get either 0 or 1 red in our 10 draws. **Getting exactly one red and getting at least one red are the same in this case.** Thus, the answer to this problem is the same as part b)

$$P(E_0') = 1 - P(E_0) = 0.1. \qquad \blacksquare$$

You should study the previous two examples and be sure you understand the difference between them. This difference is important. Notice that in example 3.18, it is possible (although unlikely) to get more than 1 defective. In example 3.19, it is impossible to get more than 1 defective.

EXAMPLE 3.20   Consider a city with a population of 4 million people. Suppose 1 million support Mr. Jones for mayor while 3 million support Ms. Smith. A polling company picks 4 people at random and asks whom they support. What is the probability of event E, that exactly 2 of the people support Mr. Jones, occurring a) if the polling is conducted in order but without replacement; and b) if the polling is conducted in order but with replacement?

Since the polling is in order, there are six ways in which exactly 2 of the 4 people polled support Mr. Jones. These are (J,J,S,S), (J,S,J,S), (J,S,S,J), (S,J,J,S), (S,J,S,J), and (S,S,J,J), where an S or J in the jth position ($j = 1, 2, 3,$ or 4) means that the jth person picked supports

Ms. Smith or Mr. Jones, respectively. Event E can be considered as the union of 6 mutually exclusive sets. Let event E_1 be that the first of these (J,J,S,S) occurs. Likewise for events $E_2, \ldots, E_6$, respectively. Therefore

$$E = E_1 \cup \cdots \cup E_6$$

where the events $E_1, \ldots, E_6$ are mutually disjoint. It should be clear that $\#E_1 = \cdots = \#E_6$ so that

$$\#E = \#E_1 + \cdots + \#E_6 = 6\#E_1,$$

and

$$P(E) = \frac{6\#E_1}{\#T}.$$

a) Let's now suppose the drawing is in order but without replacement. We need to compute $\#E_1$ and $\#T$. To compute $\#T$ we need to compute the number of ways in which four people can be chosen from a bag containing four million people, in order but without replacement. By theorem 3.11, this is

$$\#T = \frac{4\,000\,000!}{3\,999\,996!}$$
$$= 4\,000\,000 \times 3\,999\,999 \times 3\,999\,998 \times 3\,999\,997.$$

To compute the number of ways of getting 4 people such that the first 2 prefer Mr. Jones and the last 2 prefer Ms. Smith, we consider the 4 stages, pick a Jones supporter from among his million supporters, pick one of the remaining Jones supporters, pick a Smith supporter from among her 3 million supporters, pick one of the remaining Smith supporters. This can be done in

$$\#E_1 = 1\,000\,000 \times 999\,999 \times 3\,000\,000 \times 2\,999\,999.$$

Thus, drawing **without replacement** gives

$$P(E) = \frac{6 \times 1\,000\,000 \times 999\,999 \times 3\,000\,000 \times 2\,999\,999}{4\,000\,000 \times 3\,999\,999 \times 3\,999\,998 \times 3\,999\,997}$$
$$= 0.210\,937\,535.$$

b) Let's now suppose the drawing is in order and with replacement. We need to compute $\#E_1$ and $\#T$. To compute $\#T$ we need to compute the number of ways in which four people can be chosen from a bag containing four million people, in order and with replacement.

By theorem 3.7 we get that

$$\#T = 4\,000\,000^4.$$

To compute the number of ways of getting 4 people such that the first 2 prefer Mr. Jones and the last 2 prefer Ms. Smith, we consider the 4 stages, pick a Jones supporter from among his million supporters, pick another Jones supporter from his million supporters, pick a Smith supporter from among her 3 million supporters, pick another Smith supporter from her 3 million supporters. This can be done in

$$\#E_1 = 1\,000\,000^2 \times 3\,000\,000^2.$$

Note that this is just an application of theorem 3.9 twice, the first stage is to draw 2 Jones supporters, the second stage is to draw 2 Smith supporters. Thus, drawing **with replacement** gives

$$P(E) = \frac{6 \times 1\,000\,000^2 \times 3\,000\,000^2}{4\,000\,000^4} = 6 \times \left(\frac{1}{4}\right)^2 \left(\frac{3}{4}\right)^2$$
$$= 0.210\,937\,500.$$

REMARK: $P(E)$ was computed to 9 decimal places when drawing with and without replacement, above. The reason for computing this many decimal places was to compare the accuracy between drawing with and without replacement when the population is large. Note that the first 7 decimal places agree. In reality, our population size is only an estimate, so we should only compute a few decimal places.□

Note that

- the fraction of people supporting Jones is 1/4 and we want 2 of them,

- the fraction of people supporting Smith is 3/4 and we want 2 of them, and

- there are 6 cases.

Thus, in computing $P(E)$, we get terms involving each fraction of people to a power that equals the number of such people desired to be picked, and summed up a number of times (6 times) that equals the number of cases.   ∎

Note that the two probabilities are essentially the same. By comparing the two parts of the above example, it should be clear that when the number of marbles in the bag is large, and the number of marbles being drawn is small, then the probabilities when drawing with replacement and without replacement are approximately the same. This is because $999\,999/3\,999\,999$ is approximately $1/4$, $3\,000\,000/3\,999\,999$ is approximately $3/4$, and $2\,999\,999/3\,999\,997$ is approximately $3/4$. Thus we have the following.

PRINCIPLE OF LARGE POPULATIONS: When drawing a small number of marbles from a bag that contains a large number of red and blue marbles, assume that the drawing is with replacement and in order.   □

One reason for using this principle is that the answers for both experiments are approximately the same in this case, but drawing with replacement is much easier, computationally. In fact, many tables exist which give the answers to a large number of these problems. Another reason is that we often have an estimate for the fraction of "red" marbles ($p = 1/4$ in the previous example), but we do not know the total number of red marbles, $1\,000\,000$ in the previous example. In the next section, we will learn an easier method for computing probabilities when we know the **fraction** of marbles of each type.

PROBLEMS

1. A bag contains 6 red marbles and 4 blue marbles. Draw 5 marbles from the bag, with replacement and in order.
    a. What is the probability of getting exactly 4 red marbles?
    b. What is the probability of getting at least 4 red marbles?
    c. What is the probability of getting at most 1 red marble?

2. Repeat problem 1, except draw without replacement, but in order.

3. There are 4000 students enrolled at a university, of which 1500 are freshmen and 2500 are upperclassmen. What is the probability that 4 students chosen at random will include
    a. exactly 3 freshmen,
    b. exactly 2 freshman,
    c. at least 2 freshman.

4. Suppose that $1/5$ of all people like liver and onions. What is the probability that in a group of 6 people, at least one person likes liver and onions?

5. A bag contains 3 red marbles and 4 blue marbles. Two marbles are drawn from the bag.

    a. What is the probability that both marbles are red,
       i. if the drawing is with replacement and in order, and
      ii. if the drawing is in order, but without replacement?
    b. What is the probability that 1 marble is of each color,
       i. if the drawing is with replacement and in order,
      ii. if the drawing is in order, but without replacement?

6. The probability that it will rain on any given day is 0.3.
    a. What is the probability that it will rain on exactly 2 of the next 4 days?
    b. What is the probability that it will not rain in the next 4 days?
    c. What is the probability it will rain at least once in the next 4 days?

7. In Russian roulette, a gun has 5 empty chambers and one loaded chamber. A person spins the cylinder, and then pulls the trigger.
    a. What is the probability that the gun will not fire if the person pulls the trigger 3 times, given that the cylinder is spun before each attempt?
    b. What is the probability that the gun will not fire given that the person spins the cylinder once, and then fires the gun 3 times in a row, without spinning the cylinder again?

8. A certain baseball player is batting three hundred, meaning that the probability he will get a hit at any time at bat is 0.3. Suppose this player gets to bat 5 times in one game.
    a. What is the probability he will not get a hit?
    b. What is the probability he will get at least 1 hit?
    c. What is the probability he will get a hit every time he bats?

9. In a card game, you are dealt 2 aces and 3 worthless cards. You throw away the 3 worthless cards and are dealt 3 more cards. What is the probability you will get at least 1 more ace in the 3 cards? Remember that the 3 cards you are given are from the 47 remaining cards of which only 2 are aces.

10. In a card game you are dealt 5 cards: an ace, a king, a queen, a ten, and a four. You throw away the four and are dealt 1 more card. What is the probability it is a jack?

11. In a card game, you are dealt 3 spades, a heart and a diamond. You throw away the heart and the diamond and are dealt 2 more cards. What is the probability they are both spades?

12. In blackjack, you are dealt 2 cards. You are an immediate winner if one of your cards is an ace, and the other card is one of the following: a king, a queen, a jack, or a ten. What is the probability you are an immediate winner?

13. In a simplified version of Keno, a player picks 5 different numbers from 1 to 20. Five numbers from 1 to 20 are then picked at random by the casino, in order but without replacement. What is the probability that
    a. none of your numbers are included in the five numbers picked by the casino,
    b. exactly four of your numbers are among the five picked, and
    c. at least four of your numbers are among the five picked?

14. In a certain game, a die is rolled 6 times. You win if a 6 comes up on at least 1 roll of the die. What is the probability that you lose?

15. Ten men and fifteen women work for Gigantic Corporation.
    a. Each day, for 4 days in a row, an employee is chosen for special recognition, with replacement. What is the probability that women will be chosen on at least 2 of the days?
    b. Four employees are chosen for promotion. If the four are chosen at random, what is the probability that at least 2 women are chosen?

3.3 Multistage tasks

Before we continue, we need a new concept, the concept of **conditional probability**. Suppose we have 2 events, E and F. We perform a task and observe that event E has occurred. We now want to know the probability that F has also occurred? We denote this probability by $P(F|E)$, which is read as, what is the probability that event F will happen given that event E has happened?

EXAMPLE 3.21 Suppose that we roll a die. Let E be the event that a number greater than 3 occurs. Let F be the event that a 6 occurs. a) Someone looks at the die and tells us that event E has occurred. What is $P(F|E)$? b) Someone looks at the die and tells us that event F has occurred. What is $P(E|F)$?

a) Since we know that E has occurred, we know that the die is showing a 4, 5, or 6. Thus, the number of simple events that can occur is 3. Of these 3 possible results, $P(F|E)$ is the probability that the 6 occurs. This happens in just 1 of the 3 simple events, so $P(F|E) = 1/3$.

b) Since we know that F has occurred, we know that the die is showing a 6. Thus we know for certain that E has occurred, that is, the number showing is greater than 3. Thus $P(E|F) = 1$. ∎

The **intersection** of two events, E and F, is the event composed of all simple events that are in both E and F. This event is denoted by

$$E \cap F.$$

For example, we draw a card from a deck of cards. Let E be the event that the card is a spade and let F be the event that the card is an ace. Then

$$E \cap F = \{\text{the ace of spades}\},$$

and therefore it consists of 1 simple event.

Many applications involve 2-stage processes. In these applications, we are often interested in an event E_1 which involves some result on the first stage of the process, and an event E_2 which involves some result on the second stage of the process. For example, suppose we draw 2 marbles from a bag containing 2 red and 3 blue marbles, with replacement and in order. Let event E_1 be that the first marble is red (and the second marble can be any color), while event E_2 is that the second marble is red (and the first marble can be any color). Then the event $E_1 \cap E_2$ is the event that the first and second marbles are both red. By the multiplication principle, there are $2^2 = 4$ ways of getting a red marble on both draws, when drawing with replacement and in order. Thus the intersection of E_1 and E_2 contains 4 simple events, that is,

$$\#(E_1 \cap E_2) = 4.$$

Note that $\#E_1 = 2 \times 5 = 10$, since we must get a red marble on the first stage and any marble on the second stage. Similarly, $\#E_2 = 5 \times 2 = 10$, also.

EXAMPLE 3.22   Consider a bag that contains 4 red marbles and 6 blue marbles. Draw 2 marbles from the bag, in order but without replacement. Let E_1 be the event that the first marble drawn is red. Let E_2 be the event that the second marble drawn is red. Let's compute $P(E_1)$, $P(E_2)$, $P(E_1 \cap E_2)$, $P(E_2|E_1)$, and $P(E_1|E_2)$.

Since we are drawing without replacement but in order, we will use the multiplication principle to compute the total number of simple events. This is a 2-stage process. The number of ways the first stage or draw can be accomplished is 10, since there are 10 marbles. When the first stage is completed, there are only 9 marbles left in the bag, and so the second stage or draw can be accomplished in 9 ways. By the multiplication principle, $\#T = 10 \times 9 = 90$.

To compute the number of simple events in event E_1, we will again use the multiplication principle. The first stage is to draw a red marble, which can be done in 4 ways. Since one red marble has already been removed, there are 9 marbles remaining and so the second stage can be done in 9 ways, since it can be any color. Thus, the number of simple events in E_1 is $4 \times 9 = 36$. Therefore, $P(E_1) = 36/90 = 0.4$.

To compute the number of simple events in E_2, we again consider a 2-stage process. But the number of ways the second stage can be done, that is, the number of red marbles we can choose from on our second draw depends on the marble we drew on the first draw. Thus, we consider 2 cases and use the addition principle.

We could get a red marble on the first draw, case 1, or we could get a blue marble on the first draw, case 2. The number of simple events in case 1, such that the second marble is red, is, by the multiplication principle, $4 \times 3 = 12$. This is because there are 4 red marbles to choose from on the first draw, but once this has been done, there are 3 red marbles to choose from on the second draw.

Likewise, case 2 can be done in $6 \times 4 = 24$ ways, since there are 6 blue marbles to choose from on the first draw and 4 red marbles to choose from on the second draw. Thus, event E_2 is composed of $12 + 24 = 36$ simple events, and $P(E_2) = 36/90 = 0.4$.

A second way to find $P(E_2)$ is to do the second stage first and the first stage second. Specifically, we can pick anyone of the 4 red marbles on the second stage. We can then pick anyone of the remaining marbles on the first stage. Again we get that $\#E_2 = 4 \times 9 = 36$ and $P(E_2) = 36/90 = 0.4$.

There is an even easier method for finding $P(E_1)$ and $P(E_2)$. Since E_1 is only concerned with the first draw, let's restrict our attention to the first draw. The total number of simple events that may occur on the first draw is 10, that is, any one of the 10 marbles. The total number of ways event E_1 can occur is 4, that is, any one of the 4 red marbles being drawn. Thus we again get that $P(E_1) = 4/10 = 0.4$. To find $P(E_2)$, we want to compute the probability that the second marble drawn is red **before we have made the first draw.** Thus, as far as we know **now**, anyone of the 10 marbles are equally likely to be drawn on the second draw, and any of the four red marbles may be drawn on the second draw. So, $P(E_2) = 4/10 = 0.4$. This is a case of "Don't use what you don't know."

Now let's compute $P(E_1 \cap E_2)$. The event $E_1 \cap E_2$ is the event that a red marble is drawn on the first and second draws. By the multiplication principle, this can be done in $4 \times 3 = 12$ ways, so $P(E_1 \cap E_2) = 12/90 = 2/15$.

Let's find $P(E_2|E_1)$. We are given that event E_1 occurred, that is, that the first marble drawn was red, so we know that any one of $\#E_1 = 36$ simple events are equally likely to have occurred. For event E_2 to also occur, we must determine for which of these 36 events did E_2 also occur, that is, for which events in which the first marble is red, is the second marble also red. But these are the events in $E_1 \cap E_2$ and we know $\#(E_1 \cap E_2) = 4 \times 3 = 12$. Thus $P(E_2|E_1) = 12/36 = 1/3$.

Similarly, to compute $P(E_1|E_2)$, we know that one of the 36 simple events in event E_2 has occurred and we want to know the likelihood that one of the 12 events in $E_1 \cap E_2$ has also occurred. Thus, $P(E_1|E_2) = 12/36 = 1/3$.

There is another way to find these conditional probabilities. In computing $P(E_2|E_1)$, we know that the first draw is red. Thus, there are 9 marbles left to be drawn on the second draw, of which 3 are red. Thus $P(E_2|E_1) = 3/9 = 1/3$.

Likewise, for $P(E_1|E_2)$, we draw a marble, put it in our pocket, draw a second marble, look at it, and observe that it is red. What is the probability that the first marble was also red? We know that the first marble is **not** the red marble we got on the second draw. Thus, it could be any of the other 9 marbles. To be red, it would have to be one of the other 3 red marbles. Thus, $P(E_1|E_2) = 3/9 = 1/3$.

Note that $P(E_2|E_1) = 1/3 < 0.4 = P(E_2)$, that is, given that a red marble was drawn on the first draw, it is less likely that a red marble will be drawn on the second draw, since there are fewer red marbles in the bag when the second draw is made.   ■

PRINCIPLE OF CONDITIONAL PROBABILITY: To compute $P(F|E)$ we first need to compute how many simple events are possible, that is, $\#E$. Next, we need to determine the number of events in F, given that E has occurred. But this is just the number of events in $E \cap F$, that is, $\#(E \cap F)$. By the counting principle for probability, we know that

$$P(F|E) = \frac{\#(E \cap F)}{\#E}. \qquad \square$$

EXAMPLE 3.23 Consider a bag that contains 4 red marbles and 6 blue marbles. Draw 2 marbles from the bag, in order and with replacement. Let E_1 be the event that the first marble drawn is red, and let E_2 be the event that the second marble drawn is red. Let's compute $P(E_1)$, $P(E_2)$, $P(E_1 \cap E_2)$, $P(E_2|E_1)$, and $P(E_1|E_2)$.

The total number of simple events, that is, the total number of ways of drawing 2 marbles, in order and with replacement, from a bag containing 10 marbles, is $\#T = 10^2 = 100$. Let's use the multiplication principle to compute the number of simple events that compose E_1. This is a 2-stage task. For E_1 to occur, the result of the first draw or stage must be red, and there are 4 ways for this to happen. There are no restrictions on the second stage or draw, so this may occur in 10 ways, since any of the 10 marbles may be drawn. By the multiplication principle, E_1 is composed of $4 \times 10 = 40$ simple events. Thus $P(E_1) = 40/100 = 0.4$.

Alternatively, since E_1 is only concerned with the first draw, let's restrict our attention to the first draw. The total number of simple events that may occur on the first draw is 10, that is, any one of the 10 marbles. The total number of ways event E_1 can occur is 4, that is, any one of the 4 red marbles being drawn. Thus, $P(E_1) = 4/10 = 0.4$.

Similarly, E_2 is composed of $10 \times 4 = 40$ simple events, and therefore, $P(E_2) = 40/100 = 0.4$, (or there are 10 possible marbles on the second draw and there are 4 possible red marbles).

Event $E_1 \cap E_2$ is the event that a red was drawn on the first and on the second draw. By the multiplication principle, and since the drawing was with replacement, event $E_1 \cap E_2$ can be done in $4 \times 4 = 16$ ways, and thus $P(E_1 \cap E_2) = 16/100 = 0.16$.

To compute $P(E_2|E_1)$, we observe that one of the simple events in event E_1 has occurred. Thus the total number of events that we must consider is $\#E_1 = 40$, not the total of $\#T = 100$. Thus we know that in the 2 draws, the first draw was red. How many ways can E_2 occur, given that the first marble was red, that is, how many simple events are there in which the first marble is red (given) and the second marble is also red (event E_2)? These are just the simple events in $E_1 \cap E_2$, which was computed as 16. Thus, by the counting principle for conditional probability,

$$P(E_2|E_1) = \frac{\#(E_1 \cap E_2)}{\#E_1} = 16/40 = 0.4.$$

Observe that

$$P(E_1 \cap E_2) = 16/100 = (40/100) \times (16/40) = P(E_1)P(E_2|E_1).$$

Similarly, $\#E_2 = 40$ and $\#(E_1 \cap E_2) = 16$, so $P(E_1|E_2) = 16/40 = 0.4$.

The conditional probabilities could have been worked differently. For $P(E_2|E_1)$ we know that the first draw is red. We now replace that marble. What is the probability that the second marble is also red? Since there are 10 marbles in the bag when we make the second draw, and 4 of those marbles are red, $P(E_2|E_1) = 4/10 = 0.4$. Likewise, if someone draws a marble, looks at it, returns it to the bag, draws a second marble, and tells us the second marble is red, then, as far as we are concerned, there are still 10 marbles that could have drawn on the first draw of which 4 are red. Thus $P(E_1|E_2) = 4/10 = 0.4$.

Notice in this example that $P(E_2|E_1) = P(E_2)$, that is, knowing what happened at the first stage of the process does not affect what will happen at the second stage of the task. ∎

DEFINITION 3.24 *Suppose we have a 2-stage task. Let event E_1 be that a certain result happens on the first stage, (as in getting a red marble on the first draw). Let event E_2 be that a certain result happens on the second stage, (as in getting a red marble on the second draw). If*

$$P(E_2) = P(E_2|E_1),$$

that is, if knowing the result of the first stage tells us nothing new about the likelihood of results of the second stage, then we say that events E_1 and E_2 are independent events.

REMARK: In fact if E and F are **any** two events such that $P(F) = P(F|E)$, then E and F are independent events. In the applications we consider, our events tend to be the results of different stages and it is usually clear if they are independent events. ☐

Suppose we are drawing marbles, in order, from a bag containing r red marbles and b blue marbles. Let E_1 be the event that we get a red marble on the first draw and let E_2 be the event that we get a red on the second draw. If we are drawing **with replacement**, then events

E_1 and E_2 are clearly independent, since

$$P(E_2) = \frac{r}{r+b} = P(E_2|E_1).$$

If we are drawing **without replacement**, events E_1 and E_2 are not independent since

$$P(E_2) = \frac{r}{r+b}, \qquad \text{but} \qquad P(E_2|E_1) = \frac{r-1}{r-1+b}.$$

In this case, E_1 and E_2 are said to be **dependent events**.

Events E_1 and E_2 of example 3.22 are dependent, since knowing that a red marble was drawn on the first draw makes it less likely that a red marble will be drawn on the second draw. Events E_1 and E_2 of example 3.23 are independent since the marble drawn on the first draw is replaced, and thus the result of the second draw is independent of what happened on the first draw.

We now develop a method for simplifying computations involving conditional probabilities. Recall that

$$P(E_1) = \frac{\#E_1}{\#T}, \qquad P(E_1 \cap E_2) = \frac{\#(E_1 \cap E_2)}{\#T},$$

and

$$P(E_2|E_1) = \frac{\#(E_1 \cap E_2)}{\#E_1}.$$

Observe that

$$P(E_1)P(E_2|E_1) = \left(\frac{\#E_1}{\#T}\right)\left(\frac{\#(E_1 \cap E_2)}{\#E_1}\right) = \frac{\#(E_1 \cap E_2)}{\#T} = P(E_1 \cap E_2).$$

Thus we have proved the following theorem.

THEOREM 3.25 Multiplication principle for probability *Suppose that events E and F are any two events. Then*

$$P(E \cap F) = P(E)P(F|E). \tag{3.6}$$

*If E and F are **independent events**, that is, if $P(F|E) = P(F)$, then*

$$P(E \cap F) = P(E)P(F). \tag{3.7}$$

Suppose we are studying a multistage process in which E_1 and E_2 are independent events corresponding to results of the first and second stages, respectively. The multiplication principle for counting says that we multiply the number of ways of performing each stage to get the number of ways of performing both stages. Equation (3.7) says that we multiply the probability of each event occurring to get the probability that both events occur. In reality, theorem 3.25 is just the multiplication principle for counting in a disguised form.

Equation (3.6) can be used to compute probabilities for tasks in which we are **drawing in order but without replacement**, that is, when events E_1 and E_2 are dependent. Note that by equation (3.6) we have that

$$P(E_1 \cap E_2) = P(E_1)P(E_2|E_1) = P(E_2)P(E_1|E_2).$$

When using theorem 3.25, it is usually easy to see which form to use. For example, if E_1 is the result of the first stage and E_2 is the result of the second stage, then the form $P(E_1)P(E_2|E_1)$ is often the easiest way to work the problem.

EXAMPLE 3.26   Suppose we draw 2 marbles, in order but without replacement, from a bag containing 5 red and 7 blue marbles. What is the probability that both marbles are red? Let E_1 be the event that the first marble is red and let E_2 be the event that the second marble is red. We wish to compute $P(E_1 \cap E_2)$.

Applying theorem 3.25, we have that $P(E_1 \cap E_2) = P(E_1)P(E_2|E_1)$. Since there are 12 marbles in the bag of which 5 are red, we know that $P(E_1) = 5/12$. Given that the first marble drawn is red, we know there are 11 marbles left in the bag of which 4 are red, so $P(E_2|E_1) = 4/11$. Thus,

$$P(E_1 \cap E_2) = (5/12) \times (4/11) = 20/132. \qquad \blacksquare$$

Theorem 3.25 can also be helpful in computing the probability of certain events when the simple events **are not equally likely**.

EXAMPLE 3.27   Suppose we have two bags of marbles. The first bag contains 4 red marbles and 6 blue marbles. The second bag contains 4 red marbles and 1 blue marble. A fair coin is flipped. If heads comes up, we draw a marble from the first bag. If tails comes up, we draw a marble from the second bag. What is the probability that the marble we draw is red?

In this task, each event is not equally likely. Since there are fewer marbles in bag 2, any particular marble in bag 2 is more likely to be drawn then a particular marble in bag 1. Let's see how theorem 3.25 and the addition principle for probability can eliminate our problems.

Denote the event that the marble is red by R, the event that heads comes up on the coin by H and the event that tails comes up on the coin by T. The events that compose R can be broken into 2 cases. The first case is that heads comes up on the coin and then we get a red marble, which we recognize as $H \cap R$. The second case is that tails comes up on the coin and then we get a red marble, which we recognize as $T \cap R$. By the addition principle for probability.

$$P(R) = P(H \cap R) + P(T \cap R).$$

By theorem 5.12, we know $P(H \cap R) = P(H)P(R|H)$ and $P(T \cap R) = P(T)P(R|T)$. We know that $P(H) = 0.5 = P(T)$. To compute $P(R|H)$, we are given that the coin came up heads, and thus we are drawing from a bag that contains 10 marbles of which 4 are red. Thus the probability we get a red marble from this bag is 4/10, so $P(R|H) = 4/10 = 0.4$. Likewise, for $P(R|T)$ we are drawing from a bag containing 5 marbles of which 4 are red and so $P(R|T) = 4/5 = 0.8$. Combining these results, we have that

$$P(R) = 0.5 \times 0.4 + 0.5 \times 0.8 = 0.6.$$

Note that events R and T are not independent, since knowing the result of the coin flip gives you information about the likelihood of getting a red marble on the second stage, that is,

$$P(R|T) = 0.8 \neq 0.6 = P(R). \qquad \blacksquare$$

How can we use theorem 3.25 when events E_1 and E_2 are independent, that is, how can we tell if two events are independent? The answer is that it is usually obvious that the events are independent.

EXAMPLE 3.28   Suppose that a bag contains 3 red marbles and 7 blue marbles. Two marbles are drawn at random, in order and with replacement. Let event E_1 be that the first marble is red, and let event E_2 be that the second marble is blue. Find $P(E_1 \cap E_2)$, that is, find the probability that the first marble is red and the second is blue.

Since we are replacing the first marble before we draw the second marble, the color of the first marble has no effect on the color of the second marble, that is, the events are independent. Thus, by equation (3.7)

$$P(E_1 \cap E_2) = P(E_1)P(E_2) = (3/10) \times (7/10) = 0.21. \qquad \blacksquare$$

REMARK: Often two events, E_1 and E_2 are defined to be indepen-dent if $P(E_1 \cap E_2) = P(E_1)P(E_2)$. But this definition usually makes it less convenient to actually determine if events are independent. Using definition 3.24, that $P(E_2) = P(E_2|E_1)$, it is usually obvious if the events are independent. Such was the case in examples 3.27 and 3.28.   □

Many problems use a combination of the addition principle and theorem 3.25.

EXAMPLE 3.29   Let's play baseball. The score is tied, bases loaded and two outs in the bottom of the ninth inning, with the home team at bat. The man at bat has a probability of 0.3 of driving in the winning run, and thus a probability of 0.7 of getting an out. If the game goes into extra innings, the home team has a probability of 0.5 of winning the game. What is the probability the home team will win, which we will denote as $P(W)$?

There are 2 cases. The first case is that the man at bat will hit in the winning run, which we will call event H. The second case is that the batter gets an out but the home team wins in extra innings, which we will call event $H' \cap W$. By the addition principle,

$$P(W) = P(H) + P(H' \cap W).$$

We are told that $P(H) = 0.3$. The event $H' \cap W$ is the intersection of the event that the man at bat gets an out H' and the event that the team will win in extra innings W. By theorem 3.25 $P(H' \cap W) = P(H')P(W|H')$. We are told that $P(H') = 0.7$ and that $P(W|H') = 0.5$. Therefore, $P(H' \cap W) = 0.7 \times 0.5 = 0.35$. Thus,

$$P(W) = 0.3 + 0.35 = 0.65. \qquad ■$$

EXAMPLE 3.30   Tennis players usually have a fast serve and a slow serve. The fast serve is less likely to be in than the slow serve, but is more likely to win the point if it is good. Specifically, suppose Ace Johnson, a professional tennis player, has a probability of 0.5 of getting his fast serve in, and a probability of 0.9 of getting his slow serve in. Given that his fast serve is in, the probability that he will win the point is 0.8. Given that his slow serve is in, the probability he will win the point is just 0.6, since the slower serve is easier to return. Ace, as server, gets two tries to get a serve in. What is the probability that Ace will win the next point if he serves fast on his first serve and, should his first serve be out, slow on his second serve?

Let W represent the event that Ace wins the point. There are two cases to consider. Case 1 is the event that his first serve is in, which we denote by S_1, and he then goes on to win the point. Case 2 is the event that his first serve is out, which we denote by S_1', and he then goes on to win the point with his second serve.

In case 1 we must compute the probability that the first serve is in and Ace wins the point, $P(W \cap S_1)$. By theorem 3.25, we get $P(W \cap S_1) = P(S_1)P(W|S_1)$. But we were given that $P(S_1) = 0.5$ and that $P(W|S_1) = 0.8$. So

$$P(W \cap S_1) = 0.5 \times 0.8 = 0.4.$$

Case 2 is that the first serve is out, and Ace wins the point, $P(W \cap S_1')$. By theorem 3.25, $P(W \cap S_1') = P(S_1')P(W|S_1')$. We know that $P(S_1') = 0.5$, since the probability he misses his first serve is 0.5. What is the probability he wins the point given that he missed his first serve, that is, what is $P(W|S_1')$? To win the point given that he missed his first serve, his second serve must be good and then he must win the point, which we will denote $P(W \cap S_2)$, that is, $P(W|S_1') = P(W \cap S_2)$.

Again, by theorem 3.25

$$P(W|S_1') = P(W \cap S_2) = P(S_2)P(W|S_2) = 0.9 \times 0.6 = 0.54.$$

Therefore,

$$P(S_1' \cap W) = P(S_1')P(W|S_1') = 0.5 \times 0.9 \times 0.6 = 0.27.$$

Notice that this is just a 3-stage process, miss the first serve, get the second serve in, and then win the point. The 0.5 was the probability of the first stage, missing the serve. The 0.9 was the probability of the second stage, getting the second serve in given that we missed our first serve. The 0.6 was the probability of the third stage, winning the point, given that the second serve was in.

By the addition principle,

$$P(W) = P(W \cap S_1) + P(W \cap S_1') = 0.4 + 0.27 = 0.67.$$

In problems 1 and 2 you will compute the probability that Ace wins the point using different serving strategies. You will find that his best strategy is to serve fast on his first serve and slow on his second serve. This can be observed in watching professional tennis players. This is not always the best strategy though. The best strategy for a player depends on the particular probabilities for getting serves in and for winning points on those serves. For weekend players, both serves slow is often the best approach.   ■

PROBLEMS

1. In example 3.30, what is the probability that Ace wins the point if he serves fast on both his first serve and his second serve?

2. In example 3.30, what is the probability that Ace wins the point if he serves slow on both his first serve and his second serve?

3. We have two bags, the first bag contains 3 red marbles and 7 blue marbles, while the second bag contains 2 red marbles and 2 blue marbles. A marble is drawn from the first bag and put into the second bag. A marble is then drawn from the second bag. What is the probability that the second marble is red?

4. Suppose you have a bag containing 2 blue marbles. You roll a die and whatever number comes up, you put that many red marbles into the bag. Then you draw a marble from the bag. What is the probability that the marble drawn is red? (Hint: consider 6 cases, the six results of the die, and use the addition principle.)

5. If Donald Pagan is elected mayor, the probability he will raise taxes is 0.2, while if Falter Mongoose is elected mayor, the probability he will raise taxes is 0.9. Suppose the probability that Pagan gets elected is 0.6, and that Mongoose gets elected is 0.4. What is the probability taxes will be raised?

6. The probability that the Smashers will beat the Crushers on any given game is 0.6. These two teams are in a three game playoff, that is, the first team to win 2 games becomes champion. What is the probability that the Smashers will win the championship? (Hint: there are 3 cases, the Smashers can (W,W), (W,L,W), or (L,W,W), where W stands for win and L stands for lose.)

7. Answer problem 6, given that they are in a 5 game playoff, that is, the first team to win 3 games is the champion.

8. In tennis, a player must win 4 points to win a game, but must also win by 2 points. (These 4 points are called 15, 30, 40 and game.) A player with no points is at "love". Suppose two players, Ace and Spike, are equally good tennis players, that is, the probability that Ace will win a specific point is 0.5. Therefore, if the players are tied with 1 point each (15 all), 2 points each (30 all) or 3 points each (deuce), the probability is 0.5 that Ace will win the game.

   a. Suppose Ace is at 40 and Spike is at 30. What is the probability that Ace will win the game?

    b. Suppose Ace is at 40 and Spike is at 15. What is the probability that Ace will win the game? (Hint: consider 2 cases and use your answer to part a).)

    c. Suppose Ace is at 40 and Spike is at love. What is the probability that Ace will win the game?

    d. Suppose Ace is at 30 and Spike is at 15. What is the probability that Ace will win the game?

    e. Suppose Ace is at 30 and Spike is at love. What is the probability that Ace will win the game?

    f. Suppose that Ace is at 15 and Spike is at love. What is the probability that Ace will win the game?

9. Three coins are in a bag, one with heads on both sides, one with tails on both sides, and one with heads on one side and tails on the other. One coin is drawn at random from the bag, placed on the table, and it is observed that heads is showing. What is the probability that heads is also on the other side? (Be careful. The answer is not 1/2.)

3.4 An introduction to Markov chains

In this section we will study a type of probability problem which is indicative of a large class of problems. Let's start with a simple example. Consider two bags containing marbles. The first bag contains 3 red marbles and 2 blue marbles. The second bag contains 3 red marbles and 7 blue marbles. A marble is drawn from the first bag, its color is recorded, and it is returned to the bag from which it was drawn. If the first marble drawn was **red**, we draw the second marble from **bag 1**. If the first marble drawn was **blue**, we draw the second marble from **bag 2**.

We now repeat the entire process with the second marble, the third marble, and so forth. Thus, we draw the nth marble, record its color, and return it to the bag from which it was drawn. If the nth marble is **red**, we draw the $n+1$th marble from **bag 1**, but if the nth marble is **blue**, we draw the $n+1$th marble from **bag 2**. What is the probability that the kth marble drawn will be red? This process is called a **Markov chain**.

A **Markov chain** is a multistage process (drawing a marble at each stage), in which the same results (red or blue) are possible after each stage (draw). The probability of each result (red or blue) on the $n+1$th stage depends only on the result of the nth stage. For the above problem, this means that we know the probability of getting a red or blue marble on the $n + 1$th draw once we are told the color of the nth marble drawn. For example, if the nth marble is red, we draw the $n + 1$th marble from bag 1 and so the probability it is red is $3/5 = 0.6$, while if the nth marble is blue, we draw the $n + 1$th marble from bag 2 and so the probability it is red is $3/10 = 0.3$.

To study this situation, let $p(n)$ be the probability that the nth marble is red, and let $q(n) = 1 - p(n)$ be the probability that the nth marble is blue. Since the first marble is being drawn from bag 1, we know that $p(1) = 0.6$ and $q(1) = 0.4$.

What is $p(2)$, that is, what is the probability that the second marble drawn is red? There are two cases or ways in which a red marble can be drawn on the second draw. The first case is that the first marble is red and the second marble is red, denoted rr and its probability of happening is $p(rr)$, while the second case is that the first marble is blue and the second marble is red, denoted br with probability $p(br)$. To compute $p(2)$, we need to compute the probability of each of the two cases and then add them together, that is,

$$p(2) = p(rr) + p(br).$$

This is a result of theorem 3.16, the **addition principle** for probability.

What is the probability of getting rr? This is a two-stage process. The first stage is to get a red on draw 1 and the probability of this happening is $p(1) = 0.6$. The second stage is to get a red on draw 2 **given that we got a red on draw 1**, that is, given that we are drawing from bag 1. The probability of this happening is 0.6. By theorem 3.25, the multiplication principle for probability, the probability of getting a red on the first and second draw is the product of the individual probabilities, that is,

$$p(rr) = 0.6p(1) = 0.6^2 = 0.36.$$

This can be seen as the top branch of the tree in figure 3.2.

What is the probability of getting br? This is a two-stage process, also. The first stage is to get a blue on draw 1, and the probability of this happening is $q(1) = 0.4$. The second stage is to get a red on draw 2 **given that we got a blue on draw 1**, that is, given that we are drawing from bag 2. The probability of this happening is 0.3. Again by the multiplication principle, the probability of case 2 is

$$p(br) = 0.3q(1) = 0.3 \times 0.4 = 0.12.$$

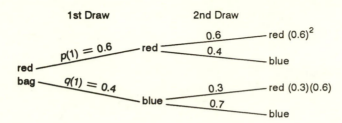

FIGURE 3.2. Tree diagram showing ways of getting a red marble on the second draw of an experiment. The sum of the two ways is $(0.6)^2 + (0.3)(0.4) = 0.48$.

This is represented as a lower branch of the tree in figure 3.2. Adding these together, we get

$$p(2) = 0.6p(1) + 0.3q(1) = 0.36 + 0.12 = 0.48.$$

We could use the same logic to compute $q(2)$, the probability of getting a blue marble on the second draw, but there is a simpler method. Since we must get a red or a blue marble on draw 2 and we know the probability that it is red equals 0.48, then the probability it is blue must equal $1 - 0.48 = 0.52$, that is,

$$q(2) = 1 - p(2) = 0.52.$$

What is $p(n + 1)$, that is, what is the probability that the $n + 1$th marble drawn is red? There are two cases. Case 1 is that the nth marble is red and the $n + 1$th marble is red, denoted rr; while case 2 is that the nth marble is blue and the $n + 1$th marble is red, denoted br. (Note that we are listing only the **last two** marbles drawn.) To compute $p(n+1)$, we need to compute the probability of each of the two cases and then add them together, that is,

$$p(n + 1) = p(rr) + p(br).$$

What is the probability of getting rr? This is a two-stage process. The first stage is to get a red on draw n, and the probability of this happening is $p(n)$. The second stage is to get a red on draw $n + 1$ **given that we got a red on draw** n, that is, given that we are drawing from bag 1. The probability of this happening is 0.6. Thus, the probability of case 1 is, using the multiplication principle,

$$p(rr) = 0.6p(n).$$

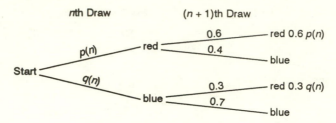

FIGURE 3.3.   Tree diagram showing ways of getting a red marble on the $n+1$th draw of an experiment. The sum of the two ways is $0.6p(n) + 0.3q(n)$.

This is represented by the upper branch in the tree diagram in figure 3.3.

Similarly, the probability of getting br, again by the multiplication principle, is

$$p(br) = 0.3q(n).$$

This is represented in a lower branch of the tree diagram in figure 3.3.

Adding these together, we get

$$p(n+1) = 0.6p(n) + 0.3q(n).$$

Since we must get a red or blue marble on the nth draw, we must have

$$p(n) + q(n) = 1, \qquad \text{or} \qquad q(n) = 1 - p(n).$$

Substitution into the above equation gives the first order affine dynamical system

$$p(n+1) = 0.6p(n) + 0.3(1 - p(n)) = 0.3p(n) + 0.3.$$

The equilibrium value for this equation is $p = 3/7$, so the general solution is

$$p(k) = c(0.3)^k + 3/7.$$

Using the fact that $0.6 = p(1) = 0.3c + 3/7$, we get $c = 4/7$ and the particular solution is

$$p(k) = (4/7)(0.3)^k + 3/7.$$

Alternatively, since the first marble is drawn from bag 1, this means an imagined marble 0 "was red", that is, $p(0) = 1$. This also gives $c = 4/7$.

Observe that

$$\lim_{k \to \infty} p(k) = 3/7,$$

that is, the probability that the 1000th marble is red is essentially 3/7. Thus, the probability that the 1000th marble is blue is 4/7. In effect, this says that approximately three-sevenths of the marbles drawn will be red and four-sevenths blue.

Suppose we keep all the rules the same for our marble drawing problem, except that we draw the first marble from bag 2 instead of bag 1, that is, $p(0) = 0$. Then the particular solution is

$$p(k) = -(3/7)(0.3)^k + 3/7,$$

but we still have that

$$\lim_{k \to \infty} p(k) = 3/7.$$

Observe that the initial condition does not affect the **long range probability**, which seems to make sense, since getting a red marble on the first draw should have little effect on our getting a red marble on the 1000th draw. To see this mathematically, we consider the general solution. In particular, we note that

$$\lim_{k \to \infty} p(k) = \lim_{k \to \infty} \left[c(0.3)^k + 3/7 \right] = 3/7.$$

Also $\lim_{k \to \infty} q(k) = \lim_{k \to \infty} [1 - p(k)] = 4/7$. We also see that, for this example, the initial value $p(0)$ plays no part in determining the long term probabilities $p(k)$ and $q(k)$.

This example is called a **regular Markov chain** with two states, red and blue. In section 8.4 we will consider the general theory of Markov chains in which there may be more than two states, that is, our bags may contain marbles of more than two colors.

PROBLEMS

1. Suppose we play our marble drawing game, but the first bag contains 4 red and 2 blue marbles and the second bag contains 3 red and 5 blue marbles.
   a. Find the affine dynamical system that models this situation.
   b. Find the general solution and the long term probabilities for $p(k)$ and $q(k)$, that is, the limit as k goes to infinity. (Note that although half the marbles are red, the long range probability of drawing a red marble is slightly greater than one-half.)
   c. Find the probability that the fourth marble is red given that we draw the first marble from the first bag.

d. Suppose we flip a fair coin and if we get heads we draw our first marble from the first bag, but if we get tails, we draw our first marble from the second bag. Find the particular solution and $p(4)$.

2. Suppose there are two competing hamburger chains, A and B. Suppose a research firm has determined that if a customer eats a hamburger from chain A, then the probability they will buy their next hamburger from chain A is 0.7 while the probability they buy their next hamburger from chain B is 0.3. Similarly, they have determined that if a customer eats a hamburger from chain B, then the probability they will buy their next hamburger from chain B is 0.6 while the probability they buy their next hamburger from chain A is 0.4. Determine the long term market share of each chain, that is, determine the fraction of hamburgers bought from chain A and the fraction bought from chain B.

3. Suppose that if it rains one day, the probability it will rain the next day is 0.5, while if it does not rain one day, the probability it will rain the next day is only 0.2.
    a. If it rains Wednesday, what is the probability it will rain Saturday?
    b. What fraction of the days does it rain?

4. If you win a game of chess against your friend, the probability you will win the next game is 0.7, while if you lose, then the probability you will win the next game is 0.4. What fraction of the games of chess do you win when playing your friend?

5. If you eat a doughnut for breakfast today, then the probability you will eat a doughnut for breakfast tomorrow is 0.1. If you do not eat a doughnut for breakfast today, then the probability you will eat a doughnut for breakfast tomorrow is 0.4.
    a. If you do not eat a doughnut for breakfast on Monday, what is the probability you will eat a doughnut for breakfast on Friday?
    b. On average, how many doughnuts do you eat in a week?

6. Suppose a bag contains 2 red marbles and 15 blue marbles. A marble is drawn from the bag, its color is noted and it is returned to the bag. If the marble drawn was blue, another marble is drawn, while if the marble drawn was red, the game ends. What is the probability that k or fewer marbles will be drawn?

CHAPTER 4

Nonhomogeneous dynamical systems

4.1 Exponential terms

Let's reconsider our study of savings accounts. The affine dynamical system

$$a(n+1) = (1+I)a(n) + b$$

describes a savings account paying $100I$ per cent interest, compounded annually, in which we add an additional b dollars each year. In reality, our salary will hopefully be increasing, so each year we will be able to add more money to our savings account than we did the previous year. To study this, we need to consider dynamical systems in which we do not add a constant b, but instead add an amount that is a function of time, that is, dynamical systems of the form

$$a(n+1) = ra(n) + g(n). \tag{4.1}$$

This is called a **first-order nonhomogeneous dynamical system.**

Assume that $a(k) = a$ is a constant solution to the nonhomogeneous dynamical system. Then, by substitution of $a(n+1) = a$ and $a(n) = a$, we get

$$a = ra + g(n),$$

that is, $g(n) = (1 - r)a$ for every value of n. But then the function g is constant and does not depend explicitly on n. The dynamical system is thus affine. Nonhomogeneous dynamical systems **do not have fixed points unless they are affine.**

Let's consider a savings account in which we increase our deposit by 10 per cent each year, that is, if at the end of the first year we deposit 100, then at the end of the second year we deposit $100 \times 1.1 = 110$, at the end of the third year we deposit $110 \times 1.1 = 121 = 100 \times 1.1^2$, and so forth. We thus have two quantities that are changing, the amount of our deposit and the amount in our account. Let $a(n)$ be the amount in our account at the beginning of year n, immediately after interest and the additional deposit from the previous year have been added. Let $b(n)$ be the amount we deposit in our account at the **end** of the nth year.

We have that $b(1) = 100$. The assumption that we increase our deposit by 10 per cent each year translates into the equations $b(2) = b(1) + 0.1b(1) = 1.1 \times 100$, $b(3) = 1.1b(2)$, ..., that is, into the dynamical system

$$b(n+1) = 1.1b(n) \qquad \text{for} \quad n = 1, 2, \ldots.$$

The particular solution to this dynamical system is

$$b(k) = 100 \times 1.1^{k-1}.$$

The amount in our account satisfies the equations

$$a(1) = (1 + I)a(0) + b(1), \quad a(2) = (1 + I)a(1) + b(2), \quad \ldots,$$

which can be summarized by the dynamical system

$$a(n+1) = (1 + I)a(n) + b(n+1), \qquad \text{for} \quad n = 0, 1, \ldots.$$

We now substitute our solution $b(n + 1) = 100 \times 1.1^n$ into this dynamical system, giving the nonhomogeneous dynamical system

$$a(n+1) = (1 + I)a(n) + 100 \times 1.1^n, \qquad \text{for} \quad n = 0, 1, \ldots.$$

In this case, $g(n) = 100 \times 1.1^n$.

REMARK: Notice that we were actually studying two quantities that were changing. One quantity $b(n)$ depended only on its previous values and we could solve for it. The second quantity $a(n)$ depended on the previous values of both of the quantities. But the $b(n)$ term could be eliminated by substituting its solution. This process is how many nonhomogeneous dynamical systems are derived. □

In this section, we want to find the general and particular solutions to first-order nonhomogeneous dynamical systems when the function $g(n)$ is an exponential, that is, when

$$g(n) = bs^n$$

where b and s are constants.

Let's rewrite the affine dynamical system $a(n+1) = ra(n) + b$ (with $r \neq 1$) as

$$a(n+1) = ra(n) + b1^n$$

and its general solution as

$$a(k) = cr^k + a1^k,$$

where a is the equilibrium value $a = b/(1-r)$. This should indicate that the general solution of the dynamical system

$$a(n+1) = ra(n) + bs^n$$

might be of the form

$$a(k) = cr^k + as^k,$$

where the constant a depends on the system and the constant c depends on the initial value.

EXAMPLE 4.1   Consider the dynamical system

$$a(n+1) = 3a(n) + 4^n.$$

In this case, $r = 3$, $s = 4$, and $b = 1$. From the previous discussion, we might guess that the general solution is of the form

$$a(k) = c3^k + a4^k,$$

where a is a constant which can be determined from the dynamical system, while c is determined from the initial value. There is nothing wrong with making intelligent guesses as long as we verify that our guess is correct.

By our guess $a(n) = c3^n + a4^n$ and $a(n + 1) = c3^{n+1} + a4^{n+1}$. Substitution into the dynamical system gives

$$c3^{n+1} + a4^{n+1} = 3(c3^n + a4^n) + 4^n.$$

Simplification gives

$$3c3^n + 4a4^n = 3c3^n + 3a4^n + 4^n.$$

After subtracting the term $3c3^n$ from both sides and then dividing by 4^n, we have

$$4a = 3a + 1, \quad \text{or} \quad a = 1.$$

Thus the general solution is

$$a(k) = c3^k + 4^k.$$

Let's find the particular solution when $a_0 = 2$. Since $2 = a(0) = c3^0 + 4^0 = c + 1$, it follows that $c = 1$, and

$$a(k) = 3^k + 4^k. \qquad \blacksquare$$

For the dynamical system

$$a(n+1) = ra(n) + bs^n \qquad (4.2)$$

we expect the solution to be of the form

$$a(k) = cr^k + as^k.$$

Substitution back into the dynamical system gives

$$cr^{n+1} + as^{n+1} = r\left(cr^n + as^n\right) + bs^n$$

or, after cancellation and division by s^n,

$$as = ra + b.$$

This gives that $a = b/(s - r)$. Thus, the general solution to the above dynamical system is, when $r \neq s$,

$$a(k) = cr^k + \left(\frac{b}{s - r}\right)s^k. \qquad (4.3)$$

The case in which $r = s$ is considered in section 4.2.

Dynamical systems involving exponentials arise in problems related to the environment. Let's see one particular example in which we study the pollution in the Great Lakes.

EXAMPLE 4.2 Most of the water flowing into Lake Ontario is from Lake Erie. Suppose that the pollution of these lakes ceased suddenly. How long would it take for the pollution level in each lake to be reduced to 10 per cent of its present level?

First, to simplify matters, let's assume that 100 per cent of the water in Lake Ontario comes from Lake Erie and all of the water that leaves Lake Erie goes into Lake Ontario. Let $a(n)$ and $b(n)$ be the total amount of pollution in Lake Ontario and Lake Erie, respectively, in year n. Since pollution has stopped, the concentration of pollution in the water coming into Lake Erie is zero. It has also been determined that, each year, the percentage of the water replaced in Lakes Erie and Ontario is approximately 38 and 13 per cent, respectively. This means that each year, 38 per cent of the water in Lake Erie flows into Lake Ontario and is replaced by rain and pure water flowing in from other sources. Also each year, 13 per cent of Lake Ontario's water flows out and is replaced by the water flowing in from Lake Erie.

Assuming the concentration of the pollution in each lake is constant throughout that lake, we have that 38 per cent of the pollution in Lake Erie is removed each year. Thus, the dynamical system

$$b(n+1) = 0.62b(n),$$

describes how the pollution is changing in Lake Erie from year to year.

Each year 13 per cent of the pollution in Lake Ontario is removed, but the pollution, $0.38b(n)$, that was removed from Lake Erie is added to Lake Ontario. This means that the pollution in Lake Ontario is described by the dynamical system

$$a(n+1) = 0.87a(n) + 0.38b(n).$$

The particular solution for the first equation is

$$b(k) = (0.62)^k.$$

For simplicity let's assume that $b_0 = 1$ unit of pollution. The first question was, how long will it take for the amount of pollution $b(k)$ in Lake Erie to be equal to 10 per cent of its present value of $b_0 = 1$? That is, letting $b(k) = 0.1$ in the formula for the solution, find k. Substitution gives

$$0.1 = (0.62)^k.$$

Taking logarithms gives

$$k \log 0.62 = \log 0.1, \quad \text{or} \quad k = 5$$

after rounding up to the next integer.

Substituting the solution for Lake Erie into the dynamical system for Lake Ontario gives

$$a(n+1) = 0.87a(n) + 0.38(0.62)^n.$$

Using solution (4.3) and $a(0) = a_0$, you should find that the particular solution is

$$a(k) = (a_0 + 1.52)(0.87)^k - 1.52(0.62)^k.$$

We cannot solve for k, the time it takes for $a(k)$ to decrease to $0.1a_0$, since there are two unknowns in this equation. To simplify matters, we observe that Lake Ontario is approximately three times the size of Lake Erie, so we assume it has three times the pollution but the same initial concentration of pollution, that is, $a_0 = 3b_0 = 3$. Substitution into the above formula gives

$$a(k) = 4.52(0.87)^k - 1.52(0.62)^k.$$

Substituting $0.1a_0 = 0.3$ for $a(k)$ gives

$$0.3 = 4.52(0.87)^k - 1.52(0.62)^k.$$

We cannot explicitly solve for k in this equation. In calculus, a technique called **Newton's method** is discussed which can be used to find the zeros of functions. This technique, or trying different values of k until finding one that sufficiently satisfies this equation, gives that $k = 19.47$. Spreadsheets or calculators could be used to find that $19 < k < 20$. So it will take over 20 years for Lake Ontario to become as unpolluted as Lake Erie will become in 5 years.

Among some of the unrealistic assumptions that we were making in this example are that 1) all pollution is soluble; 2) no rain goes into Lake Ontario; and 3) no water evaporates. As we have said before, successful modeling requires making simplifying assumptions. Even with the omission of numerous important components of this problem, it was still somewhat difficult to study. ∎

In section 2.1, we discussed radioactive decay. To review, we assumed that some fixed proportion p of the atoms of material b decayed each time period. This led to the dynamical system

$$b(n + 1) = b(n) - pb(n)$$

where $b(n)$ was the amount of material b after n time periods. The solution to this problem is

$$b(k) = (1 - p)^k b(0)$$

which gives the amount of b after k time periods.

For some radioactive materials, each atom of one material that decays, decays into an atom of a second radioactive material. These two radioactive materials form what is called a **radioactive chain.**

EXAMPLE 4.3 Assume that 10 per cent of radioactive material b and 60 per cent of radioactive material a decay each time period, respectively. Also assume that each atom of material b decays into an atom of material a. Thus, the number of atoms of material b is described by the dynamical system

$$b(n+1) = b(n) - 0.1b(n).$$

Since each atom of b changes into an atom of a, the loss to b becomes a gain for a giving the dynamical system

$$a(n+1) = a(n) - 0.6a(n) + 0.1b(n)$$

describing how material a changes.

The equation for $b(n)$ can be simplified to $b(n+1) = 0.9b(n)$ which has the solution $b(k) = (0.9)^k b_0$. Substituting this into the equation for $a(n)$ gives the nonhomogeneous dynamical system

$$a(n+1) = 0.4a(n) + 0.1(0.9)^n b_0.$$

By equation (4.3), the solution to this dynamical system is

$$a(k) = c(0.4)^k + 0.2b_0(0.9)^k \qquad (4.4)$$

(since $s - r = 0.5$ so that $b/(s - r) = 0.1b_0/0.5 = 0.2b_0$).

Since 0.9 and 0.4 are less than 1, both $a(n)$ and $b(n)$ must decrease to zero. But observe that

$$\frac{a(k)}{b(k)} = \frac{c(0.4)^k + 0.2b_0(0.9)^k}{(0.9)^k b_0} = \frac{c}{b_0}\left(\frac{4}{9}\right)^k + 0.2.$$

Since $4/9 < 1$, $(4/9)^k$ goes to zero as k goes to infinity and so

$$\frac{a(k)}{b(k)} \rightarrow 0.2$$

as k goes to infinity. This means that after a long period of time, the amount of material a will be about 20 per cent of the amount of material b. In other words, there will be about 5 atoms of b for each atom of a. ■

Many radioactive materials have very short half-lives, but they can still be found in nature. One might think that after millions of years, all of these materials would have decayed. Why haven't they? In the previous example, material a decayed relatively quickly, but the total amount of it, $a(n)$, went to zero slowly because of the $(0.9)^k$ term. In fact, the amount of material a became about one-fifth of material b. This shows that when a radioactive material is in a chain, and it is the daughter of another radioactive material that decays more slowly, the materials actually decay at essentially proportional rates.

Let's now consider a financial application.

EXAMPLE 4.4   Suppose when we retire, we have one million dollars in an annuity that pays 10 per cent interest, compounded annually. Suppose we withdraw b_0 dollars from this account today to live on for the next year. Due to a 5 per cent inflation rate, we will need to withdraw 5 per cent more each year than the previous year to live at the same level. How much can we withdraw today so that our savings will last 20 years?

Let $b(n)$ be the amount that we withdraw at the **end** of the nth year. Since $b(0) = b_0$ and

$$b(n+1) = 1.05b(n),$$

then $b(k) = 1.05^k b_0$.

Let $a(n)$ be the amount in our account at the **end** of the nth year, just following the addition of interest and the withdrawal for that year. Thus, to compute $a(1)$, the amount in our account at the end of the first year, we add the year's interest $0.1a(0)$ and subtract our withdrawal. But this withdrawal is occurring after one year and is thus $b(1)$. This gives the equation $a(1) = 1.1a(0) - b(1) = 1.1a(0) - 1.05b_0$. Also note that we first withdraw b_0 dollars today, so that the amount $a(0)$ in our account at the beginning is $1\,000\,000 - b_0$ dollars. Generalizing, the amount in our account at the end of the $n + 1$th year is the amount at the end of the nth year plus interest minus a withdrawal giving the dynamical system

$$a(n+1) = 1.1a(n) - b(n+1).$$

After substituting $b(n+1) = 1.05^{n+1} b_0 = 1.05b_0 \times 1.05^n$, this gives

$$a(n+1) = 1.1a(n) - 1.05b_0 \times 1.05^n$$

with $a(0) = 1\,000\,000 - b_0$. Using solution (4.3) with $r = 1.1$, $s = 1.05$, $b = -1.05b_0$, and $b/(s-r) = 21b_0$, this gives the general solution of

$$a(k) = c1.1^k + 21b_0 1.05^k.$$

Using $1\,000\,000 - b_0 = a(0) = c + 21 b_0$, we find that $c = 1\,000\,000 - 22 b_0$. The original question was to find b_0 so that $a(20) = 0$. Solving

$$0 = (1\,000\,000 - 22 b_0)1.1^{20} + 21 b_0 1.05^{20}$$

gives that

$$b_0 = \frac{1\,000\,000 \times 1.1^{20}}{22 \times 1.1^{20} - 21 \times 1.05^{20}} = 72\,898.58.$$

Thus, we can withdraw $72\,898.58$ dollars now and can increase the amount we withdraw by 5 per cent each year, and our account will last for 20 years. ■

As another example of dynamical systems in which $r \neq s$ let's again consider the geometric series

$$1 + s^1 + s^2 + \cdots + s^n,$$

which we first studied in section 2.4. Here, s is a fixed constant. Note that when $s = 1$, the sum is just $n + 1$, so we will assume that $s \neq 1$ in the following discussion. By letting

$$a(n) = 1 + s^1 + s^2 + \cdots + s^n,$$

we have

$$a(n + 1) = a(n) + s^{n+1} = a(n) + ss^n,$$

with $a_0 = 1$. Note that this is of the form of dynamical system (4.2) with $r = 1$ and $b = s$. Since $s \neq 1$, the general solution is, by equation (4.3),

$$a(k) = c + \left(\frac{s}{s-1}\right) s^k.$$

Since $a_0 = 1 = c + s/(s - 1)$, we have that

$$c = 1 - \frac{s}{s-1} = \frac{s-1}{s-1} - \frac{s}{s-1} = \frac{-1}{s-1},$$

so the particular solution is

$$a(k) = \frac{-1}{s-1} + \frac{s^{k+1}}{s-1} = \frac{s^{k+1} - 1}{s-1}.$$

DEFINITION 4.5   *We call the infinite sum*

$$g(0) + g(1) + g(2) + \cdots$$

*an **infinite series**. Suppose a(k) is the solution to the dynamical system*

$$a(n+1) = a(n) + g(n+1),$$

with $a_0 = g(0)$, that is,

$$a(n) = g(0) + g(1) + \cdots + g(n).$$

If

$$\lim_{k \to \infty} a(k) = a,$$

*then we say that the infinite series **converges** to a, and we write*

$$g(0) + g(1) + g(2) + \cdots = a.$$

From our discussion above, we have

$$a(k) = 1 + s + \cdots + s^k = \frac{s^{k+1} - 1}{s - 1}.$$

We note that if $-1 < s < 1$, then s^{k+1} goes to zero as k goes to infinity so that

$$\lim_{k \to \infty} a(k) = \frac{-1}{s - 1} = \frac{1}{1 - s}.$$

Thus, if $|s| < 1$, then we have the geometric series

$$1 + s + s^2 + \cdots = \frac{1}{1 - s}.$$

In particular, we have, for $s = 1/2$,

$$1 + \frac{1}{2} + \left(\frac{1}{2}\right)^2 + \cdots = \frac{1}{1 - 1/2} = 2.$$

EXAMPLE 4.6   Geometric series can be used to rewrite numbers with an infinitely repeating decimal (including all zeros, such as $0.5 = 0.500\ldots$) as fractions. For example, the number $a = 1.11\ldots$ can be written as the geometric series

$$a = 1 + 0.1 + 0.1^2 + \cdots.$$

Thus, it follows that $\lim_{k \to \infty} a(k) = a$, where $a(k)$ is the solution to the dynamical system

$$a(n+1) = a(n) + 0.1^{n+1}$$

with $a(0) = 1$. In this case, $s = 0.1$ so we can write

$$a = \frac{1}{1 - 0.1} = \frac{10}{9}.$$ ■

We now have a method for rewriting any number with an infinitely repeating decimal as a fraction, thus showing that such numbers are rational numbers. Let's illustrate this method by way of another example. You should be able to use this method on any other repeating decimal.

EXAMPLE 4.7   Write the number

$$2.361\,616\,1\ldots = 2.3\overline{61}$$

as a fraction. The line over the 61 means that 61 keeps repeating.

1. Rewrite the number as the sum of two numbers: the first part, possibly zero, which does not repeat and the infinitely repeating decimal. Thus, we rewrite our number as

    $$2.3\overline{61} = 2.3 + 0.0\overline{61}.$$

2. Factor a power of 10 out of the repeating part so that the first repetition occurs before the decimal place:

    $$2.3\overline{61} = 2.3 + 10^{-3}(61.\overline{61}).$$

3. Factor out the number that repeats:

    $$2.3\overline{61} = 2.3 + 10^{-3} \times 61 \times 1.\overline{01}.$$

4. Write the repeating part as a geometric series, and then replace the geometric series with its sum:

    $$2.3\overline{61} = 2.3 + 10^{-3} \times 61\,(1 + 0.01 + 0.01^2 + \cdots)$$

    $$= 2.3 + 10^{-3} \times 61\left(\frac{1}{1 - 0.01}\right)$$

    $$= 2.3 + 10^{-3} \times 61\left(\frac{100}{99}\right)$$

5. Simplify:

    $$2.3\overline{61} = \frac{23}{10} + \left(\frac{61}{1000}\right)\left(\frac{100}{99}\right) = \frac{1169}{495}.$$ ■

From this discussion, it is clear that any number ending in a repeating sequence of digits can be rewritten as a fraction. It is also easy to show that any fraction can be rewritten as a number with an eventually repeating decimal expansion. These numbers are called **rational numbers**. All other numbers are called **irrational numbers**. The study of properties of rational and irrational numbers is an interesting area of mathematics, but is outside the scope of this text.

PROBLEMS

1. Find the general solution to each of the following dynamical systems.
   a. $a(n+1) = 2a(n) + 3^n$
   b. $a(n+1) = -3a(n) + 15 \times 2^n$
   c. $a(n+1) = a(n) + (-1)^n$
   d. $a(n+1) = 2a(n) + 3^n + 4^n$
   e. $a(n+1) = a(n) + 2^n - 5^n$
   f. $a(n+1) = 3a(n) + 2 \times 4^n - 6$

2. Find the particular solution to each equation in problem 1 above, given that $a(0) = 1$ in each case.

3. Find the general solution to the dynamical system

$$a(n+1) = 2a(n) + n3^n - 5 \times 3^n.$$

4. What should be the form of the general solution to the dynamical system

$$a(n+1) = ra(n) + b_1 ns^n + b_2 s^n$$

when $r \neq s$?

5. Consider the infinite series

$$0(0.2)^0 + 1(0.2)^1 + 2(0.2)^2 + \cdots.$$

Determine that this series converges and find to what value it converges. Do this by finding the general solution to the dynamical system

$$a(n+1) = a(n) + (n+1)(0.2)^{n+1}$$

and computing

$$\lim_{k \to \infty} a(k).$$

(Hint: $\lim_{k \to \infty} ks^k = 0$ when $|s| < 1$.)

6. Suppose your money is in the bank earning 8 per cent interest. Your initial deposit is 1000 dollars. Because you earn more

money each year, your deposit each year will be 10 per cent more than the preceding year. That is, in year 1 your deposit will be $1000 + (0.1)1000 = 1.1 \times 1000$ dollars. Develop a dynamical system for the amount in your account, find the particular solution, and use the solution to find $a(20)$ the amount in your account after 20 years.

7. Suppose we have a savings account initially containing a_0 dollars, which collects 10 per cent interest, compounded annually. We have determined that, to live comfortably, we presently need 40 000 dollars per year. We also assume that inflation is going to be 5 per cent per year, compounded annually, so next year, the first year we will withdraw money from our account, we will need to withdraw $1.05 \times 40\,000$ dollars from our account. The following year, we need to withdraw $1.05^2 \times 40\,000$ from our account, etc.
    a. Find the dynamical system that models the amount in our account.
    b. Find the general solution to this dynamical system.
    c. Find the particular solution, given that $a(0) = a_0$.
    d. If we want this account to last for 20 years, that is, $a(20) = 0$, find a_0.

8. Suppose we initially deposit a_0 dollars into a bank account paying 10 per cent interest, compounded annually. We increase our deposit by 8 per cent per year. What must a_0 be so that we have 1 000 000 dollars in our account in 30 years?

9. You manage to land a fantastic job, and you start a savings account with 20 000 dollars invested at 10 per cent interest compounded annually. Next year, you will deposit an additional 10 000 dollars, and then each year thereafter 20 per cent more than the year before. You intend to retire in 24 years. Unknown to you, there is an evil embezzler at the bank who decides to steal 2000 dollars from your account immediately, and then each year thereafter, 30 per cent more than the year before. The embezzler is a master at fixing the books so that no one suspects anything. The very day that you retire and want your money, the embezzler safely skips the country. Alas, the bank claims their records show that you made the withdrawals yourself, and a nightmare lawsuit ensues.
    a. How much money did you expect to have for your retirement?

    b. How much money did the villain take out of the country?

    c. How much money did the villain considerately leave in your account so that you can pay your lawsuit?

10. Suppose each day, 1 per cent of radioactive material b decays into radioactive material a and that 15 per cent of radioactive material a decays into lead.

    a. Find the solution for $a(k)$ and $b(k)$, the amounts of each radioactive material after k days.

    b. After a long period of time, find the proportion of the amount of material a to material b.

    c. Suppose that after several years, there are 10 grams of material a left? How much of material b is left?

    d. Suppose that after several years, there are 20 kilograms of material b left. How much of material a is left?

11. Suppose we have 3 radioactive materials, a, b, and c. Suppose that radioactive material c decays into material b and that material b decays into material a. Suppose also that each day 1 per cent of c, 10 per cent of b, and 5 per cent of a decays, respectively. Suppose that initially there are 2.5, 11, and 9 grams of materials a, b, and c, respectively. (This is a chain of length three. There are naturally occurring chains of lengths 13, 17, and 19.)

    a. Find the solutions for $a(k)$, $b(k)$, and $c(k)$, the amounts of a, b, and c, respectively, after k days.

    b. Find the proportion of $b(k)$ to $c(k)$ for large k.

    c. Find the proportion of $a(k)$ to $c(k)$ for large k.

12. Experimental evidence indicates that each time a ball bounces, it rebounds some fixed proportion of the distance from which it fell. The particular proportion depends on the ball and the surface on which it is dropped. Suppose a ball is dropped from a height of 5 meters and that on each bounce it rebounds 90 per cent of the height of the previous rebound. For example, on the first bounce it rebounds $0.9 \times 5 = 4.5$ meters, and on the second bounce it rebounds $0.9 \times 4.5 = 4.05$ meters.

    a. Develop a dynamical system for $b(n)$, the height the ball rebounds on the nth bounce.

    b. Develop a dynamical system for $a(n)$, the total distance the ball has traveled when it strikes the ground for the nth time. For example, when the ball strikes the ground the first time it has traveled $a(1) = 5$ meters and when it strikes for the second time, it has traveled $a(2) = 5 + 2 \times 4.5 = 14$ meters.

    c. How far will the ball have traveled when it stops bouncing?

13. The following infinite series converge to what numbers?
    a. $1 + (2/3) + (2/3)^2 + (2/3)^3 + \cdots$
    b. $1 - (1/2) + (1/2)^2 - (1/2)^3 + \cdots$
    c. $(1/4)^3 + (1/4)^4 + (1/4)^5 + \cdots$
    d. $1 + (2/5)^2 + (2/5)^4 + (2/5)^6 + \cdots$

14. Express the following repeating decimals as fractions.
    a. $0.55\overline{5}$
    b. $0.13\overline{13}$
    c. $1.254\overline{54}$

4.2   Exponential terms, a special case

Let's consider exponentials in which $r = s$, that is, the dynamical system

$$a(n + 1) = ra(n) + br^n. \tag{4.5}$$

We expect that $a(k)$ involves an r^k term, but it is not clear what other terms it might involve. One possible approach is to factor out the r^k term and see what is left, that is, try to determine $a'(k)$ when

$$a(k) = r^k a'(k).$$

To do this, we substitute $a(n+1) = r^{n+1} a'(n+1)$ and $a(n) = r^n a'(n)$ into equation (4.5), giving

$$r^{n+1} a'(n + 1) = r[r^n a'(n)] + br^n.$$

Dividing both sides by r^{n+1} gives that $a'(n)$ satisfies the dynamical system

$$a'(n + 1) = a'(n) + (b/r).$$

But the general solution to this first-order affine dynamical system is

$$a'(k) = c + (b/r)k.$$

Thus, the general solution to dynamical system (4.5) is

$$a(k) = r^k a'(k) = r^k \left(c + \frac{bk}{r} \right). \tag{4.6}$$

EXAMPLE 4.8   Find the general solution to the dynamical system

$$a(n + 1) = 3a(n) + 6 \times 3^n.$$

In this case, $r = s = 3$ and $b = 6$. Applying solution (4.6) gives that the general solution is

$$a(k) = (c + 2k)3^k$$

since $b/r = 2$.

Now let's find the particular solution when $a(0) = 5$. Set $a(0) = [c + 2(0)]3^0 = c = 5$. So the particular solution is

$$a(k) = (5 + 2k)3^k. \qquad \blacksquare$$

Let's summarize the results of this section and section 4.1.

THEOREM 4.9   Nonhomogeneous dynamical system in which $g(n)$ is an exponential   *Consider the dynamical system*

$$a(n + 1) = ra(n) + bs^n.$$

- *If $r \neq s$, then the general solution is*

$$a(k) = cr^k + \frac{bs^k}{s - r}.$$

- *If $r = s$, then the general solution is*

$$a(k) = \left(c + \frac{bk}{r}\right)r^k.$$

In both cases the constant c depends on the initial value.

Let's apply our results to a financial problem.

EXAMPLE 4.10   Suppose we deposit a_0 dollars into a savings account that pays 8 per cent interest, compounded annually. Each succeeding year, we deposit 8 per cent more that we did the previous year. At the end of 30 years, we want to have 1 000 000 dollars in our account. What should our initial deposit a_0 be so that we can accomplish this?

Let $b(n)$ be our deposit after n years. Then $b(0) = a_0$ and

$$b(n + 1) = 1.08b(n)$$

since our deposits **increase** by 8 per cent. The solution to this dynamical system is $b(k) = 1.08^k a_0$.

Let $a(n)$ be the amount in our account after n years, immediately after the interest is added and the deposit is made. Then $a(0) = a_0$ and $a(n + 1) = 1.08a(n) + b(n + 1)$ or, after substitution

$$a(n + 1) = 1.08a(n) + 1.08a_0 \times 1.08^n.$$

Since $b/r = a_0$, the general solution is

$$a(k) = 1.08^k (c + a_0 k).$$

Since $a_0 = a(0) = 1.08^0 (c + a \times 0) = c$, the particular solution is

$$a(k) = 1.08^k (1 + k) a_0.$$

To have one million dollars after 30 years, we must solve

$$1.08^{30}(1 + 30) a_0 = 1\,000\,000 \qquad \text{or} \qquad a_0 = \frac{1\,000\,000}{1.08^{30} \times 31} = 3205.72.$$

Thus, if we initially deposit 3205.72 dollars and increase our deposit by 8 per cent per year, we will have a million dollars in 30 years. ∎

The techniques of this section and section 4.1 can be combined to find the solution for more general nonhomogeneous dynamical systems.

EXAMPLE 4.11   Let's find the general solution to the dynamical system

$$a(n + 1) = 3a(n) + 6 \times 3^n + 3 \times 4^n \qquad (4.7)$$

If we did not have the 6×3^n term in dynamical system (4.7), then $b/(s - r) = 3$ and the general solution would be

$$a(k) = c3^k + 3 \times 4^k$$

where c depends on the initial value. If we did not have the 3×4^n term in dynamical system (4.7), then we would have the case in which $s = r$, $b/r = 2$, and the general solution would be

$$a(k) = c3^k + 2k3^k,$$

again with c depending on the initial value. Let's try a combination of these two solutions and see if it works. Specifically, let's try

$$a(k) = c3^k + 2k3^k + 3 \times 4^k. \qquad (4.8)$$

Using both $n+1$ and n for k in equation (4.8), and then substituting the result into dynamical system (4.7) gives

$$c3^{n+1} + 2(n+1)3^{n+1} + 3 \times 4^{n+1} = 3(c3^n + 2n3^n + 3 \times 4^n) + 6 \times 3^n + 3 \times 4^n.$$

Multiplying through by the 3 on the right, simplifying the exponents on the left, and then canceling the $3c(3)^n$ terms gives

$$6(n + 1)3^n + 12 \times 4^n = 6n3^n + 9 \times 4^n + 6 \times 3^n + 3 \times 4^n.$$

Canceling the $6n3^n$ terms gives

$$6 \times 3^n + 12 \times 4^n = 9 \times 4^n + 6 \times 3^n + 3 \times 4^n.$$

Clearly the 3^n and 4^n terms cancel giving equality. Thus, equation (4.8) satisfies dynamical system (4.7) and is thus the general solution.

Let's find the particular solution when $a(0) = 1$. Solving

$$a(0) = c3^0 + 2 \times 0 \times 3^0 + 3 \times 4^0 = c + 3 = 1$$

gives that $c = -2$ and the particular solution is

$$a(k) = (-2 + 2k)3^k + 3 \times 4^k. \qquad \blacksquare$$

PROBLEMS

1. Find the general solution to each of the following dynamical systems.
   a. $a(n+1) = 2a(n) - 2^n$
   b. $a(n+1) = -3a(n) + 15(-3)^n$
   c. $a(n+1) = -a(n) + (-1)^n$
   d. $a(n+1) = 3a(n) + 2 \times 3^n + 2^n$
   e. $a(n+1) = 3a(n) + 3^n + 8$
   f. $a(n+1) = 3a(n) + 9 \times 3^n + 6n3^n$

2. Find the particular solution to each equation in problem 1 given that $a(0) = 1$ in each case.

3. What should be the form of the general solution to the dynamical system

$$a(n+1) = ra(n) + b_1 r^n + b_2 nr^n?$$

4. Suppose your money is in the bank earning 10 per cent interest. Your initial deposit is 1000 dollars. Because you earn more money each year, your deposit each year will be 10 per cent more than the preceding year.
   a. Develop a dynamical system for $a(n)$, the amount of money in your account after n years.
   b. Find the particular solution to the dynamical system in the previous part.
   c. Find the amount in your account after 20 years, that is, find $a(20)$.

5. Suppose we have a savings account which now contains a_0 dollars, and that this account earns 7 per cent interest, compounded

annually. Next year, we will withdraw 40 000 dollars from our account. Due to inflation and our increased spending habits, we will increase the amount of our withdrawal by 7 per cent each year. How much will we need to have in our account initially, that is, what is a_0, if we want our account to last for 20 years?

4.3   Fractal geometry

In this section, we will study 2-dimensional objects that are defined recursively. We will learn how to construct such objects and will then use the recursive nature of how they were defined to find their perimeter and area.

The first object we will construct is called the Sierpinski triangle. The first step is to construct an equilateral triangle whose sides are all of length one. This can be seen in figure 4.1 a).   To compute the first iteration, the midpoints of each of the line segments are connected and then the middle triangle is discarded, giving the shaded portion of figure 4.1 b). To compute the second iteration, the midpoints of the sides of each of the remaining triangles are connected and the middle (down pointing) triangles are discarded, giving the shaded portion of figure 4.1 c).

Call the original triangle $t(0)$. Denote the shaded portion of figure 4.1 b) by $t(1)$. Similarly, the shaded portion of figure 4.1 c) is $t(2)$. Continue constructing $t(3), t(4), \ldots$.

Let $a(n)$ be the area of the nth iteration, $t(n)$. The area of an equilateral triangle is $\sqrt{3}s^2/4$ where s is the length of a side. Therefore, since each side of $t(0)$ is one unit in length, then $a(0) = \sqrt{3}/4$. Note that all 4 triangles in figure 4.1 b) are congruent. Since one of them is discarded,

$$a(1) = a(0) - 0.25a(0) \qquad \text{or} \qquad a(1) = 0.75a(0).$$

In figure 4.1 c), one-fourth of each of the remaining triangles is discarded, so that $a(2) = 0.75a(1)$. Thus, the areas satisfy the dynamical system $a(n+1) = 0.75a(n)$, and thus the area of the kth iteration is

$$a(k) = 0.75^k a(0) = \frac{\sqrt{3}}{4} \left(\frac{3}{4}\right)^k.$$

Let $p(n)$ be the total perimeter of the nth iteration, that is, the total perimeter of all of the shaded triangles in $t(n)$. One approach is to first start with the perimeter of $t(0)$. Since each side is length

a)

b)

c)

d)

FIGURE 4.1. Figure a) is the original triangle, figure b) is the first iteration, and figure c) is the second iteration. Figure d) is the limit of the iterations, the Sierpinski triangle.

one, $p(0) = 3$. To get $p(1)$, we add to $p(0)$ the perimeter of the unshaded triangle in figure 4.1 b). Each side of this triangle is of length one-half and there are three of them, so $p(1) = p(0) + 1.5$. To get $p(2)$, we add the perimeter of the 3 new unshaded triangles in figure 4.1 c) to $p(1)$. Each side of one of these triangles is of length $1/4$, so each of these triangles has perimeter $3/4$. Since there are 3 of these triangles, the total perimeter added is $9/4$ so

$$p(2) = p(1) + \frac{9}{4}.$$

To compute $p(n+1)$ we add $p(n)$ and the perimeter of the new unshaded triangles. We must compute the number of new unshaded triangles and the perimeter of each of these triangles. Let $s(n)$ be the length of one side of one of the new unshaded triangles on the nth iteration. As seen in figure 4.1 b), $s(1) = 0.5$. It should also be clear that the side of each new unshaded triangle is half the length of a side of the previous unshaded triangles, that is,

$$s(n+1) = 0.5s(n).$$

The solution to this dynamical system is $s(k) = 0.5^k c$. Since $s(1) = 0.5^1 c = 0.5$, then $c = 1$ and $s(k) = 0.5^k$. Thus, the perimeter of **each** of the new unshaded triangles added on the kth iteration is $3s(k) = 3 \times 0.5^k$.

The number of unshaded triangles added at each iteration is 3 times the number added on the previous iteration. Thus, letting $N(n)$ be the number of new unshaded triangles on the nth iteration, we have that $N(n+1) = 3N(n)$ so that $N(k) = 3^k c$. Since from figure 4.1 b), $N(1) = 1 = 3^1 c$, we have that $c = 1/3$ and $N(k) = 3^k/3 = 3^{k-1}$.

Putting this all together, we have the total perimeter $p(n+1)$ on the $n+1$th iteration is the previous total perimeter $p(n)$ plus the number of new unshaded triangles $N(n+1)$ times the perimeter of each of these triangles $3s(n+1)$. After substituting $N(n+1) = 3^n$ and $3s(n+1) = 3 \times 0.5^{n+1}$, this translates into the dynamical system

$$p(n+1) = p(n) + 3^n \times 3 \times 0.5^{n+1} = p(n) + 1.5 \times \left(\frac{3}{2}\right)^n.$$

By theorem 4.9, the general solution to this dynamical system is

$$p(k) = c + 3 \times \left(\frac{3}{2}\right)^k.$$

Since $p(0) = c + 3 = 3$, we have that $c = 0$ and the particular solution is

$$p(k) = 3 \times \left(\frac{3}{2}\right)^k.$$

Notice that the perimeter of the nth iteration increases toward infinity while the area decreases to zero. Thus, if we constructed the 50th iteration, the resulting shaded part would have an area less than one-millionth, but the total length of lines needed to draw it would be almost 2 billion units in length. This figure has negligible area but it would require many pencils to draw it accurately.

The Sierpinski triangle 4.1 d) is an example of a **fractal**. A fractal is a figure which is self-similar under a change in scale. For the

Sierpinski triangle, any triangular shaped portion is a replica of the entire figure, but on a smaller scale.

Another example of a fractal is the Koch snowflake seen in figure 4.2 d). The construction of the Koch snowflake is also recursive. Again start with the equilateral triangle in figure 4.2 a) in which each side is of length one. Construct an equilateral triangle on the middle third of each line segment and remove the common side. This is given in figure 4.2 b). For the second iteration, again construct equilateral triangles on the middle third of each of the line segments of figure 4.2 b) to get figure 4.2 c). The limit figure to this process is the Koch snowflake of figure 4.2 d).

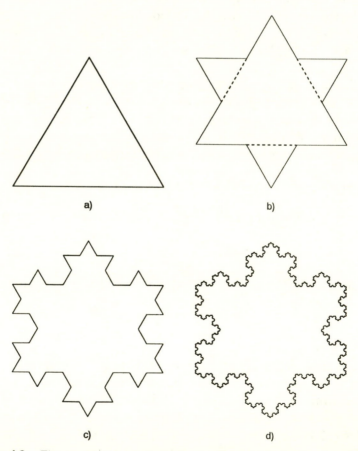

a)   b)

c)   d)

FIGURE 4.2. Figure a) is the original triangle, figure b) is the first iteration, and figure c) is the second iteration. Figure d) is the limit of the iterations, the Koch snowflake.

Our goal is to find the perimeter and area of the Koch snowflake, or the perimeter and area of the kth iteration which approximates the Koch snowflake.

The perimeter is relatively easy. On each iteration, each line segment is divided into 3 equal parts and the middle third is then replaced with 2 line segments of the same length. This can be seen in figure 4.2 b) in which the dotted lines are replaced by the other 2 sides of the equilateral triangle constructed on the dotted lines. Thus, each original line segment is replaced by 4 segments of one-third the length. So the perimeter satisfies the dynamical system

$$p(n+1) = \frac{4}{3}p(n),$$

where $p(n)$ is the perimeter of the nth iteration. Since $p(0) = 3$, the particular solution is

$$p(k) = 3\left(\frac{4}{3}\right)^k.$$

The total area is a little more complex. We will find its area in stages. Let $A(n)$ be the total area of the nth iteration. We have that

$$A(n+1) = A(n) + (\text{area of triangles added on } n+1\text{th iteration}).$$

Let $N(n)$ be the number of equilateral triangles and $a(n)$ be the area of one of the triangles added on the nth iteration. Thus,

$$A(n+1) = A(n) + N(n+1)a(n+1). \tag{4.9}$$

The area of the original equilateral triangle is $a(0) = \sqrt{3}/4$. Since the lengths of the sides of each new triangle constructed on the $n+1$th iteration are $1/3$ the lengths of the sides constructed on the nth iteration, then the areas of the triangles constructed on the $n+1$th iteration are $(1/3)^2 = 1/9$ of the areas of the triangles constructed on the nth iteration, that is,

$$a(n+1) = (1/9)a(n).$$

Thus, $a(k) = c(1/9)^k$. Since $\sqrt{3}/4 = a(0) = c$, we have

$$a(k) = \frac{\sqrt{3}}{4}\left(\frac{1}{9}\right)^k. \tag{4.10}$$

At each iteration, a triangle is constructed on the middle third of each edge of the preceding figure, which replaces each edge with 4 smaller edges. Thus, since there are 4 times as many edges after each iteration, there will be 4 times as many triangles constructed one iteration as there were on the previous iteration. Thus, $N(n+1) = 4N(n)$ and so $N(k) = c4^k$. There are 3 triangles constructed on the first iteration, that is, $N(1) = 3$. Using this initial value, we find that $3 = N(1) = 4c$ so $c = 3/4$ and

$$N(k) = 3 \times 4^{k-1}. \tag{4.11}$$

Substituting solutions (4.10) and (4.11) into dynamical system (4.9) gives, after a little simplification

$$A(n+1) = A(n) + \frac{\sqrt{3}}{12} \left(\frac{4}{9}\right)^n. \tag{4.12}$$

Using theorem 4.9 with $r = 1$, $s = 4/9$ and $b = \sqrt{3}/12$, gives that

$$A(k) = c - \frac{3\sqrt{3}}{20} \left(\frac{4}{9}\right)^k.$$

Using the fact that $\sqrt{3}/4 = A(0) = c - 3\sqrt{3}/20$ gives that $c = 2\sqrt{3}/5$ and the solution to dynamical system (4.12) is

$$A(k) = \frac{2\sqrt{3}}{5} - \frac{3\sqrt{3}}{20} \left(\frac{4}{9}\right)^k. \tag{4.13}$$

Note that as k goes to infinity, $(4/9)^k$ goes to zero and

$$A(k) \to \frac{2\sqrt{3}}{5}.$$

The Koch snowflake has finite area but infinite boundary. It is also a fractal in that any small piece of its boundary is a replica of any larger piece.

For our last construction, we start with a square with each side of length one. The first iteration is to connect the midpoints of the sides of this square to form an inscribed square. See figure 4.3 a).

At each step, the midpoints of the sides of the innermost square are connected to form a new inscribed square. See figure 4.3 b) to see the second iteration.

Let $s(n)$ equal the length of the side of the nth inscribed square, where $s(0) = 1$. Note that the side of the $n + 1$th inscribed square is also the hypotenuse of a right triangle whose legs are half of two

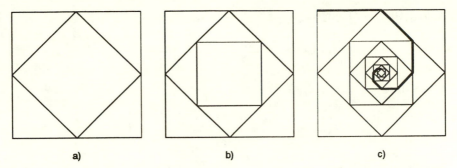

FIGURE 4.3. Figure a) is the first iteration. Figure b) is the second iteration. Figure c) is the limit of the iterations with the spiral highlighted.

sides of the nth inscribed square, that is, each leg is of length $s(n)/2$. By the Pythagorean theorem

$$s(n+1) = \sqrt{\left(\frac{s(n)}{2}\right)^2 + \left(\frac{s(n)}{2}\right)^2} = (\sqrt{2}/2)s(n).$$

Thus, $s(k) = (\sqrt{2}/2)^k c$. Since $s(0) = 1 = c$, we have

$$s(k) = (\sqrt{2}/2)^k.$$

Right triangles are formed at the corners of each square. One leg of each right triangle at each level (chosen in a clockwise manner) is highlighted in figure 4.3 c) to form a spiral. We wish to compute the length of the spiral. Let $\ell(n)$ be the length of the first n segments of the spiral. Then $\ell(1) = 0.5$ since the first leg is half the length of a side of the original square. It is also clear that

$$\ell(n+1) = \ell(n) + (\text{the length of the } n+1\text{th leg}).$$

But the length of the $n+1$th leg is half the length of a side of the nth **inscribed** square. Thus

$$\ell(n+1) = \ell(n) + \frac{s(n)}{2} = \ell(n) + \frac{1}{2}\left(\frac{\sqrt{2}}{2}\right)^n \qquad (4.14)$$

Using theorem 4.9 with $r = 1$, $s = \sqrt{2}/2$, and $b = 1/2$ gives that the general solution to dynamical system (4.14) is

$$\ell(k) = c + \frac{1}{\sqrt{2}-2}\left(\frac{\sqrt{2}}{2}\right)^k = c - \frac{2+\sqrt{2}}{2}\left(\frac{\sqrt{2}}{2}\right)^k.$$

Using the fact that

$$0.5 = \ell(1) = c - \frac{2 + \sqrt{2}}{2}\left(\frac{\sqrt{2}}{2}\right) = c - \frac{1 + \sqrt{2}}{2}$$

gives that $c = (2 + \sqrt{2})/2$. Thus,

$$\ell(k) = \left(\frac{2 + \sqrt{2}}{2}\right)\left[1 - \left(\frac{\sqrt{2}}{2}\right)^k\right].$$

Since $(\sqrt{2}/2)^k$ goes to zero, the length of the spiral is

$$\lim_{k\to\infty} \ell(k) = \frac{2 + \sqrt{2}}{2}.$$

It is interesting that the length of the spiral is the same as the perimeter of the outermost right triangle.

PROBLEMS

1. Construct a square with sides of length one. Construct squares on the middle half of the top of this square as the first iteration. For the second iteration, construct a square on the middle half of the top of the new square. Continue by constructing new squares on the middle half of the top of the last square constructed. The figure should resemble a tower.
   a. Construct the first 3 iterations.
   b. Find the external perimeter of the kth iteration and find the perimeter of the limit figure.
   c. Find the area of the kth iteration and find the area of the limit figure.

2. Construct an equilateral triangle with sides of length one. For the first iteration, connect the midpoints of the three sides to get an inscribed equilateral triangle. On each iteration, connect the midpoints of the previous inscribed triangle to get a new inscribed triangle.
   a. Construct the first 5 iterations.
   b. Draw a spiral by starting at a corner of the outer triangle and go in a clockwise direction to a corner of the next inscribed triangle. Keep going by connecting a corner of one triangle to a corner of the next inscribed triangle in a clockwise direction around each triangle. How long is the limit spiral?

3. Construct an equilateral triangle with sides of length one. At each iteration, mark points one-third of the way from one corner

to the next, going in a clockwise rotation. Connect these three marks to form a new inscribed equilateral triangle. Note that the 3 corner triangles are all 30-60-90 degree right triangles.

  a. Draw the first five iterations (for a total of six triangles).

  b. Draw a spiral by going from the corner of one triangle to a corner of the next inscribed triangle, going in a clockwise rotation. Note that you are going 1/3rd of a side of each triangle. What is the length of the limit spiral?

  c. Draw a spiral by going from the corner of one triangle to a corner of the next inscribed triangle, going in a counterclockwise rotation. What is the length of the limit spiral?

4. Draw a square with sides of length one. Make a mark three-sevenths unit from each corner, using a clockwise rotation. Connect the four marks to form a new inscribed square. Again, make a mark three-sevenths of the way from each corner of this inscribed square to the next corner, in a clockwise direction. Connect these four marks forming another inscribed square. Continue in this manner.

  a. Construct the first six iterations (six inscribed squares).

  b. Draw a spiral by starting at an outer corner and go clockwise to a corner of the next inscribed square. Continue going in this clockwise manner. Note that you are going three-sevenths of the length of the side of each square. What is the length of the limit spiral?

  c. Draw a spiral by going counterclockwise from a corner of one square to a corner of the next inscribed square. Note that you are going 4/7ths of the length of the side of each square. What is the length of the limit spiral?

  d. Shade one triangle formed by the corner of one square and a side of the next inscribed square going in a clockwise direction. This will give you a shaded spiral. Develop a dynamical system for $a(n)$, the total area of the spiral using the first n inscribed squares. Find the solution to this dynamical system and find the limit of $a(k)$ as k goes to infinity to find the area of the limit spiral. Does the answer make sense?

5. Draw a regular pentagon with sides of length one. At each iteration, connect the midpoints of the previously drawn regular pentagon to form a new inscribed pentagon. Draw a spiral by going from the corner of one pentagon to a corner of the next

inscribed pentagon, going in a clockwise rotation and starting at a corner of the outer pentagon.

    a. Draw the first four iterations.

    b. Using the fact that triangles formed by the midpoints of two adjacent sides of a regular pentagon and the corner at which the two sides meet has angles of 108, 36, and 36 degrees, find the length of the limit spiral.

  6. Construct a shaded square with sides of length one. For each iteration, construct shaded squares on the middle third of each and every line segment of the previous figure's boundary.

    a. Construct the figure that results after 3 iterations.

    b. Let $s(n)$ be the length of each line segment after n iterations. Find a dynamical system to describe $s(n + 1)$ in terms of $s(n)$ and then find the particular solution given that $s(0) = 1$.

    c. Let $b(n)$ be the number of line segments after n iterations. Find a dynamical system for $b(n)$ and find its particular solution given that $b(0) = 4$.

    d. Let $a(n)$ be the total area of the nth iteration. Find a dynamical system for $a(n+1)$ in terms of $a(n)$, $b(n)$, and $s(n)$. Substitute in the solutions for $b(n)$ and $s(n)$ and then find the particular solution, $a(k)$ given that $a(0) = 1$.

    e. Find the area of the limit figure by letting k go to infinity in the solution for $a(k)$. Note that the area is the area of a square with sides of length $\sqrt{2}$. Does this make sense from your results of part a)?

4.4   Polynomial terms

In section 4.1 we considered savings accounts in which we increased our deposit by a fixed per cent each year. Now let's consider a savings account in which we increase our deposit by a fixed amount each year. Suppose we make an initial deposit of 1000 dollars at 8 per cent interest compounded annually. At the end of the first year we deposit 100 dollars and increase our deposit by 100 dollars for each year thereafter. Let's find a formula for $a(k)$, the amount in our account at the end of the kth year.

To begin with, let's study our deposit. Let $b(n)$ be our deposit at the end of the nth year. Since our deposit at the end of the first year is 100 dollars, then $b(1) = 100$. Since we increase our deposit by 100 dollars per year, we have that $b(n+1) = b(n) + 100$. The general solution to this dynamical system is $b(k) = c + 100k$. Since $100 = b(1) = c + 100$, we have that $c = 0$ and the particular solution is $b(k) = 100k$.

Let $a(n)$ be the amount in our account at the end of the nth year, immediately after interest and our deposit have been added. Then $a(n+1) = 1.08a(n) + b(n+1)$ or, since $b(n+1) = 100(n+1) = 100n + 100$,

$$a(n+1) = 1.08a(n) + 100n + 100. \qquad (4.15)$$

This leads us to the point of this section. What is the solution to a nonhomogeneous dynamical system

$$a(n+1) = ra(n) + g(n)$$

in which $g(n)$ is a polynomial in n?

Recall that the general solutions of the dynamical systems (with $r \neq 1$)

$$a(n+1) = ra(n) \qquad \text{and} \qquad a(n+1) = ra(n) + b$$

are

$$a(k) = cr^k \qquad \text{and} \qquad a(k) = cr^k + c_0,$$

respectively, where the constant c_0 is determined from the dynamical system. We might then suspect that the general solution of the dynamical system

$$a(n+1) = ra(n) + b_1 n + b_0$$

is

$$a(k) = cr^k + c_1 k + c_0,$$

where c_1 and c_0 are two constants that are determined from the dynamical system. Let's test this hypothesis.

EXAMPLE 4.12 Let's solve dynamical system (4.15), $a(n+1) = 1.08a(n) + 100n + 100$. Our hypothesis is that the solution has the form

$$a(k) = 1.08^k c + c_1 k + c_0.$$

This would mean that $a(n+1) = 1.08^{n+1} c + c_1(n+1) + c_0$ and $a(n) = 1.08^n c + c_1 n + c_0$. Substitution into dynamical system (4.15) gives

$$1.08^{n+1} c + c_1(n+1) + c_0 = 1.08(1.08^n c + c_1 n + c_0) + 100n + 100$$

or, after multiplying through by 1.08 on the right and canceling the $1.08^{n+1}c$ terms,

$$c_1 n + c_1 + c_0 = 1.08c_1 n + 1.08c_0 + 100n + 100.$$

Moving all terms to the left and collecting all terms involving an n gives

$$(c_1 - 1.08c_1 - 100)n + (c_1 + c_0 - 1.08c_0 - 100) = 0.$$

If the two expressions within parentheses are both zero, then we have equality for all values of n. Thus, we have a solution if we pick c_1 and c_0 so that

$$c_1 - 1.08c_1 - 100 = 0 \qquad \text{and} \qquad c_1 + c_0 - 1.08c_0 - 100 = 0.$$

The first of these equations is solved when $c_1 = -1250$. Substitution into the second of these equations gives $-1250 - 0.08c_0 - 100 = 0$ or $c_0 = -16\,875$. Thus we have that

$$a(k) = c(1.08)^k - 1250k - 16\,875$$

is a general solution to dynamical system (4.15). Using the fact that $1000 = a(0) = 1.08^0 c - 1250 \times 0 - 16\,875$ gives that $c = 17\,875$ and the particular solution is

$$a(k) = 17\,875(1.08)^k - 1250k - 16\,875.$$

Thus after 10 years we have $a(10) = 17\,875(1.08)^{10} - 1250 \times 10 - 16\,875 = 9215.78$ dollars in our account. ∎

EXAMPLE 4.13   As a second example, let's consider the dynamical system

$$a(n + 1) = 3a(n) + 8n,$$

in which $g(n) = 8n + 0$. We assume the general solution is of the form

$$a(k) = 3^k c + c_1 k + c_0,$$

where c_1 and c_0 are determined from the dynamical system. It is important that $a(k)$ must have a term equal to the highest power of n in $g(n)$ and **all powers less than the highest power, down to the constant.** In this case, we have

$$a(n + 1) = c3^{n+1} + c_1(n + 1) + c_0 \qquad \text{and} \qquad a(n) = c3^n + c_1 n + c_0.$$

Substitution into the dynamical system gives

$$3^{n+1}c + c_1(n+1) + c_0 = 3(3^n c + c_1 n + c_0) + 8n.$$

Multiplying through by 3 on the right side and subtracting $3^{n+1}c$ from both sides gives

$$c_1 n + c_1 + c_0 = 3c_1 n + 3c_0 + 8n.$$

Remember that c_1 and c_0 are constants which are to be determined, while n can be any integer.

Collect all terms on the left giving

$$c_1 n + c_1 + c_0 - 3c_1 n - 3c_0 - 8n = 0.$$

Collect all terms involving an n, giving

$$(-2c_1 - 8)n + (c_1 - 2c_0) = 0$$

after simplification.

Since this equality must hold for **all** integer values n, it must hold when $n = 0$, that is,

$$c_1 - 2c_0 = 0.$$

But if this term equals zero, then we must have $(-2c_1 - 8)n = 0$ for $n = 1, 2, \dots$. For this to happen, we must have

$$-2c_1 - 8 = 0.$$

But this means that $c_1 = -4$. Substituting -4 for c_1 into $c_1 - 2c_0 = 0$ gives $-4 - 2c_0 = 0$ or $c_0 = -2$. Therefore, it follows that

$$a(k) = 3^k c - 4k - 2$$

satisfies the dynamical system and is thus a solution.

This solution is the general solution since, if we are given $a(0)$, then solving

$$a(0) = c3^0 - 0 - 2 = c - 2$$

for c gives $c = a(0) + 2$, and we now have the particular solution. For example, given that $a(0) = 8$, we then have $c = 10$, and the particular solution to the dynamical system is

$$a(k) = 10 \times 3^k - 4k - 2. \qquad \blacksquare$$

EXAMPLE 4.14 As a third example, let's consider the dynamical system

$$a(n+1) = a(n) + 2n + 1,$$

in which $g(n) = 2n + 1$. We assume the general solution is of the form

$$a(k) = 1^k c + c_1 k + c_0,$$

where c_1 and c_0 are determined from the dynamical system. In this case, we have

$$a(n+1) = c + c_1(n+1) + c_0 \qquad \text{and} \qquad a(n) = c + c_1 n + c_0.$$

Substitution into the dynamical system gives

$$c + c_1(n+1) + c_0 = (c + c_1 n + c_0) + 2n + 1.$$

Cancellation gives

$$c_1 = 2n + 1.$$

Remember that c_1 is a constant while n can vary. Thus, this equation cannot be satisfied for all values of n. The coefficient of $a(n)$ being 1 in the dynamical system is a special case and will be studied in section 4.5. ∎

All that was required to find the general solution in the above examples was

- to make a good guess at the solution,
- substitute that guess into the dynamical system,
- solve two linear equations by setting the sum of the terms not involving n equal to zero, and the sum of the terms involving n equal to zero.

We can summarize the results of this section in a theorem.

THEOREM 4.15   Nonhomogeneous dynamical systems in which $g(n)$ is a polynomial   *Consider the dynamical system*

$$a(n+1) = ra(n) + g(n),$$

where

$$g(n) = b_m n^m + b_{m-1} n^{m-1} + \cdots + b_1 n + b_0.$$

Here, m is a fixed constant and $g(n)$ is an mth-degree polynomial in n. If $r \neq 1$, then the general solution is of the form

$$a(k) = cr^k + c_m k^m + c_{m-1} k^{m-1} + \cdots + c_1 k + c_0.$$

The constant c depends on the initial value and the constants c_m, c_{m-1}, ..., c_1, and c_0 can be found by substituting into the dynamical system and then solving $m + 1$ equations for the $m + 1$ unknowns by setting the coefficient of the n^m term, the coefficient of the n^{m-1} term, ..., the coefficient of the n term, and the constant term equal to zero.

The techniques of this section can be combined with the techniques of the previous sections to find the solution when the nonhomogeneous term $g(n)$ is more complex.

EXAMPLE 4.16   Find the general solution to the dynamical system

$$a(n+1) = 2a(n) - 3^n - 2n + 1.$$

In this case, you should be able to guess that the solution is of the form

$$a(k) = c2^k + c_2 3^k + c_1 k + c_0.$$

Substitution into the dynamical system gives

$$c2^{n+1} + c_2 3^{n+1} + c_1(n+1) + c_0 = 2(c2^n + c_2 3^n + c_1 n + c_0) - 3^n - 2n + 1.$$

After simplifying, bringing all terms to the left, and grouping the 3^n terms, the n terms, and the constants, we have

$$3^n(c_2 + 1) + n(-c_1 + 2) + (c_1 - c_0 - 1) = 0.$$

Setting each of the terms in parentheses equal to zero and solving gives

$$c_2 = -1, \quad c_1 = 2, \quad c_0 = 1.$$

The general solution is then

$$a(k) = c2^k - 3^k + 2k + 1. \qquad\qquad \blacksquare$$

PROBLEMS

1. Find the general solution to each of the following dynamical systems.
   a. $a(n+1) = 2a(n) - n$
   b. $a(n+1) = 4a(n) + 6n + 5$
   c. $a(n+1) = -a(n) + 4n^2 - 2$
   d. $a(n+1) = 2a(n) + 3^n - n$

2. Find the particular solution to each equation in problem 1 above, given that $a(0) = 1$ in each case.

3. Suppose your savings account has 1000 dollars. Suppose you deposit an additional 100 dollars at the end of the first year, an additional 300 at the end of the second year, an additional 500 at the end of the third year, etc. If your bank pays 7 per cent interest compounded annually, how much will be in your account after 20 years?

4.5 Polynomial terms, a special case

We saw in example 4.14 that $r = 1$ in the dynamical system

$$a(n+1) = ra(n) + g(n),$$

was a special case. We tried to find a solution that was a polynomial of the same degree as $g(n)$, and failed. Recall that the solution of the first-order affine dynamical system

$$a(n+1) = a(n) + b \quad \text{is} \quad a(k) = c + bk.$$

In other words, when $r = 1$ and $g(n) = b$ is a zeroth-degree polynomial, the solution involves a first-degree polynomial $c + bk$. Thus, we might suspect that the solution to the first-order nonhomogeneous dynamical system

$$a(n+1) = a(n) + b_1 n + b_0,$$

is

$$a(k) = c + c_1 k^2 + c_0 k.$$

EXAMPLE 4.17 Consider the dynamical system

$$a(n+1) = a(n) + n + 1,$$

with $a(0) = 0$. Note that $a(1) = a(0) + 1 = 0 + 1$, $a(2) = a(1) + 2 = 0 + 1 + 2$ and in general, $a(k) = 0 + 1 + \cdots + k$. Since $r = 1$, we expect the general solution to be

$$a(k) = c + c_1 k^2 + c_0 k$$

for some value of c_0 and c_1. Since

$$a(n+1) = c + c_1 (n+1)^2 + c_0 (n+1) = c + c_1 n^2 + 2c_1 n + c_1 + c_0 n + c_0,$$

substitution into the dynamical system gives

$$c + c_1 n^2 + 2c_1 n + c_1 + c_0 n + c_0 = c + c_1 n^2 + c_0 n + n + 1.$$

Cancellation gives

$$2c_1 n + c_1 + c_0 = n + 1, \qquad \text{or} \qquad (2c_1 - 1)n + (c_1 + c_0 - 1) = 0.$$

Setting the coefficient of n equal to zero and the constant term equal to zero gives the two equations $2c_1 - 1 = 0$ and $c_1 + c_0 - 1 = 0$. Solving these two equations gives $c_1 = 1/2$ and $c_0 = 1/2$. The general solution is therefore

$$a(k) = c + \frac{k^2}{2} + \frac{k}{2}.$$

Given that $a_0 = 0$ we find that $0 = a(0) = c + (1/2)0^2 + (1/2)0$, and the particular solution is

$$a(k) = (0 + 1 + \cdots + k) = \frac{k^2}{2} + \frac{k}{2} = \frac{k(k+1)}{2}.$$

Observe that we have developed a formula for the sum of the first k integers. In particular, letting $k = 100$ we have that

$$a(100) = 0 + 1 + 2 + \cdots + 100 = \frac{100 \times 101}{2} = 5050.$$

We can also use this solution to find

$$200 + 201 + 202 + \cdots + 1000.$$

Note that this is just

$$a(1000) - a(199) = \frac{1000 \times 1001}{2} - \frac{199 \times 200}{2} = 480\,600. \quad \blacksquare$$

If we had tried to find a solution to the above dynamical system, of the form

$$a(k) = c + c_1 k + c_0,$$

then, after substitution into the dynamical system, canceling terms, and simplifying, we would have

$$c_1 = n + 1.$$

But c_1 is a constant and $n + 1$ changes as n changes. Thus, we cannot pick an c_1 and c_0 to satisfy this equation.

This method can be extended in an obvious fashion to polynomials of higher degree.

THEOREM 4.18   *Consider the dynamical system*

$$a(n + 1) = a(n) + g(n),$$

where

$$g(n) = b_m n^m + b_{m-1} n^{m-1} + \cdots + b_1 n + b_0.$$

Here, m is a fixed constant and $g(n)$ is an mth-degree polynomial. The general solution is of the form

$$a(k) = c + c_m k^{m+1} + c_{m-1} k^m + \cdots + c_1 k^2 + c_0 k.$$

The constant c depends on the initial value, and the constants c_m, c_{m-1}, ..., c_1, and c_0 can be found by substituting into the dynamical system and then solving $m + 1$ equations for the $m + 1$ unknowns by setting the coefficient of the n^m term, the coefficient of the n^{m-1} term, ..., the coefficient of the n term, and the constant term equal to zero.

As our first application, let's use theorem 4.18 to sum finite series of polynomials. In example 4.17 we summed the first k integers by using $g(x) = x + 1$. By letting $g(x) = (x + 1)^2$, we have the sum of the first k integers squared, that is,

$$0^2 + 1^2 + 2^2 + \cdots + k^2 = a(0) + g(0) + g(1) + \cdots + g(k - 1).$$

In fact, we have a method for finding the sum of the mth powers of the first k integers, that is,

$$0^m + 1^m + 2^m + \cdots + k^m.$$

EXAMPLE 4.19   Suppose we want to find the sum of the squares of the first k integers, that is,

$$a(k) = 0^2 + 1^2 + 2^2 + \cdots + k^2.$$

We only need to find the solution of the dynamical system

$$a(n + 1) = a(n) + (n + 1)^2 = a(n) + n^2 + 2n + 1,$$

with $a(0) = 0$. Note that the general solution is of the form

$$a(k) = c + c_2 k^3 + c_1 k^2 + c_0 k.$$

Substitution into the dynamical system gives

$$c + c_2 (n + 1)^3 + c_1 (n + 1)^2 + c_0 (n + 1) = c + c_2 n^3 + c_1 n^2 + c_0 n + n^2 + 2n + 1.$$

Multiplication, cancellation, and simplification gives

$$(3c_2 - 1) n^2 + (3c_2 + 2c_1 - 2) n + (c_2 + c_1 + c_0 - 1) = 0.$$

Solving the three equations

$$3c_2 - 1 = 0, \quad 3c_2 + 2c_1 - 2 = 0, \quad c_2 + c_1 + c_0 - 1 = 0$$

gives $c_2 = 1/3$, $c_1 = 1/2$, and $c_0 = 1/6$. Thus, the general solution is

$$a(k) = c + \frac{k^3}{3} + \frac{k^2}{2} + \frac{k}{6} = c + \frac{k(2k+1)(k+1)}{6}.$$

Using the fact that $a(0) = 0$ gives $c = 0$ and we have the identity

$$0^2 + 1^2 + 2^2 + \cdots + k^2 = \frac{k(2k+1)(k+1)}{6}. \qquad \blacksquare$$

Let's now use this method to study the motion of a falling object. The first record of someone conducting this study is Galileo, who dropped objects off the Leaning Tower of Pisa. In one experiment he is said to have dropped his assistant off the tower. His assistant marked the side of the tower every second while falling. From this, Galileo discovered the distance traveled by his assistant was $d(0) = 0$, $d(1) = 16$, $d(2) = 64$, $d(3) = 144$, and $d(4) = 256$ where $d(n)$ is the **distance** traveled in n seconds, where distance is measured in feet. Somehow, I suspect the story is somewhat apocryphal, since the Leaning Tower of Pisa is only 179 feet tall.

Letting $v(n)$ be the average velocity during the nth second, we get that $v(1) = (d(1) - d(0))/1 = d(1) - d(0) = 16$, $v(2) = d(2) - d(1) = 48, \ldots$, $v(n+1) = d(n+1) - d(n)$. In particular, $v(3) = 80$ and $v(4) = 112$. The average acceleration during any two-second interval is given by

$$(v(n+1) - v(n))/1 = v(n+1) - v(n).$$

From our computations, $v(2) - v(1) = 32$, $v(3) - v(2) = 32$, and $v(4) - v(3) = 32$. From this, we might conclude that acceleration is constant.

The preferred method for introducing this material is to conduct an experiment on the acceleration of a falling object using an inclined airtrack, if you have access to one. In this experiment, a metal slide moves freely down the track, burning a hole in a paper strip every tenth of a second. In this case, the numbers will not work exactly, but a linear regression analysis on the acceleration should indicate that it is approximately constant.

Since our accelerations are 32, we will assume that **acceleration is a constant 32 feet per second squared**. We now go back and compute our average velocity using the formula

$$v(n+1) - v(n) = 32 \qquad \text{or} \qquad v(n+1) = v(n) + 32.$$

From this, we get

$$v(k) = 32k + c.$$

Since $v(1) = 16 = 32(1) + c$, we get $c = -16$.

We can now compute our distance traveled after $n+1$ seconds using the formula

$$d(n+1) - d(n) = v(n+1) \qquad \text{or} \qquad d(n+1) = d(n) + v(n+1).$$

Since $v(n+1) = 32(n+1) - 16 = 32n + 16$, we get the dynamical system

$$d(n+1) = d(n) + 32n + 16.$$

Since the solution is of the form $d(k) = c_2 k^2 + c_1 k + c$, substitution gives

$$c_2(n+1)^2 + c_1(n+1) + c = c_2 n^2 + c_1 n + c + 32n + 16.$$

Simplifying and collecting terms on the left gives

$$(2c_2 - 32)n + (c_2 + c_1 - 16) = 0,$$

or that $c_2 = 16$ and $c_1 = 0$. Thus the general solution is $d(k) = 16k^2 + c$. Since $d(0) = 0$ we get that $c = 0$ and the position of the object after k seconds is

$$d(k) = 16k^2.$$

We have thus discovered the quadratic nature of the formula for the position of a falling object, that is,

$$d(k) = c_1 k^2 + c_2 k + c_3$$

for some values c_1, c_2, and c_3. If you review this work, you find that these constants depend on $d(0)$, $v(1)$ (which is in terms of $d(0)$ and $d(1)$), and the acceleration due to gravity. We note that if the object is on an inclined rail, the acceleration due to gravity will depend on the angle of inclination.

PROBLEMS

1. Find the general solution to each of the following dynamical systems.
   a. $a(n+1) = a(n) + 6n + 1$
   b. $a(n+1) = a(n) + 4n + 5$
   c. $a(n+1) = a(n) - 4n$

   d. $a(n+1) = a(n) + 3n^2 - n - 1$

2. Find the particular solution to the dynamical systems in problem 1 given that $a(0) = 2$.

3. Find the particular solution to the dynamical systems in problem 1 given that $a(1) = 2$.

4. Using dynamical systems, find a formula for the sum

$$0(1) + 1(2) + 2(3) + \cdots + n(n+1),$$

and use that formula to compute

$$0(1) + 1(2) + 2(3) + \cdots + 100(101).$$

5. On the kth day of Christmas, my true love gave to me, a partridge in a pear tree, 2 turtle doves, 3 French hens, ..., and k thingamajigs.
   a. How many presents did I get on the kth day of Christmas?
   b. How many presents did I get, total, in the first k days of Christmas?

6. What should be the form of the general solution to the dynamical system

$$a(n+1) = ra(n) + nr^n?$$

7. Use your answer to problem 6 to find the particular solution to the dynamical system

$$a(n+1) = 2a(n) + n2^n$$

with $a(0) = 0$. Check your answer by computing $a(4)$ using the dynamical system and by using the solution.

8. Use your answer to problem 6 to find the particular solution to the dynamical system

$$a(n+1) = -a(n) + n(-1)^n$$

with $a(0) = 0$. Check your answer by computing $a(4)$ using the dynamical system and by using the solution.

9. Suppose you have a roll of paper, such as paper towels. Let the radius of the inner core (that the paper is wrapped about) be $r_0 = 1$ inch. Suppose that the paper is 0.002 inches thick, so when the paper is wrapped n times about the core, the radius of the entire roll is $r_1 = r_0 + 0.002n = 1 + 0.002n$ inches. Let $a(n)$

be the total length of paper when it is wrapped about the core n times.

    a. Develop a dynamical system that gives the relationship between $a(n+1)$ and $a(n)$ by computing the length of paper when it is wrapped about the roll one more time.

    b. Solve that dynamical system with $a(0) = 0$.

    c. When the outer radius is $r_1 = 2$ inches, what is the total length of paper about the roll?

    d. What is the outer radius r_1 of the roll when the length of paper left is 500 inches?

Higher-order linear dynamical systems

5.1 An introduction to second-order linear equations

In example 4.2 we considered a dynamical system that models the amount of pollution in two of the Great Lakes. Specifically, letting $a(n)$ and $b(n)$ be the amount of pollution in Lake Ontario and Lake Erie, respectively, in year n, we developed the equations

$$a(n+1) = 0.87a(n) + 0.38b(n) \qquad (5.1)$$

$$b(n+1) = 0.62b(n). \qquad (5.2)$$

The method we used to solve this system of equations was to find that $b(k) = c(0.62)^k$, and then to substitute that into equation (5.1) giving the nonhomogeneous equation

$$a(n+1) = 0.87a(n) + 0.38c \times 0.62^n,$$

whose solution is of the form

$$a(k) = c_1 0.87^k + c_2 0.62^k.$$

An alternate approach is to solve for $b(n)$ in equation (5.1), giving

$$b(n) = \frac{a(n+1) - 0.87a(n)}{0.38}$$

and consequently that

$$b(n+1) = \frac{a(n+2) - 0.87a(n+1)}{0.38}.$$

Substituting this into equation (5.2) gives

$$\frac{a(n+2) - 0.87a(n+1)}{0.38} = 0.62 \left(\frac{a(n+1) - 0.87a(n)}{0.38} \right).$$

Simplification gives the **second-order linear** dynamical system

$$a(n+2) = (0.87 + 0.62)\, a(n+1) - 0.87(0.62)\, a(n).$$

The solution still must be of the form

$$a(k) = c_1 0.87^k + c_2 0.62^k.$$

Observe that the numbers 0.87 and 0.62 are roots of the equation

$$x^2 = (0.87 + 0.62)\, x - 0.87(0.62).$$

We are led to the idea that the general solution to a second-order linear dynamical system

$$a(n+2) = b_1 a(n+1) + b_2 a(n)$$

might be of the form

$$a(k) = c_1 r^k + c_2 s^k$$

for some constants r and s, where c_1 and c_2 depend on the initial values $a(0)$ and $a(1)$.

EXAMPLE 5.1   Let's see if the solution to the dynamical system,

$$a(n+2) = -3.5a(n+1) + 2a(n)$$

is of the form

$$a(k) = c_1 r^k + c_2 s^k.$$

Substituting our guess at a solution into the dynamical system gives

$$c_1 r^{n+2} + c_2 s^{n+2} = -3.5(c_1 r^{n+1} + c_2 s^{n+1}) + 2(c_1 r^n + c_2 s^n).$$

Bringing all terms to the left, collecting terms involving r and terms involving s, and factoring gives

$$c_1 r^n (r^2 + 3.5r - 2) + c_2 s^n (s^2 + 3.5s - 2) = 0.$$

We have equality if r and s are both roots of the equation

$$x^2 + 3.5x - 2 = 0.$$

Since the two roots of this quadratic are $x = 0.5$ and $x = -4$, which can be found by factoring, we will let $r = 0.5$ and $s = -4$. This gives the general solution

$$a(k) = c_1 (0.5)^k + c_2 (-4)^k.$$

Let's find the particular solution when $a(0) = 3$ and $a(1) = -3$. In this case, we must solve the system of equations

$$a(0) = c_1 (0.5)^0 + c_2 (-4)^0 = c_1 + c_2 = 3$$
$$a(1) = c_1 (0.5)^1 + c_2 (-4)^1 = 0.5c_1 - 4c_2 = -3.$$

In this case, $c_1 = 2$ and $c_2 = 1$, and the particular solution is

$$a(k) = 2(0.5)^k + (-4)^k. \qquad \blacksquare$$

Let's now develop a method for finding both general and particular solutions to arbitrary second-order linear dynamical systems.

DEFINITION 5.2   *The **characteristic equation** for the second-order linear dynamical system*

$$a(n+2) = b_1 a(n+1) + b_2 a(n)$$

is

$$x^2 = b_1 x + b_2.$$

As we saw in the discussion of the Great Lakes and in example 5.1, the roots of the characteristic equation help determine the general solution to a second-order linear dynamical system.

THEOREM 5.3   *Consider the second-order dynamical system*

$$a(n+2) = b_1 a(n+1) + b_2 a(n).$$

Suppose that the roots of the corresponding characteristic equation

$$x^2 - b_1 x - b_2 = 0$$

are $x = r$ and $x = s$, where $r \neq s$. Then the general solution of the second-order dynamical system is

$$a(k) = c_1 r^k + c_2 s^k.$$

The numbers c_1 and c_2 depend on the initial values. In particular, if we are given $a(0)$ and $a(1)$ then

$$c_1 = \frac{a(1) - sa(0)}{r - s} \quad \text{and} \quad c_2 = \frac{a(1) - ra(0)}{s - r}.$$

REMARK: Note that to get the formula for c_2, you just replace all the s's with r's and the r with an s in the formula for c_1. Computationally, it may be easier to use the formula

$$c_2 = \frac{ra(0) - a(1)}{r - s}. \qquad\qquad \square$$

PROOF: Consider the dynamical system

$$a(n+2) = b_1 a(n+1) + b_2 a(n),$$

with characteristic equation

$$x^2 = b_1 x + b_2.$$

Suppose that r and s are two distinct roots of this equation, that is,

$$r^2 = b_1 r + b_2 \quad \text{and} \quad s^2 = b_1 s + b_2.$$

We need to show that

$$a(k) = c_1 r^k + c_2 s^k$$

is a solution to the dynamical system. In this case,

$$a(n+1) = c_1 r^{n+1} + c_2 s^{n+1} \quad \text{and} \quad a(n+2) = c_1 r^{n+2} + c_2 s^{n+2}.$$

Substitution into the dynamical system gives

$$c_1 r^{n+2} + c_2 s^{n+2} = b_1 (c_1 r^{n+1} + c_2 s^{n+1}) + b_2 (c_1 r^n + c_2 s^n).$$

If this equation is balanced, then our formula for $a(k)$ is a solution.

Bring all the terms involving r to the left of the equation and all the terms involving s to the right. This gives

$$c_1 r^{n+2} - b_1 c_1 r^{n+1} - b_2 c_1 r^n = -c_2 s^{n+2} + b_1 c_2 s^{n+1} + b_2 c_2 s^n.$$

Factoring $c_1 r^n$ out of the left side and $-c_2 s^n$ out of the right side, gives

$$c_1 r^n (r^2 - b_1 r - b_2) = -c_2 s^n (s^2 - b_1 s - b_2).$$

Our assumption that r and s were roots of the characteristic equation gives

$$c_1 r''(0) = -c_2 s''(0), \quad \text{or} \quad 0 = 0.$$

Thus,

$$a(k) = c_1 r^k + c_2 s^k$$

is a solution to the dynamical system.

If we are given $a(0)$ and $a(1)$, then solving the two equations

$$a(0) = c_1 r^0 + c_2 s^0 = c_1 + c_2 \quad \text{and} \quad a(1) = c_1 r + c_2 s$$

for the two unknowns c_1 and c_2 gives

$$c_1 = \frac{a(1) - sa(0)}{r - s} \quad \text{and} \quad c_2 = \frac{a(1) - ra(0)}{s - r}.$$

For any given $a(0)$ and $a(1)$, a particular solution to the dynamical system satisfying this set of initial values is given by this choice of c_1 and c_2. Also, given $a(0)$ and $a(1)$, it is clear that $a(2), a(3), \ldots$, are uniquely determined, so there can be only one particular solution, and this choice of c_1 and c_2 must be it. Since

$$a(k) = c_1 r^k + c_2 s^k$$

gives every particular solution for appropriate choices of c_1 and c_2, it must be the general solution, and the proof is complete. ∎

(For the reader with some knowledge of calculus, the problem of unique solutions is easy for discrete dynamical systems, but is nontrivial for differential equations.)

EXAMPLE 5.4   Consider the second-order dynamical system

$$a(n+2) = -a(n+1) + 6a(n),$$

with $a(0) = 7$ and $a(1) = -6$. The characteristic equation is

$$x^2 = -x + 6.$$

Bringing all the terms to the left and factoring gives

$$x^2 + x - 6 = (x+3)(x-2) = 0,$$

so the roots are $x = 2$ and $x = -3$. The general solution is then

$$a(k) = c_1 2^k + c_2 (-3)^k.$$

Using the formulas in theorem 5.3 with $r = 2$ and $s = -3$, we find that

$$c_1 = \frac{-6 - (-3)7}{2 - (-3)} = 3 \quad \text{and} \quad c_2 = \frac{-6 - (2)7}{-3 - 2} = 4,$$

and so the particular solution is

$$a(k) = 3 \times 2^k + 4(-3)^k.$$

Instead of using the above formulas, let's solve for c_1 and c_2 using the initial conditions. We get the two equations

$$a(0) = c_1 2^0 + c_2 (-3)^0 = c_1 + c_2 = 7 \quad \text{and} \quad a(1) = 2c_1 - 3c_2 = -6.$$

From the first equation, $c_1 = 7 - c_2$. Substitution into the second equation gives $2(7 - c_2) - 3c_2 = -6$, or, after simplifying, $c_2 = 4$. It follows that $c_1 = 7 - c_2 = 3$, giving the same answers as before. ∎

EXAMPLE 5.5   Consider the second-order dynamical system

$$a(n + 2) = 0.5a(n + 1) + 0.5a(n).$$

The characteristic equation is, after bringing all the terms to the left,

$$x^2 - 0.5x - 0.5 = (x + 0.5)(x - 1) = 0,$$

so the roots are $x = 1$ and $x = -0.5$. Since $1^k = 1$, the general solution is then

$$a(k) = c_1 + c_2(-0.5)^k.$$

Suppose we are given that $a(0) = 8$ and $a(1) = -1$. From the above formulas with $r = 1$ and $s = -0.5$, we get $c_1 = 2$ and $c_2 = 6$. Therefore, the particular solution is

$$a(k) = 2 + 6(0.5)^k.$$

Observe that, as k goes to infinity, $a(k)$ goes to 2. ∎

Let's see how a second-order dynamical system might come up in "real life". Actually, this is an interesting but artificial application. More realistic applications will be considered in later sections.

Consider a child who has a row of n blocks. The child starts at the left of the row and picks up one or two blocks and puts them into a toy box. The child continues picking up one or two blocks at a time until all the blocks have been picked up. How many ways can the child pick up the n blocks?

If the child has three blocks, there are three ways. Way 1 is to pick up the first block and then the next two, denoted $(1,2)$; way 2 is to pick up the first two and then the third, denoted $(2,1)$; and way 3 is to pick up the first then the second then the third, denoted $(1,1,1)$. If $n = 4$ then there are five ways, denoted $(1,1,1,1)$, $(1,1,2)$, $(1,2,1)$, $(2,1,1)$, and $(2,2)$.

Let $a(n)$ be the number of ways to pick up n blocks, one or two at a time. Note that $a(1) = 1$, $a(2) = 2$, $a(3) = 3$, and $a(4) = 5$. We see for the numbers computed that

$$a(3) = a(2) + a(1) \qquad \text{and} \qquad a(4) = a(3) + a(2).$$

We then might think that

$$a(n+2) = a(n+1) + a(n), \qquad \text{for} \quad n = 1, 2, \dots. \tag{5.3}$$

Let us first justify our answer for $n = 5$, that is, $a(5) = 8$. We consider two cases. The first case is that the child picks up one block initially, denoted $(1, \dots)$. The second case is that the child picks up two blocks initially, denoted $(2, \dots)$.

To count the number of ways to accomplish the first case, that is, to pick up a total of five blocks and pick up one block first, we consider two stages. The first stage is to pick up the first block which can be done one way. The second stage is to pick up the remaining four blocks which can be done in $a(4) = 5$ ways. Thus, the first case can be done in $(1)a(4) = a(4) = 5$ ways. These ways are $(1,\mathbf{1,1,1,1})$, $(1,\mathbf{1,1,2})$, $(1,\mathbf{1,2,1})$, $(1,\mathbf{2,1,1})$, and $(1,\mathbf{2,2})$. Note that the numbers in boldface are the five ways listed above that four blocks can be picked up.

To count the number of ways to accomplish the second case, that is, to pick up a total of five blocks and pick up two blocks first, we also consider two stages. The first stage is to pick up the first two blocks which can be done one way. The second stage is to pick up the remaining three blocks which can be done in $a(3) = 3$ ways. Thus, the second case can be done in $(1)a(3) = a(3) = 3$ ways. These three ways are $(2,\mathbf{1,2})$, $(2,\mathbf{2,1})$, and $(2,\mathbf{1,1,1})$.

Thus, since every way of picking up five blocks must be one of these ways, we have

$$a(5) = a(4) + a(3).$$

Now we see the general argument. Suppose there are $n+2$ blocks to pick up. We have two cases. The first case is that the child picks up one block first and then picks up the remaining $n+1$ blocks. Since there are $a(n+1)$ ways to pick up the $n+1$ remaining blocks, there are $a(n+1)$ ways to pick up $n+2$ blocks by picking up one block first. The second case is that the child picks up two blocks first and then picks up the remaining n blocks. There are $a(n)$ ways to do this.

Thus the $a(n+2)$ ways of picking up $n+2$ blocks must be one of the $a(n+1)$ ways in the first case or one of the $a(n)$ ways in the second case, giving the second-order linear dynamical system (5.3),

$$a(n+2) = a(n+1) + a(n).$$

The numbers $a(1)$, $a(2)$, ... are 1, 2, 3, 5, 8, 13, ... This sequence of numbers is called a Fibonacci sequence and dynamical system (5.3) is called a Fibonacci relation. These numbers arise both in applied mathematics and in biology.

We now attempt to find the general solution to the Fibonacci relation. Our first step is to find the roots of the characteristic equation

$$x^2 = x + 1.$$

From the quadratic formula, we obtain the roots

$$r = \frac{1 + \sqrt{5}}{2} \qquad \text{and} \qquad s = \frac{1 - \sqrt{5}}{2}.$$

Thus the general solution to the above second-order dynamical system is

$$a(k) = c_1 \left(\frac{1 + \sqrt{5}}{2} \right)^k + c_2 \left(\frac{1 - \sqrt{5}}{2} \right)^k.$$

This is amazing! The **integer** $a(k)$, which gives the number of ways to do something (and also forms a Fibonacci sequence), is given in terms of the irrational number $\sqrt{5}$.

We could solve for c_1 and c_2 by using the initial conditions that $a(1) = 1$ and $a(2) = 2$. If we say that there is 1 way of picking up 0 toys (which is not to pick them up), then we can use $a(0) = 1$ and $a(1) = 1$ as our initial values. This simplifies the algebra. Note that, using this convention, we do get $a(2) = a(1) + a(0)$. Using these initial conditions gives, after a little algebra,

$$c_1 = 0.5 + 0.1\sqrt{5} \qquad \text{and} \qquad c_2 = 0.5 - 0.1\sqrt{5}.$$

The particular solution, which gives the Fibonacci numbers, is

$$a(k) = \frac{5+\sqrt{5}}{10} \left(\frac{1+\sqrt{5}}{2} \right)^k + \frac{5-\sqrt{5}}{10} \left(\frac{1-\sqrt{5}}{2} \right)^k .$$

This can be simplified to

$$a(k) = \frac{1}{\sqrt{5}} \left(\frac{1+\sqrt{5}}{2} \right)^{k+1} - \frac{1}{\sqrt{5}} \left(\frac{1-\sqrt{5}}{2} \right)^{k+1} .$$

This is quite a complicated formula for what appears to be a simple sequence of integers. Since

$$\frac{1}{\sqrt{5}} \left(\frac{1-\sqrt{5}}{2} \right)^{k+1} < 0.5$$

for $k \geq 0$, then the term

$$\frac{1}{\sqrt{5}} \left(\frac{1+\sqrt{5}}{2} \right)^{k+1}$$

must be less than 0.5 from the kth Fibonacci number. Thus, round this number off to the nearest integer to get the Fibonacci number.

Let's consider an example that appears to be even more amazing.

EXAMPLE 5.6 Find the particular solution to the dynamical system

$$a(n+2) = 2a(n+1) - 5a(n),$$

with $a(0) = 2$ and $a(1) = 6$. Note that since the initial values are integers, $a(k)$ must be an integer for every value of k.

The characteristic equation is

$$x^2 - 2x + 5 = 0.$$

From the quadratic formula, we get that the two roots are

$$x = \frac{2 \pm \sqrt{4-20}}{2} = 1 \pm 2\sqrt{-1} = 1 \pm 2i,$$

where $i = \sqrt{-1}$. Thus, the general solution is

$$a(k) = c_1 (1 + 2i)^k + c_2 (1 - 2i)^k .$$

We are using complex numbers to compute the real solution to the above dynamical system.

To find the particular solution, we have

$$c_1 = \frac{a(1) - sa(0)}{r - s} = \frac{6 - (1 - 2i)2}{(1 + 2i) - (1 - 2i)} = \frac{4 + 4i}{4i}.$$

Rationalizing the denominator by multiplying numerator and denominator by $i = \sqrt{-1}$ gives

$$c_1 = \frac{4i - 4}{-4} = 1 - i.$$

Similarly, $c_2 = 1 + i$, and the particular solution is

$$a(k) = (1 - i)(1 + 2i)^k + (1 + i)(1 - 2i)^k.$$

It seems amazing that the solution which gives the integers, $a(0) = 2$, $a(1) = 6$, $a(2) = 2$, $a(3) = -24$, $a(4) = -58$, $a(5) = 4, \ldots$, involves complex numbers. Later, we will see "real" applications that involve complex numbers. ∎

PROBLEMS

1. For the following dynamical systems, find the characteristic equation, find the roots of the characteristic equation, give the general solution of the dynamical system, and give the particular solution of the dynamical system for the given values of $a(0)$ and $a(1)$.

    a. $a(n + 2) = a(n + 1) + 2a(n)$, with $a(0) = 3$ and $a(1) = 3$.
    b. $a(n + 2) = 2a(n + 1) + a(n)$, with $a(0) = 2$ and $a(1) = 8$.
    c. $a(n+2) = -2a(n+1) + 15a(n)$, with $a(0) = 1$ and $a(1) = 11$.
    d. $a(n+2) = -6a(n+1) - 16a(n)$, with $a(0) = 2$ and $a(1) = 4$.
    e. $a(n + 2) = 4a(n + 1) - 5a(n)$, with $a(0) = 2$ and $a(1) = 4$.
    f. $a(n+2) = -4a(n+1) - 13a(n)$, with $a(0) = 6$ and $a(1) = -6$.

2. Suppose you have a pile of bubble gum which cost 1 cent each, a pile of licorice sticks which cost 1 cent each, and a pile of mints which cost 2 cents each. You pick n cents worth of candy, one piece at a time. Let $a(n)$ be the number of ways this can be done. For example, $a(1) = 2$ (bubble gum or licorice, denoted b and l), and $a(2) = 5$ denoted (b, b), (b, l), (l, b), (l, l), and (m). Notice that the order in which the candy is chosen is important.

    a. Find a dynamical system to model this problem.
    b. Find the general solution to this dynamical system.
    c. Using $a(0) = 1$ and $a(1) = 2$, find the particular solution to this dynamical system.

3. Suppose that the general solution to a dynamical system is

$$a(k) = c_1 r^k + c_2 s^k,$$

where $r \neq s$.

a. Suppose that $a(1)$ and $a(2)$ are given. Find a formula for c_1 and c_2 in terms of $a(1)$ and $a(2)$.

b. Suppose that $a(0)$ and $a(N)$ are given, where N is a given integer greater than 0. Find a formula for c_1 and c_2 in terms of $a(0)$ and $a(N)$.

5.2 Multiple roots

Suppose we have a car that gets 20 miles per gallon of gas and that we start driving with 15 gallons of gas in the tank. Let $a(n)$ be the amount of gas we have left after driving n miles. Observe that $a(0) = 15$. Since we use 0.05 gallons of gas in driving 1 mile, we have

$$a(n+1) = a(n) - 0.05,$$

that is, if we drive one more mile, we have 0.05 fewer gallons of gas. We know that the particular solution to this first-order affine dynamical system is

$$a(k) = 15 - 0.05k.$$

Notice that $a(k)$ depends linearly on k, that is, if we plotted the points $(k, a(k))$, they would lie on a straight line.

More generally, the dynamical system that describes the amount of gas left in a car after going n miles is

$$a(n+1) = a(n) - b,$$

where b is the amount of gas used in going 1 mile. The solution is

$$a(k) = c_1 + c_2 k,$$

where c_1 depends on the amount of gas we start with and $c_2 = -b$.

Another way to approach this problem is to write the dynamical system

$$a(n+2) - a(n+1) = a(n+1) - a(n).$$

This system says that the gas used in going any one mile (from $n+1$ to $n+2$ miles) is the same as the gas used in going any other mile (from n to $n+1$ miles). Simplification gives

$$a(n+2) = 2a(n+1) - a(n).$$

The solution must still be

$$a(k) = c_1 + c_2 k$$

since we already know that this describes the amount of gas left after going k miles.

The characteristic equation for this second-order dynamical system is

$$x^2 = 2x - 1, \qquad \text{or} \qquad (x-1)^2 = 0.$$

Thus, $x = 1$ is a double root. We thus see that when $x = 1$ is a double root, the solution is of the form

$$a(k) = c_1 + c_2 k.$$

Let's check that $a(k) = c_1 + c_2 k$ is a solution to the dynamical system

$$a(n+2) = 2a(n+1) - a(n)$$

for every choice of c_1 and c_2. Substitution of $a(n+2) = c_1 + c_2(n+2)$, $a(n+1) = c_1 + c_2(n+1)$, and $a(n) = c_1 + c_2 n$ gives

$$c_1 + c_2(n+2) = 2[c_1 + c_2(n+1)] - (c_1 + c_2 n).$$

All the terms cancel and we get equality, so this is a solution. If we are given $a(0)$ and $a(1)$, then solving the equations

$$a(0) = c_1 + c_2(0) \qquad \text{and} \qquad a(1) = c_1 + c_2$$

gives $c_1 = a(0)$ and $c_2 = a(1) - a(0)$ (the amount of gas used in going 1 mile) so the particular solution is

$$a(k) = a(0) + [a(1) - a(0)]k.$$

We thus see that if $x = 1$ is a double root, that is, when the dynamical system is

$$a(n+2) = 2a(n+1) - a(n),$$

then the general solution is

$$a(k) = c_1 + c_2 k.$$

What happens if $x = r$ is a double root of the characteristic equation, that is, when the characteristic equation is

$$(x-r)^2 = 0 \qquad \text{or} \qquad x^2 = 2rx - r^2?$$

The dynamical system that has $x^2 = 2rx - r^2$ as its characteristic equation must be

$$a(n+2) = 2ra(n+1) - r^2 a(n). \tag{5.4}$$

There is a clever technique for solving this dynamical system and that is to substitute

$$a(n) = r^n a'(n)$$

into the dynamical system. (We first used this technique in section 4.2 to solve dynamical system (4.5).) This substitution gives

$$r^{n+2} a'(n+2) = 2r[r^{n+1} a'(n+1)] - r^2 [r^n a'(n)],$$

or, after dividing both sides by r^{n+2},

$$a'(n+2) = 2a'(n+1) - a'(n). \tag{5.5}$$

From the above discussion, we know that the general solution to dynamical system (5.5) is

$$a'(k) = c_1 + c_2 k.$$

By substitution, it follows that

$$a(k) = r^k a'(k) = r^k (c_1 + c_2 k)$$

is the general solution to dynamical system (5.4). We summarize this result with the following theorem.

THEOREM 5.7 *Consider the dynamical system*

$$a(n+2) = 2ra(n+1) - r^2 a(n)$$

which has $x = r$ as a double root of the corresponding characteristic equation. The general solution is

$$a(k) = (c_1 + c_2 k) r^k.$$

If $a(0)$ and $a(1)$ are known, then the particular solution is given when

$$c_1 = a(0) \qquad and \qquad c_2 = \frac{a(1) - ra(0)}{r}.$$

EXAMPLE 5.8   Find the particular solution to the dynamical system

$$a(n+2) = 4a(n+1) - 4a(n),$$

given that $a(0) = 3$ and $a(1) = 4$. Factoring the characteristic equation

$$x^2 = 4x - 4$$

gives

$$(x - 2)^2 = 0,$$

so that $x = 2$ is a double root of the equation. Thus, the general solution is of the form

$$a(k) = (c_1 + c_2 k)2^k.$$

The initial values give $c_1 = 3$ and $c_2 = -1$, so the particular solution is

$$a(k) = (3 - k)2^k. \qquad \blacksquare$$

PROBLEMS

1. Find the particular solution to each of the following dynamical systems.
   a. $a(n+2) = 2a(n+1) - a(n)$, with $a(0) = 5$ and $a(1) = 4$.
   b. $a(n+2) = -2a(n+1) - a(n)$, with $a(0) = 2$ and $a(1) = 5$.
   c. $a(n+2) = -a(n+1) + 2a(n)$, with $a(0) = 1$ and $a(1) = 0$.
   d. $a(n+2) = -a(n+1) - 0.25a(n)$, with $a(0) = 1$ and $a(1) = 1$.

2. Find the particular solution to each of the dynamical systems in problem 1 given that $a(1) = 1$ and $a(2) = 4$.

3. Find the particular solution to each of the dynamical systems in problem 1 given that $a(0) = 2$ and $a(3) = 12$.

4. Suppose that the general solution to a dynamical system is

$$a(k) = (c_1 + c_2 k)r^k.$$

   a. Suppose that $a(1)$ and $a(2)$ are given. Find a formula for c_1 and c_2 in terms of $a(1)$ and $a(2)$.
   b. Suppose that $a(0)$ and $a(N)$ are given, where N is a given integer greater than 0. Find a formula for c_1 and c_2 in terms of $a(0)$ and $a(N)$.

5.3   The gambler's ruin

We wish to consider a problem which is of interest to anyone who is planning to visit Las Vegas, Atlantic City, or Monte Carlo to gamble, namely the **gambler's ruin**.

Being a sensible person, you set aside a certain amount of money which you can afford to lose. Once this money is gone, you will quit gambling. Some people would just continue gambling until the allocated money is gone. Being more intelligent than that, you set a goal for yourself of, say, N dollars. If you reach this total, you will take your winnings and go home.

You have decided always to bet **1 dollar** on the same game, in which your probability of winning a dollar is p while the probability of losing your dollar is $1 - p = q$. The question is, what is your probability of quitting a winner, that is, quitting with a total of N? Conversely, what is your probability of being ruined, that is, of losing all your money?

Suppose you start the above game with n dollars, where $0 \leq n \leq N$. The question is, "what is the probability that you will eventually go broke under the conditions just described, **given** that you presently have n dollars?" We will denote the answer to this question by $a(n)$.

Suppose the game is the tossing of a fair coin. For each toss, the bet is 1 dollar, that is, you win 1 dollar if heads comes up and you lose 1 dollar if tails comes up. You will quit the game if you have a total of either 0 or 3 dollars. What is your probability of quitting broke if you started with 0, 1, 2, or 3 dollars?

Let $a(n)$ be the probability you will eventually go broke given that you presently have n dollars. Clearly $a(0) = 1$ and $a(3) = 0$ since you quit when you have 0 or 3 dollars. What are $a(1)$ and $a(2)$?

To compute $a(1)$, you have to compute the probability of quitting a loser, given that you presently have 1 dollar? There are two **cases**: case 1 is that you lose your dollar on this flip and thus are ruined; and case 2 is that you win a dollar on this flip and then eventually lose all your money. Since one of these cases must occur for you to be ruined, we will compute the probability of each case happening and then, by the addition principle for probability, $a(1)$ must be the sum of these two probabilities, that is,

$$a(1) = \text{prob(case 1} + \text{prob(case 2}.$$

In case 1 for $a(1)$, the probability that you lose your dollar is the probability that tails comes up, which is 0.5. You are then broke, so $\text{prob(case 1)} = 0.5$.

To compute the probability of case 2 for $a(1)$ happening, we notice that two things must happen: first we must flip a coin and get heads, and second we must then **eventually** go broke. The probability that we flip heads is $\mathrm{prob}(H) = 0.5$. Since we have flipped heads, we have won 1 dollar and now have 2 dollars. Thus, the probability that we now eventually go broke is, by our notation, $a(2)$. The probability that both of these occur (flipping heads then going broke) is, by the multiplication principle (see theorem 3.25 of section 3.3), the product of each of these probabilities, that is,

$$\mathrm{prob}(\text{case } 2) = \mathrm{prob}(H)\,a(2) = 0.5a(2).$$

So by the addition principle,

$$a(1) = 0.5 + 0.5a(2). \qquad (5.6)$$

To compute $a(2)$, we must again consider two cases. In this part, we are assuming we have 2 dollars and want to know the probability we will eventually go broke. Case 1 now is that we get heads and eventually go broke, while case 2 is that we get tails and eventually go broke.

In case 1, if we get heads, we win a dollar, and thus quit playing and cannot go broke. Thus the probability of flipping heads and going broke is 0, since we cannot flip heads and go broke. So $\mathrm{prob}(\text{case } 1) = 0$.

Case 2 for $a(2)$ is computed similarly to case 2 for $a(1)$ above, that is, we consider a two-stage event in which the first stage is that we flip tails T, and the second stage is that we eventually go broke. Again, the probability of the first stage is $\mathrm{prob}(T) = 0.5$. Since we have now lost 1 dollar and only have 1 dollar left, the probability of the second stage is the probability we eventually go broke given we presently have 1 dollar, $a(1)$. By the multiplication principle,

$$\mathrm{prob}(\text{case } 2) = 0.5a(1).$$

Now by the addition principle,

$$a(2) = \mathrm{prob}(\text{case } 1) + \mathrm{prob}(\text{case } 2) = 0 + 0.5a(1) = 0.5a(1). \quad (5.7)$$

We now have two equations involving $a(1)$ and $a(2)$. From equation (5.7), we can substitute $0.5a(1)$ into equation (5.6) for $a(2)$, giving

$$a(1) = 0.5 + 0.5(0.5a(1)), \qquad \text{or} \qquad a(1) = 2/3.$$

Consequently,

$$a(2) = 1/3.$$

This was relatively easy because our goal of $N = 3$ dollars was relatively low. But if we decide to quit with either 0 or 1000 dollars, we would have 999 equations to solve simultaneously. A simpler method is to use the theory of second-order linear dynamical systems.

We now consider the general gambler's ruin, that is, when the probability we win a dollar is p, the probability we lose a dollar is $q = 1 - p$, and we quit when we have a total of 0 or N dollars. (We are assuming that the coin may be unfair, that is, the probability p of heads occurring is not necessarily one-half.

Let $a(n)$ be the probability we will **eventually** quit with 0 dollars if we presently have n dollars. Then $a(0) = 1$ and $a(N) = 0$ are our two "initial" values. Note that the initial conditions are a bit unusual. Usually, we are given $a(0)$ and $a(1)$. But we will soon see that this causes no additional problems.

Let's now try to compute $a(n+1)$, that is, the probability that we eventually go broke given that we presently have $n + 1$ dollars. We assume that $0 < n+1 < N$, that is, $n = 0, 1, \ldots, N - 2$. There are two cases. Case 1 is the case in which we win a dollar on the next flip and eventually go broke, while case 2 is the case in which we lose a dollar on the next flip and eventually go broke. By the addition principle,

$$a(n+1) = \text{prob(case 1)} + \text{prob(case 2)}.$$

Let H be the event we win a dollar on our next flip and let T be the event we lose a dollar on our next flip.

Case 1 is a two-stage process. We are given that the probability of the first stage occurring is $\text{prob}(H) = p$. Now that we have won a dollar, we have a total of $n+2$ dollars, so the probability of the second stage is the probability we eventually go broke given that we have $n+2$ dollars, that is, $a(n+2)$. Thus, by the multiplication principle,

$$\text{prob(case 1)} = p\, a(n+2).$$

Case 2 is also a two-stage process. The probability of the first stage occurring is $\text{prob}(T) = q$. Since we have now lost a dollar, the second stage is the event that we eventually go broke given that we have n dollars, $a(n)$. By the multiplication principle,

$$\text{prob(case 2)} = qa(n).$$

By the addition principle,

$$a(n+1) = p\,a(n+2) + q\,a(n), \qquad \text{for} \quad n = 0, 1, \ldots, N-2.$$

Rearranging and dividing by p gives

$$a(n+2) = (1/p)\,a(n+1) - (q/p)\,a(n).$$

The characteristic equation is

$$x^2 = (1/p)x - q/p.$$

The two roots are $x = 1$ and $x = q/p$. Note that if $p \neq q$ then we have two distinct roots and can use theorem 5.3 to find a solution. But if $p = q = 1/2$, then we have a double root and we have to use theorem 5.7 to find a solution.

Suppose $p \neq 0.5$, that is, $p \neq q$. Then $q/p \neq 1$ and we have two roots. For simplicity, let $q/p = r$. Thus

$$a(k) = c_1\,(1)^k + c_2\,r^k = c_1 + c_2\,r^k.$$

Since

$$a(0) = c_1 + c_2 = 1 \qquad \text{and} \qquad a(N) = c_1 + c_2\,r^N = 0,$$

we can solve these equations for c_1 and c_2, giving

$$c_2 = \frac{1}{1 - r^N} \qquad \text{and} \qquad c_1 = \frac{-r^N}{1 - r^N}.$$

Thus

$$a(k) = \frac{-r^N}{1 - r^N} + \frac{r^k}{1 - r^N} = \frac{r^k - r^N}{1 - r^N}.$$

For example, suppose you start with $k = 20$ dollars, the probability you will win a dollar is $p = 0.48$, the probability you lose a dollar is $q = 0.52$, and you will quit when you go broke or have a total of $N = 40$. Then $r = 0.52/0.48 = 1.083$, and the probability you will go broke is

$$a(20) = \frac{r^{20} - r^{40}}{1 - r^{40}} = 0.832.$$

Observe that, **on each flip**, your probability of losing a dollar is only slightly more than 0.5. Also observe that you are playing "double or nothing", that is, since $N = 40$ and $n = 20$ you quit if you win or lose a net of 20 dollars. Intuition might suggest that your probability of going broke would be only slightly greater than 0.5. But instead, your probability of going broke is 0.832, while your probability of quitting a winner with a profit of 20 dollars is only $1 - 0.832 = 0.168$. This does not look good for you.

A more detailed study makes it look even worse. Suppose you set an unrealistically high goal for N. This is about the same as the strategy of playing until you are broke. Mathematically, we solve this problem by computing

$$\lim_{N \to \infty} a(k) = \lim_{N \to \infty} \frac{r^k - r^N}{1 - r^N}.$$

Note that r^k and 1 remain fixed, while r^N goes to infinity. Let $x = r^N$. We can then rewrite the above as

$$\lim_{x \to \infty} \frac{c_1 - x}{1 - x} = 1.$$

Thus, we see that the higher we set our goal, the more likely we are to go broke.

Let's now compute the formula for $a(k)$ under the condition that we are playing a fair game, that is, $p = q = 0.5$. In this case $q/p = 1$ and 1 is a double root. Thus

$$a(k) = (c_1 + c_2 k)\, 1^k = c_1 + c_2\, k.$$

Since $1 = a(0) = c_1$ and $0 = a(N) = c_1 + Nc_2$, it follows that $c_1 = 1$, $c_2 = -1/N$, and

$$a(k) = 1 - \frac{k}{N} = \frac{N - k}{N}.$$

For example, suppose you flip a fair coin and with heads you win a dollar and with tails you lose a dollar. Thus $p = q = 0.5$. Suppose you start with $k = 20$ and quit when you are ruined or when you have $N = 40$. Then your probability of ruin is $a(20) = (40 - 20)/40 = 0.5$, a much better result then when $p = 0.48$.

This also brings up the issue of the instability of the value $p = 05$. If the probability of winning any one bet is **exactly** 0.5, then the game is fair and the probability of doubling our money is 0.5. But if the actual probability of winning any one bet is off by a little, say $p = 0.499$, then when we are trying to double a large amount of money (start with n and quit with $2n$ dollars betting one dollar at a time), the probability of ruin is quite large. Thus, a small error in our estimate of p near $p = 0.5$ can cause a large error in our predictions.

MORAL: Don't gamble when $p < 0.5$. In the above, we used the same amounts of money but the probabilities were even money $p = 0.5$, and slightly against us $p = 0.48$. A difference of 0.02 in a game wouldn't seem to make much difference, but in reality it changed our chances of being ruined from 0.5 to 0.832. If you try these computations for larger amounts of money the results are even more dramatic. If $k = 100$, $N = 200$, and $p = 0.5$, our chances of ruin are 0.5, while if $p = 0.48$, our chance of ruin is over 0.999. $\square$

The result is that the more you gamble (when the odds are against you), the more likely it is that you will be ruined. In fact if you must gamble, your best chance to double your money is to place all your money on one bet. If you win, quit. If you lose, don't say you weren't warned.

PROBLEMS

1. Suppose you are in a game in which the probability of winning any particular bet is 0.49, while the probability of losing is 0.51. You start with 50 dollars and will quit when you have 100 dollars. What is your probability of ruin,
   a. if you make 1-dollar bets,
   b. if you make 5-dollar bets (Hint: Let 5 dollars equal one unit. Thus, you begin with 10 units, and quit when you have 20 units.),
   c. if you make 10-dollar bets,
   d. if you make 25-dollar bets, and
   e. if you make 50-dollar bets?

2. Suppose you are in a game in which the probability of winning any particular bet is 0.49, while the probability of losing is 0.51. You will quit when you have 100 dollars. Assume 1-dollar bets.
   a. What amount of money k do you need to start with so that you have at least a 50 per cent chance of quitting a winner?
   b. Same question as part (a) except that the probability of winning any one bet is 0.48.

3. Suppose you are in a game in which, on each play, there is a probability of 0.1 that you will win 2 dollars, there is a probability of 0.3 that you will win 1 dollar, and there is a probability of 0.6 that you will lose 1 dollar. You will quit when you are broke or when you have at least N dollars.
   a. Suppose you quit if you have 3 or more dollars. What is $a(0)$, $a(1)$, $a(2)$, $a(3)$, and $a(4)$, where $a(n)$ is the probability of eventually going broke if you presently have n dollars?

    b. Suppose you will quit when you are broke or when you have at least N dollars. Write a third-order dynamical system to model this situation. What are the three "initial" values?

5.4 Sex-linked genes

As was discussed earlier, many inherited traits of individuals are determined by a pair of alleles. If one allele A is dominant, while another allele a is recessive, then an individual has the recessive trait only if the pair of alleles are both recessive, that is, (a, a). For certain traits, women have a pair of alleles, but men have only one allele, which they inherit from their mother. Such traits are called **sex-linked traits**. Therefore, if a man inherits an A-allele, he will have the dominant trait, but if he inherits an a-allele, he will have the recessive trait. Color blindness and hemophilia are two such traits. In these cases, the alleles which cause problems are recessive. In the following, we will show why it is much more common for a man to exhibit a sex-linked trait than a woman.

    Consider a sex-linked trait (in which women have a pair of alleles, but men have a single allele). Suppose that a woman has the pair of alleles (A, a), and a man has the allele A. Thus, both the man and the woman exhibit the dominant trait. Suppose these two individuals have an offspring. If the child is a boy, he will inherit either the A- or a-allele from his mother. Thus the boy has a probability of 0.5 of inheriting the a-allele and thus having the recessive trait. But if the child is a girl, she will inherit the A-allele from her father, and either the A- or a-allele from her mother. Therefore, the girl will have either the pair of alleles (A, a) or (A, A), and so she is certain to exhibit the dominant trait.

    To study this problem, we must keep the alleles of the males separate from the alleles of the females. Thus, we will let $p(n)$ and $q(n)$ represent the proportion of A-alleles and a-alleles among the women of generation n, respectively. Likewise, we will let $P(n)$ and $Q(n)$ represent the proportion of A-alleles and a-alleles among the men of generation n, respectively.

    Let u, v, and w represent the proportion of dominant homozygotes, heterozygotes, and recessive homozygotes, respectively, among the population of women in generation $n + 1$. To be a dominant homozygote, a girl must inherit an A-allele from her mother and an A-allele from her father. To compute the probability of this happening, we consider a two-stage process. The first stage is to draw an A-allele from the population of women of generation n, and the probability of this happening is $p(n)$. The second stage is to draw an A-allele from the population of men of generation n, and the probability of this happening is $P(n)$. Thus, by the multiplication principle for probability,

$$u = p(n)P(n).$$

To compute the probability that a woman is a heterozygote, we must consider two cases. The first case is that she inherits an A-allele from her mother and an a-allele from her father. The probability of this happening is $p(n)Q(n)$. The second case is that she inherits an a-allele from her mother and an A-allele from her father, which by the multiplication principle is $q(n)P(n)$. Adding the probabilities of these two cases, we find that the probability a woman is a heterozygote is

$$v = p(n)Q(n) + P(n)q(n).$$

Similarly, the probability that a woman is a recessive homozygote is

$$w = q(n)Q(n).$$

Suppose the total number of women in generation $n+1$ is W. Then the total number of alleles in the population of women of generation $n + 1$ is $2W$, since each woman has two alleles. The total number of A-alleles in the population of women of generation $n + 1$ is: the total number of dominant homozygotes (which is uW) times 2 (since each dominant homozygote has two A-alleles) plus the total number of heterozygotes (which is vW) times 1 (since each heterozygote has one A-allele). Recall that the recessive women have no A-alleles. Thus the total number of A-alleles in the women of generation $n + 1$ is

$$W(2u + v) = W[2p(n)P(n) + p(n)Q(n) + P(n)q(n)]$$
$$= W[p(n)P(n) + p(n)Q(n) + p(n)P(n) + P(n)q(n)],$$

after rearranging terms. Since $p(n) + q(n) = 1$ and $P(n) + Q(n) = 1$, it follows that

$$p(n)P(n) + p(n)Q(n) = p(n)[P(n) + Q(n)] = p(n),$$

and

$$p(n)P(n) + P(n)q(n) = P(n).$$

Thus, it follows that the total number of A-alleles in the women of generation $n+1$ is

$$W(2u+v) = W[p(n)+P(n)].$$

The proportion of A-alleles in the women of generation $n+1$ is the total number of the A-alleles divided by the total number of alleles, which gives

$$p(n+1) = \frac{W[p(n)+P(n)]}{2W} = 0.5p(n)+0.5P(n).$$

We now turn our attention to the men of generation n. Since men only have one allele, they are either dominant or recessive, and the proportion of men that are of each of these genotypes is the same as the proportion of alleles of type A and a among the men of generation n. Thus, the proportion of dominant men in generation n is $P(n)$, while the proportion of recessive men in generation n is $Q(n)$.

To compute the proportion of dominant men in generation $n+1$, we must compute the probability that a man will inherit an A-allele. Since a man in generation $n+1$ inherits his allele from a woman of generation n, the probability that a boy in generation $n+1$ inherits an A-allele is

$$P(n+1) = p(n).$$

Likewise, $Q(n+1) = q(n)$.

To review, the proportions of A-alleles among women and among men satisfy the dynamical systems

$$p(n+1) = 0.5p(n)+0.5P(n) \qquad \text{and} \qquad P(n+1) = p(n),$$

respectively. Let us substitute $n+1$ for n in the first of these dynamical systems to get

$$p(n+2) = 0.5p(n+1)+0.5P(n+1).$$

Since we know from above that $P(n+1) = p(n)$, we can substitute $p(n)$ for $P(n+1)$ in the previous equation to get

$$p(n+2) = 0.5p(n+1)+0.5p(n)$$

as the second-order linear dynamical system that gives the relationship for the proportion of A-alleles among the women of generation n. We first studied this dynamical system in example 5.5.

As we found in example 5.5, the characteristic equation for this dynamical system is

$$x^2 = 0.5x + 0.5,$$

the roots are $x = 1$ and $x = -0.5$, and the general solution is

$$p(k) = c_1 + c_2 (-0.5)^k,$$

where c_1 and c_2 depend on the initial values.

As k goes to infinity, $(-0.5)^k$ goes to zero and so $p(k)$ goes to c_1. Since the current population of women today is obviously the result of many previous generations, we can assume that k is large, that is, we can assume that $p(k) = c_1$, where $p(k)$ is the proportion of a-alleles among women today.

Since $P(k) = p(k-1) = c_1 + c_2 (-0.5)^{k-1}$, which is also close to c_1, we can assume that the proportion of A-alleles in the male population is $P(k) = c_1$ also.

To review, the proportion of A-alleles in the male and in the female populations is stable at some level $c_1 = p$, and thus the proportion of a-alleles in the male and female populations is $q = 1 - p$. Thus, the proportion of dominant homozygotes, heterozygotes, and recessive homozygotes among the population of women in any generation is

$$u = p^2, \qquad v = 2pq, \qquad w = q^2,$$

respectively, while the proportion of dominant and recessive men in any generation is

$$p \qquad \text{and} \qquad q,$$

respectively.

Color blindness is a sex-linked trait in which the proportion of men that are color blind is $q = 0.01$, which, as we have seen above, is also the proportion of a-alleles. Thus, we expect the proportion of women that are color blind to be $w = (0.01)^2 = 0.0001$, so that only 1 out of 10 000 women will exhibit this recessive trait.

PROBLEMS

1. Suppose that for a certain sex-linked trait we know that $p = 0.7$. What proportion of women and of men will exhibit the recessive trait? What proportion of the women are heterozygotes?

2. Answer problem 1 with $p = 0.98$.

3. Suppose a recessive sex-linked allele, a, has recently appeared in the population, and so has not reached equilibrium. At present, we estimate that the proportion of a-alleles in women is $q(0) = 0.1$, and that the proportion of this allele in men is $Q(0) = 0.2$. This means that $p(0) = 0.9$ and $P(0) = 0.8$.

   a. Using the dynamical system

   $$p(n+1) = 0.5p(n) + 0.5P(n),$$

   find $p(1)$.

   b. Using $p(1)$ from part (a), find the particular solution to this dynamical system.

   c. What is the equilibrium value for the dynamical system in part (a)?

4. Repeat problem 3 using $q(0) = 0.98$ and $Q(0) = 0.99$.

5. Suppose that a trait is sex-linked. Let $p(n)$ and $P(n)$ represent the proportion of A-alleles among the adult women and men, respectively, of generation n. Suppose that a fraction μ of the A-alleles mutate to a-alleles each generation. Develop a second-order linear dynamical system for $p(n+2)$ and find the solution to that system. (Hint: It is easiest to assume that the A-alleles mutate to a-alleles in the children.)

6. Suppose we have a fatal sex-linked trait in which all recessive individuals die before adulthood. Thus, all adult males contain one A-allele. Let $p(n)$ and $q(n)$ be the proportion of A- and a-alleles in the adult women of generation n.

   a. Compute $p(n+1)$, the proportion of A-alleles among the females of generation $n+1$?

   b. Solve the dynamical system in part (a) and find the stable equilibrium value.

   c. Suppose that the fraction of males being born with this trait is 10^{-2} telling us that now the proportion of a-alleles in the women today is $q(0) = 10^{-2}$. Use this information to find the particular solution to the dynamical system in part (a).

5.5　Stability for second-order affine equations

In this section, we are going to study the stability of second-order linear dynamical systems.

First assume that **the two roots of the characteristic equation, r and s, are distinct (r is not a double root) and are not equal to 1.** Thus, we are assuming that the solution to the dynamical system

$$a(n+2) = b_1 a(n+1) + b_2 a(n)$$

(where b_1 and b_2 are given constants) is of the form

$$a(k) = c_1 r^k + c_2 s^k,$$

where c_1 and c_2 depend on the initial values. We now want to find under what conditions does $a(k)$ go to zero.

Suppose, in addition to the fact that $r \neq s$, that they satisfy

$$|r| < 1 \qquad \text{and} \qquad |s| < 1.$$

Then, as k goes to infinity, $c_1 r^k$ and $c_2 s^k$ both go to zero, and so the solution satisfies

$$\lim_{k \to \infty} a(k) = \lim_{k \to \infty} (c_1 r^k + c_2 s^k) = 0.$$

In this case, zero is an attracting fixed point or stable equilibrium value.

If either one of the roots, say r, is greater than 1 (in absolute value) then $a(k)$ does not go to zero since $|r^k|$ goes to infinity. We note that $|a(k)|$ may not go to infinity in a strict sense since the terms $c_1 r^k$ and $c_2 s^k$ may cancel each other for some, but not all, values of k.

Now suppose that r is a double root of the characteristic equation and that $|r| < 1$. Then the general solution is of the form

$$a(k) = (c_1 + c_2 k) r^k.$$

We know that the term $c_1 r^k$ goes to zero as k goes to infinity. But the term $c_2 k r^k$ also goes to zero as k goes to infinity. A brief sketch of the proof of this fact is as follows.

PROOF: Let $b(k) = k r^k$. Then $b(k)$ satisfies the dynamical system

$$b(n+1) = \frac{(n+1)r}{n} b(n).$$

You should substitute $b(n+1) = (n+1)r^{n+1}$ and $b(n) = nr^n$ into this dynamical system to see that it works.

Thus each number, $b(n)$, is a nonconstant multiple of the previous number, that multiple being $(n+1)r/n$. Since $|r| < 1$, if n is large enough, specifically if $n > r/(1-r)$, then

$$\left| \frac{(n+1)r}{n} \right| \leq r_0$$

for some constant $r_0 \leq 1$. So from that point m on, each number is less than a constant fraction of the previous number, that is,

$$|b(n+m)| \leq r_0|b(n+m-1)| \leq r_0^2|b(n+m-2)| \leq \cdots \leq r_0^n|b(m)|.$$

Since $r_0 < 1$, it follows that

$$\lim_{n \to \infty} (n+m)r^{n+m} = \lim_{n \to \infty} |b(n+m)| \leq \lim_{n \to \infty} r_0^n|b(m)| = 0.$$

Thus,

$$\lim_{k \to \infty} kr^k = 0$$

and the proof is complete.   ∎

We have shown the following.

THEOREM 5.9   *Suppose that the dynamical system*

$$a(n+2) = b_1 a(n+1) + b_2 a(n)$$

*has two real roots r and s. If $|r| < 1$ and $|s| < 1$, with $r = s$ being possible, then zero is an **attracting fixed point**.*

Suppose the roots of the dynamical system are complex. Does $|a(k)|$ go to zero, infinity, or does it do something else?

Before continuing, let us review complex numbers. Let x, y, w, and z be four real (not complex) numbers. Then

$$(x+yi)(w+zi) = (xw - yz) + (xz + yw)i.$$

Using this formula, we get

$$(3+2i)(-1+5i) = (-3-10) + (15-2)i = -13 + 13i.$$

You can **visualize** a complex number, $x + yi$, as the point (x, y) in the plane. Note that the real part x is the horizontal coordinate and the imaginary part y is the vertical coordinate. Observe that, by the Pythagorean theorem, the distance between the point (x, y) and the origin is given by $\sqrt{x^2 + y^2}$. From this idea, we can define the absolute value of a complex number as its distance from the origin.

DEFINITION 5.10   *The absolute value of a complex number $x + yi$ is*

$$|x+yi| = \sqrt{x^2 + y^2}.$$

For example,

$$|3 + 4i| = \sqrt{9 + 16} = 5,$$

and

$$\left| \frac{1}{2} + \frac{\sqrt{3}i}{2} \right| = \sqrt{\frac{1}{4} + \frac{3}{4}} = 1.$$

Notice that $(1/2, \sqrt{3}/2)$ is a point on the unit circle.

Suppose that the root r to the characteristic equation for a second-order dynamical system is complex, that is, $r = x + yi$. What we will see is that, if $|r| = |x + yi| < 1$, then $|r^k|$, the distance of the complex number r^k from the origin, will be "small" for large k, that is, the real part of r^k will be close to zero and the complex part of r^k will be close to zero. We say that

$$\lim_{k \to \infty} r^k = 0.$$

So if both roots r and s satisfy $|r| < 1$ and $|s| < 1$, then r^k and s^k will both go to zero and so $a(k)$ goes to zero. Thus, we can restate theorem 5.9 without the word "real".

THEOREM 5.11   *Suppose that the characteristic equation for the dynamical system*

$$a(n + 2) = b_1 a(n + 1) + b_2 a(n)$$

has two, not necessarily distinct, roots r and s. If $|r| < 1$ and $|s| < 1$, then the fixed point zero is **attracting***. If either $|r| > 1$ or $|s| > 1$, or both, then the fixed point zero is* **repelling***.*

PROOF: For the attracting part of this theorem, all we need to show is that, if $|r| = |x + yi| < 1$, then $|r^k|$ goes to zero. We **know**, since $|r| < 1$, that $|r|^k$ goes to zero. To prove the theorem, we will show that

$$|r^k| = |r|^k.$$

First we will show that if r and s are two complex numbers then $|rs| = |r||s|$, that is, you can multiply the numbers and then take the absolute value, or you can take the absolute value of each number and then multiply.

$$|rs| = |(x + yi)(w + zi)| \qquad \text{and} \qquad |r||s| = |x + yi||w + zi|.$$

We will see that they are equal. Since

$$(x + yi)(w + zi) = (xw - yz) + (yw + xz)i,$$

it follows that

$$|(x + yi)(w + zi)| = \sqrt{(xw - yz)^2 + (yw + xz)^2}$$

$$= \sqrt{(xw)^2 + (yz)^2 + (yw)^2 + (xz)^2}$$

after multiplying out and canceling the $2xywz$ terms. Also,

$$|x + yi||w + zi| = \sqrt{x^2 + y^2}\sqrt{w^2 + z^2}$$

$$= \sqrt{(x^2 + y^2)(w^2 + z^2)}$$

$$= \sqrt{(xw)^2 + (yz)^2 + (yw)^2 + (xz)^2}.$$

So we see that $|rs| = |r||s|$.

Now, let $a(n) = |r^n|$, where r is a complex number. Note that $a(0) = |r^0| = 1$. We have just shown in the above argument that

$$a(2) = |r^2| = |rr| = |r||r| = |r|a(1)$$

and that

$$a(3) = |r^3| = |rr^2| = |r||r^2| = |r|a(2).$$

In fact,

$$a(n + 1) = |r^{n+1}| = |rr^n| = |r||r^n| = |r|a(n).$$

The solution to this first-order linear dynamical system is

$$a(k) = |r|^k a(0) = |r|^k.$$

We have just shown that $a(k) = |r^k| = |r|^k$. Thus, if $|r| < 1$, then

$$\lim_{k \to \infty} a(k) = \lim_{k \to \infty} |r^k| = \lim_{k \to \infty} |r|^k = 0$$

and zero is attracting. ∎

Suppose one of the roots satisfies $|r| > 1$. If $|r| > |s|$, it is not difficult to show that $|a(k)| \to \infty$. If $|r| = |s|$, then the solution may oscillate with increasing amplitude. This means that there will be k-values for which $|a(k)|$ is quite large, but there may be other k-values for which $|a(k)|$ is small or even zero. Zero is still repelling, because

some $a(k)$-values are large. We will not go into the details of this part of the theorem.

EXAMPLE 5.12   Consider the dynamical system

$$a(n+2) = -0.25a(n). \tag{5.8}$$

The roots of the characteristic equation are $r = 0.5i$, and $s = -0.5i$. Therefore, the general solution is

$$a(k) = c_1(0.5i)^k + c_2(-0.5i)^k,$$

where c_1 and c_2 depend on the initial values. Since $|\pm 0.5i| = 0.5 < 1$, it follows that $c_1(0.5i)^k$ and $c_2(-0.5i)^k$ go to zero as k goes to infinity, so that $a(k)$ goes to 0. See figure 5.1, in which the solution to this dynamical system is plotted with $a(0) = -4$ and $a(1) = -8$.   ■

EXAMPLE 5.13   Consider the second-order dynamical system

$$a(n+2) = 2a(n+1) - 2a(n). \tag{5.9}$$

From the quadratic formula, the roots of the characteristic equation are

$$r = 1 + i \qquad \text{and} \qquad s = 1 - i,$$

and the general solution is

$$a(k) = c_1(1+i)^k + c_2(1-i)^k.$$

Since

$$|1 \pm i| = \sqrt{1^2 + 1^2} = \sqrt{2} > 1,$$

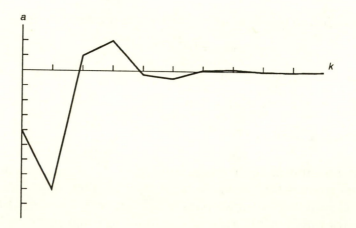

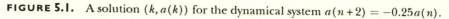

FIGURE 5.1.   A solution $(k, a(k))$ for the dynamical system $a(n+2) = -0.25a(n)$.

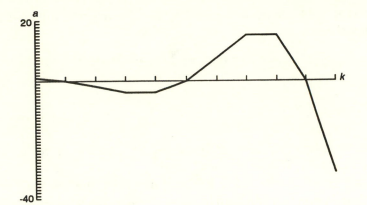

FIGURE 5.2. A solution $(k, a(k))$ for the dynamical system $a(n+2) = 2a(n+1) - 2a(n)$.

zero is a repelling fixed point. This can be seen in figure 5.2, where the particular solution is plotted with $a(0) = 1$ and $a(1) = 0$. Although it may be difficult to see, $a(k)$ actually oscillates with increasing amplitude.   ■

There is an easy method for computing $|x + iy|$ when $r = x + iy$ is a root of the characteristic equation for the second-order dynamical system

$$a(n+2) + b_1 a(n+1) + b_2 a(n) = 0,$$

where b_1 and b_2 are real numbers. First, the characteristic equation is

$$r^2 + b_1 r + b_2 = 0.$$

Suppose it can be factored into

$$r^2 + b_1 r + b_2 = (r - r_1)(r - r_2) = 0.$$

Note that $r_1 r_2 = b_2$, that is, **the product of the roots of a polynomial equals the constant term in the polynomial.**

Second, if one of the roots of the characteristic equation is $r_1 = x + iy$, then the other root must be $r_2 = x - iy$ so that their product, b_2, is real. But $b_2 = r_1 r_2 = (x + iy)(x - iy) = x^2 + y^2 = |x \pm iy|^2$.

Putting these facts together, we get the following theorem.

THEOREM 5.14   *Suppose the roots of the characteristic equation for the dynamical system*

$$a(n+2) + b_1 a(n+1) + b_2 a(n) = 0,$$

are $x \pm iy$. *Then*

$$|x + iy| = \sqrt{b_2}.$$

This theorem will save quite a bit of algebra in some of the applications that follow in this chapter.

If $|a \pm bi| = 1$, then $|(a \pm bi)^k| = 1$ for all k, and so the "size" of the solution remains constant in the sense that the solution oscillates. But we will see that the solution can behave in quite a strange manner.

EXAMPLE 5.15   Consider the dynamical system

$$a(n+2) = a(n+1) - a(n). \tag{5.10}$$

The characteristic equation is

$$r^2 - r + 1 = 0.$$

From the quadratic formula, the roots are

$$r = \frac{1 \pm \sqrt{3}i}{2}$$

The general solution is

$$a(k) = c_1 \left(\frac{1 + \sqrt{3}i}{2} \right)^k + c_2 \left(\frac{1 - \sqrt{3}i}{2} \right)^k.$$

By theorem 5.14, $|1/2 \pm (\sqrt{3}/2)i| = 1$, so the fixed point zero is neither attracting nor repelling. This can be seen in figure 5.3, in which the particular solution is plotted for $a(0) = 6$ and $a(1) = -1$. If you compute several values of $a(k)$, you will find that $a(2) = -7$, $a(3) = -6$, $a(4) = 1$, and $a(5) = 7$. You then find that these six numbers keep repeating, that is, $a(k)$ forms a **6-cycle**.

In fact, for any initial values $a(0)$ and $a(1)$, the solution to this dynamical system will form a 6-cycle.   ■

Let's see one more example in which the roots of the characteristic equation equal 1 in absolute value. The behavior of the solution will be quite interesting.

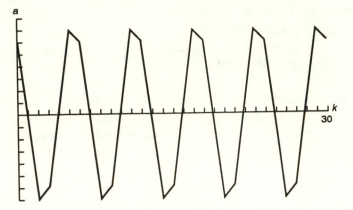

FIGURE 5.3. A solution $(k, a(k))$ for the dynamical system $a(n+2) = a(n+1) - a(n)$.

EXAMPLE 5.16 Consider

$$a(n+2) = 1.2a(n+1) - a(n). \qquad (5.11)$$

The roots of the characteristic equation are

$$r = 0.6 \pm 0.8i,$$

which satisfy $|0.6 \pm 0.8i| = 1$, by theorem 5.14. In figure 5.4, $a(k)$ is plotted with $a(0) = 1$ and $a(1) = -2$. Is it periodic? It doesn't appear to be, although it almost is (the peaks are slightly different).

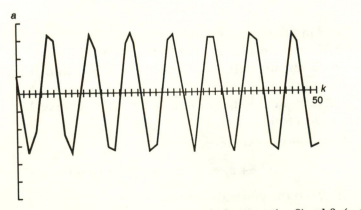

FIGURE 5.4. A solution $(k, a(k))$ for the dynamical system $a(n+2) = 1.2a(n+1) - a(n)$.

This solution is what mathematicians call "almost periodic". We will not give a precise definition of almost period. Intuitively, a solution is almost period if it looks periodic (as this solution does), but it never repeats itself exactly.   ∎

You might wonder why solutions involving complex numbers tend to oscillate. The reason is that complex numbers are closely connected with trigonometric functions, specifically the sine and cosine functions. This will be discussed further in section 5.11.

PROBLEMS

1. Compute the following:
    a. $(2+3i)(7-2i)$
    b. $(-1+3i)(5+4i)$
    c. $(2-7i)^2$
    d. $|2-7i|$
    e. $|2+i|$

2. Consider the dynamical system

$$a(n+2) = 4a(n+1) - 5a(n).$$

   a. Find the general solution.
   b. Is zero an attracting or repelling fixed point?
   c. Given that $a(0) = 2$ and $a(1) = 6$, plot the particular solution from $(0, a(0))$ through $(5, a(5))$.

3. Consider the dynamical system

$$a(n+2) = -a(n+1) - 0.5a(n).$$

   a. Find the general solution.
   b. Is zero an attracting or repelling fixed point?
   c. Given that $a(0) = 0$ and $a(1) = -1$, find the particular solution.
   d. Compute $a(2)$ through $a(12)$ and observe how $|a(k)|$ is decreasing.

4. Consider the dynamical system

$$a(n+2) = 0.8a(n+1) - 0.16a(n)$$

   a. Find the general solution.
   b. Determine if zero is attracting or repelling.
   c. Find the particular solution when $a(0) = 5$ and $a(1) = 6$.

d. When $a(0) = 5$ and $a(1) = 6$, determine the first value of k for which $|a(k)| < 1$.

5. Consider the dynamical system

$$a(n + 2) = 6a(n + 1) - 8a(n) + 6.$$

a. Find the general solution. This requires using your knowledge of second-order linear dynamical systems and your knowledge of first-order affine dynamical systems.
b. Find the fixed point and determine if it is attracting or repelling.

6. Consider the dynamical system

$$a(n + 2) = 0.25a(n) + 3.$$

a. Find the general solution.
b. Find the fixed point and determine if it is attracting or repelling.

7. Consider the dynamical system

$$a(n + 2) = 4a(n + 1) - 5a(n) + 6.$$

a. Find the general solution.
b. Find the fixed point and determine if it is attracting or repelling.

8. Consider the dynamical system

$$a(n + 2) = -a(n + 1) - 0.5a(n) - 5.$$

a. Find the general solution.
b. Find the fixed point and determine if it is attracting or repelling.

5.6 Modeling a vibrating spring

In section 4.5 we modeled the motion of a falling object. In particular, we assumed that the acceleration of a falling object is constant. Since gravity acting on a falling object is also constant, we might assume that **the acceleration of an object is proportional to the force acting on that object.**

An experiment can be run using a spring and several objects of the same weight. Add n weights of constant mass to the bottom of the spring and measure the displacement $d(n)$ of the spring. In this experiment, you should find that the addition of each weight causes the spring to be displaced by the same amount, that is,

$$d(n+1) = d(n) + b \tag{5.12}$$

for some value b which depends on the spring and the mass of the weights being used. Since $d(0) = 0$, the solution to dynamical system (5.12) is

$$d(k) = bk. \tag{5.13}$$

We have observed Hooke's law, that **the force exerted by a spring is proportional to its displacement**.

We now come to the main point of this section. We add a weight to a spring and find its rest position with the weight added, which we denote as position zero. We then pull the weight down slightly and let go. The spring will oscillate. Our goal is to model this behavior mathematically. To do this, let $d(n)$ be the **displacement** from rest (position zero) of the weighted spring at time n, where one conveniently short period of time is chosen as our unit, say 0.01 seconds. Positions below point zero will be denoted as negative positions, while points above zero will be denoted as positive positions.

The trick is to compute the average acceleration of the spring during 2 consecutive time periods. Acceleration is the change in velocity divided by time. We will divide by 1 for one unit of time, giving

$$v(n+2) - v(n+1).$$

We picked this particular point in time for later convenience. Since average velocity is given by $v(n+1) = d(n+1) - d(n)$ and $v(n+2) = d(n+2) - d(n+1)$ (using the same unit of time) substitution gives that **average acceleration** is

$$[d(n+2) - d(n+1)] - [d(n+1) - d(n)] = d(n+2) - 2d(n+1) + d(n). \tag{5.14}$$

Our assumption is that acceleration $d(n+2) - 2d(n+1) + d(n)$ is proportional to force, and we saw that the force of the spring is proportional to the displacement of the spring $d(m)$ at some point in time m. This leads to the relationship

$$d(n+2) - 2d(n+1) + d(n) = -pd(m)$$

where p is a positive constant of proportionality. The negative sign in front of p is because the force is in the opposite direction from the displacement, that is, if the end of the spring is below point zero (negative $d(m)$) then the force exerted by the spring is in the positive (upward) direction, and if the end of the spring is above point zero (positive $d(m)$) then the force acting on the object is in the negative (downward) direction.

One problem is that the force varies during the 2 units of time, so we have to decide what point m in time to use for the displacement. Two possibilities are, (1) the displacement at the beginning of the time period, $d(n)$ or (2) the displacement at the middle of the time period, $d(n+1)$. Let's try each in the equation, analyze the solutions, and then decide. A third possibility of $d(n+2)$ will be treated in the problems at the end of the section.

Let us try $d(n+1)$ and an appropriately small value for p, say $p = 0.02$. The equation then becomes $d(n+2) - 2d(n+1) + d(n) = -0.02d(n+1)$ or

$$d(n+2) - 1.98d(n+1) + d(n) = 0.$$

This is a linear, second-order dynamical system. The characteristic equation is

$$x^2 - 1.98x + 1 = 0.$$

The roots of this equation are

$$r = \frac{1.98 \pm \sqrt{-0.0796}}{2} = 0.99 \pm \sqrt{0.0199}\, i.$$

Thus, the general solution is

$$d(k) = c_1 (0.99 + \sqrt{0.0199}\, i)^k + c_2 (0.99 - \sqrt{0.0199}\, i)^k.$$

Since

$$|0.99 \pm \sqrt{0.0199}\, i|^2 = 1$$

by theorem 5.14, we have that $|d(k)|$ does not go to zero or infinity. Instead, the solution oscillates. This can be seen in figure 5.5 in which the particular solution $(k, d(k))$ is plotted for $k = 1, \ldots, 100$, with $d(0) = 4$ and $d(1) = 5$. Notice the oscillatory behavior that we would expect from a vibrating spring.

Now we try the same equation except we use $-pd(n)$. This gives the dynamical system $d(n+2) - 2d(n+1) + d(n) = -0.02d(n)$ or

$$d(n+2) - 2d(n+1) + 1.02d(n) = 0.$$

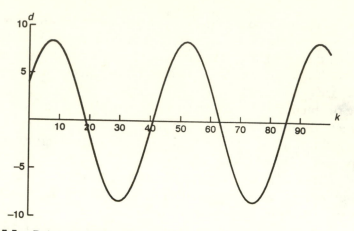

FIGURE 5.5.   Points $(k, d(k))$ using $-pd(n+1)$. Note how the spring oscillates with constant amplitude.

The characteristic equation is

$$x^2 - 2x + 1.02 = 0$$

which has the roots

$$r = \frac{2 \pm \sqrt{0.08}i}{2} = 1 \pm \sqrt{0.02}i.$$

Observe that by theorem 5.14, $|r| = \sqrt{1.02} > 1$, so the solution

$$d(k) = c_1(1 + \sqrt{0.02}i)^k + c_2(1 - \sqrt{0.02}i)^k$$

oscillates with increasing amplitude. This can be seen in figure 5.6 in which the points $(k, d(k))$ are plotted for $k = 0, \dots, 200$. Here we used $d(0) = 0.2$ and $d(1) = 0.1$.

It is clear from our analysis and from figure 5.6 that the fixed point zero is repelling, so $-pd(n)$ gives unreasonable results. It appears that the best choice is the middle of the time period being considered, that is, $-pd(n+1)$. Thus, a dynamical system that describes our vibrating spring is

$$d(n+2) - 2d(n+1) + d(n) = -pd(n+1). \qquad (5.15)$$

Our analysis describes an oscillating spring well, if friction is ignored. We all know that the spring motion will eventually die out. To model this, we need to **add** a term to dynamical system (5.15) to model the force of friction. One possible assumption is that the force of friction is proportional to velocity, but in the opposite direction. In other words, the faster you move, the more you feel the wind

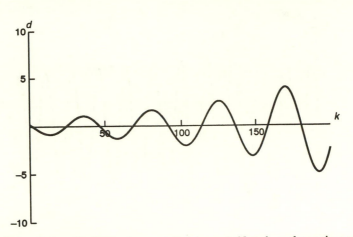

FIGURE 5.6.   Points $(k, d(k))$ when $-pd(n)$ is used. Note how the spring oscillates
with increasing amplitude.

pushing you back. Adding this to dynamical system (5.15) gives

$$d(n+2) - 2d(n+1) + d(n) = -pd(n+1) - qv(m),$$

for some m. As before, we can use $v(n)$, $v(n+1)$, or $v(n+2)$.

Using $v(n) = d(n) - d(n-1)$ would lead to a third-order equation,
so we will not consider this. Instead, let us try $v(n+1)$. The remaining
case will be encountered in the problems. For simplicity, let $p = 0.02$
as before, and assume that the force of friction is much less than the
force of the spring, that is, $q = 0.002$. This gives

$$d(n+2) - 2d(n+1) + d(n) = -0.02d(n+1) - 0.002v(n+1). \quad (5.16)$$

Substitution of $v(n+1) = d(n+1) - d(n)$ and simplification gives

$$d(n+2) - 1.978d(n+1) + 0.998d(n) = 0.$$

The roots of the characteristic equation

$$x^2 - 1.978x + 0.998 = 0$$

are

$$x = 0.989 \pm \sqrt{0.019879}\, i.$$

The solution is then

$$d(k) = c_1 (0.989 + \sqrt{0.019879}\, i)^k + c_2(0.989 - \sqrt{0.019879}\, i)^k.$$

Since by theorem 5.14,

$$|x| = \sqrt{0.998} < 1,$$

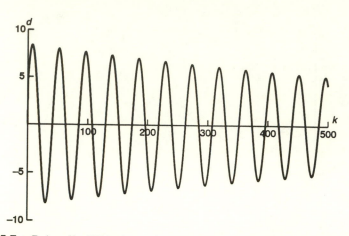

FIGURE 5.7.   Points $(k, d(k))$ when $-0.002v(n+1)$ is used for friction, and $d(0) = 4$ and $d(1) = 5$. Note how the spring oscillates with decreasing amplitude.

the solution $d(k)$ oscillates to zero. Thus, the friction dampens out the oscillations. This can be seen in figure 5.7.

Let's consider the general case of dynamical system (5.16), that is,

$$d(n+2) - 2d(n+1) + d(n) = -pd(n+1) - qv(n+1),$$

or

$$d(n+2) + (p+q-2)d(n+1) + (1-q)d(n) = 0$$

after substituting for $v(n+1)$ and simplifying. Applying the quadratic formula to the characteristic equation

$$x^2 + (p+q-2)x + (1-q) = 0$$

gives that the roots are

$$x = \frac{2-p-q \pm \sqrt{(p+q-2)^2 - 4 + 4q}}{2}.$$

We have that

$$(p+q-2)^2 - 4 + 4q = (p+q)^2 - 4(p+q) + 4 - 4 + 4q = (p+q)^2 - 4p.$$

Assume p is small, say $p < 1$, which it usually is. Also, assume that the friction constant q is smaller than the spring constant p. Then

$$(p+q)^2 - 4p < (p+p)^2 - 4p = 4p^2 - 4p = 4p(p-1) < 0.$$

Thus, the term under the root is negative and the roots are

$$x = \frac{2 - p - q \pm \sqrt{4p - (p+q)^2}\, i}{2}.$$

We have by theorem 5.14 that

$$|x| = \sqrt{1 - q} < 1.$$

Theorem 5.14 saved quite a bit of algebra, here.

Note that when we did this analysis using $p = 0.02$ and $q = 0.002$, that $|x| = \sqrt{0.998} = \sqrt{1 - q}$. Since the roots are complex, the position $d(k)$ oscillates. Since $|x| < 1$, the solutions oscillate toward zero as was seen in figure 5.7.

An important lesson comes out of our analysis in this section. Consider the dynamical system

$$d(n+2) - 2d(n+1) + d(n) = -pd(m). \tag{5.17}$$

We discovered that if the force of the spring acts at time $m = n$, then the solution oscillates with increasing amplitude, that is, is unstable. If the force acts at time $m = n+1$, the solution oscillates with constant amplitude, being neither stable nor unstable. In problem 1 you will discover that if the force acts at time $m = n + 2$, then the solution oscillates to zero and is stable.

Suppose that dynamical system (5.17) modeled an economy instead of spring motion. The equation states that the present economy depends on the economy in the last two time periods. Let the present be time period 2 and the relationship is then

$$d(2) - 2d(1) + d(0) = -pd(m).$$

Instead of the spring applying the force $-pd(m)$, it is now the government that is applying the force. Unlike the spring, the government can apply the force at any point in time. Suppose that economic advisers suggest that if an input proportional to the present economy is applied, that the economy will stabilize. This means we should have tax credits or government expenditures or something that subtracts $pd(2)$ from the economy, giving the dynamical system

$$d(2) - 2d(1) + d(0) = -pd(2)$$

and similarly for each of the following time periods. We observe that this results in stability.

But suppose that bureaucratic delays cause the input to occur in the next time period, that is, the force is not applied in time period 2, but in time period 3. This gives the equation

$$d(3) - 2d(2) + d(1) = -pd(2)$$

which from our above analysis causes constant oscillations, not stability.

Worse yet, suppose the delays cause the input to occur two time periods later, giving the equation

$$d(4) - 2d(3) + d(2) = -pd(2).$$

This dynamical system is unstable, having increasing oscillations.

The moral is that it is not enough to know what to do, but you must know when to do it. The right medicine at the wrong time can actually cause things to get worse instead of better. While not proved, it is generally considered that a delay in applying a force causes instability.

PROBLEMS

1. Suppose in modeling our spring motion, we assume that acceleration is proportional to the force of the spring at time $n + 2$ and that $p = 0.04$, that is, the model is

   $$d(n+2) - 2d(n+1) + d(n) = -0.04d(n+2).$$

   Solve this dynamical system and describe the behavior it predicts by finding $|x|$, where x is a root of the characteristic equation. Since we are neglecting friction, why does this behavior seem incorrect?

2. Find the absolute values of the roots of the characteristic equation, $|x|$ for the dynamical system

   $$d(n+2) - 2d(n+1) + d(n) = -pd(n+2).$$

   Use this to predict the behavior of the solution.

3. Assume that the force of the spring acts at time $n + 1$, but the friction term acts at time $n + 2$, that is, we have the dynamical system

   $$d(n+2) - 2d(n+1) + d(n) = -pd(n+1) - qv(n+2).$$

   a. Let $p = 0.02$ and $q = 0.01$. Find the absolute value of the roots of the characteristic equation, $|x|$, and use this to predict the behavior of the solution.

    b. Assume that $1 > p > q$. (This will assure that the roots of the characteristic equation are complex.) Find the absolute value of the roots of the characteristic equation, $|x|$, and use this to predict the behavior of the solution.

4. Instead of using the velocity at time $n + 1$ or $n + 2$ for the friction term, use the average of these velocities. This gives the dynamical system

$$d(n+2) - 2d(n+1) + d(n) = -pd(n+1) - (q/2)[v(n+2) + v(n+1)].$$

    a. Find the absolute value of the roots of the characteristic equation, $|x|$, and use this to predict the behavior of the solution, when $p = 0.04$ and $q = 0.02$.

    b. Assume that $1 > p > q$. Find the absolute value of the roots of the characteristic equation, $|x|$, and use this to predict the behavior of the solution.

5. Let's discover what can happen if our object is in a viscous material, such as molasses, where the force of friction is relatively large compared to the force of the spring. For example, let $p = 0.03$ and let $q = 0.37$, where the force of the spring is given by $-pd(n+1)$ and the force of friction if given by $-qv(n+1)$.

    a. Give a dynamical system to describe this process.

    b. Find the general solution to this dynamical system.

    c. Find the particular solution to this dynamical system given that $d(0) = 4$ and $v(1) = -1$. Compute $d(2)$ through $d(10)$. Describe the behavior in this case.

    d. Find the particular solution to this dynamical system given that $d(0) = 4$ and $v(1) = -2$. Compute $d(2)$ through $d(15)$. Describe the behavior in this case. (Note that depending on the initial velocity, the object either goes directly toward zero or can pass zero once and then go toward zero from the other side. It does not oscillate.)

5.7   Second-order nonhomogeneous equations

In this section, we study equations of the form

$$a(n+2) = b_1 a(n+1) + b_2 a(n) + f(n),$$

where b_1 and b_2 are given constants and f is a function of n. The results are similar to those of first-order nonhomogeneous equations, discussed in Chapter 4.

EXAMPLE 5.17   Suppose we wish to find the general solution to the (affine) nonhomogeneous equation

$$a(n+2) = 5a(n+1) - 6a(n) + 4. \tag{5.18}$$

(From now on, we will consider an affine dynamical system to be a type of nonhomogeneous equation.)   First, consider the corresponding linear dynamical system

$$a(n+2) = 5a(n+1) - 6a(n).$$

The two roots of the characteristic equation are $x = 2$ and $x = 3$. Therefore, the general solution of the linear dynamical system is

$$a(k) = c_1 2^k + c_2 3^k,$$

where c_1 and c_2 depend on the initial values.

Hence, we might expect the general solution to the original nonhomogeneous dynamical system (5.18) to be of the form

$$a(k) = c_1 2^k + c_2 3^k + c_3,$$

where c_1 and c_2 depend on the initial values, and c_3 depends on the term added on to the linear equation, that is, on $f(n) = 4$. To determine the number c_3, we can substitute

$$a(n) = c_1 2^n + c_2 3^n + c_3$$
$$a(n+1) = c_1 2^{n+1} + c_2 3^{n+1} + c_3$$
$$a(n+2) = c_1 2^{n+2} + c_2 3^{n+2} + c_3$$

into the dynamical system. After canceling the 2^n and 3^n terms via the characteristic equation and solving for the unknown c_3, we get $c_3 = 2$. Thus, the general solution to the above nonhomogeneous equation is

$$a(k) = c_1 2^k + c_2 3^k + 2.$$

The computation can be simplified by omitting the terms $c_1 2^n$ and $c_2 3^n$. This can be done because these terms depend only on the initial conditions. Therefore we only need to substitute

$$a(n) = c_3, \quad a(n+1) = c_3, \quad a(n+2) = c_3$$

into the dynamical system, giving

$$c_3 = 5c_3 - 6c_3 + 4,$$

again getting $c_3 = 2$.

You can check that the above value is a solution by substituting it back into the nonhomogeneous dynamical system. You will see that everything cancels, giving a balanced equation.

You can see that the above is the general solution by solving

$$a(0) = c_1 + c_2 + 2 \qquad \text{and} \qquad a(1) = 2c_1 + 3c_2 + 2$$

for c_1 and c_2.   ■

EXAMPLE 5.18   Now consider the equation

$$a(n+2) = 5a(n+1) - 6a(n) + 2n + 1. \tag{5.19}$$

From the previous example, we know that the roots of the characteristic equation corresponding to the linear dynamical system

$$a(n+2) = 5a(n+1) - 6a(n)$$

are $x = 2$ and $x = 3$. Therefore, we expect the general solution to the nonhomogeneous dynamical system (5.19) to be of the form

$$a(k) = c_1 2^k + c_2 3^k + c_3 + c_4 k,$$

where c_1 and c_2 depend on the initial values, while c_3 and c_4 depend on the function $f(n) = 2n + 1$. As in the previous example we omit the 2^n and 3^n terms, since they will cancel each other out, anyway, via the characteristic equation. To find c_3 and c_4, we substitute

$$a(n) = c_3 + c_4 n$$
$$a(n+1) = c_3 + c_4(n+1) = c_3 + c_4 n + c_4$$
$$a(n+2) = c_3 + c_4(n+2) = c_3 + c_4 n + 2c_4$$

into the nonhomogeneous dynamical system, giving

$$c_3 + 2c_4 + c_4 n = 5c_3 + 5c_4 + 5c_4 n - 6c_3 - 6c_4 n + 2n + 1,$$

or

$$2c_3 - 3c_4 + 2c_4 n = 2n + 1.$$

Now collect all the terms on the left, group the terms involving n together and factor out the n, and group the constant terms (not involving n) together, giving

$$(2c_4 - 2)n + (2c_3 - 3c_4 - 1) = 0.$$

We want to choose c_3 and c_4 so that this equation is satisfied **for every value of** n. Notice that if $2c_4 - 2$ (the coefficient of n) equals zero, and if the constant term, $2c_3 - 3c_4 - 1$, also equals zero, then we have $0 = 0$ and the equation is satisfied. Thus we want c_3 and c_4 to satisfy the equations

$$2c_4 - 2 = 0 \quad \text{and} \quad 2c_3 - 3c_4 - 1 = 0.$$

Solving the first equation gives $c_4 = 1$. Substituting this into the second equation gives $c_3 = 2$. Therefore the general solution to the above nonhomogeneous dynamical system is

$$a(k) = c_1 2^k + c_2 3^k + 2 + k,$$

where c_1 and c_2 depend on the initial values.   ∎

Let's summarize the rule for finding the general solution to a non-homogeneous second-order linear dynamical system in which the nonhomogeneous term is a polynomial. Recall that, for first-order nonhomogeneous equations, there was a special case when $r = 1$. For second-order equations there are two roots of the characteristic equation, so there will be two special cases, exactly one root equals 1 and 1 is a double root.

THEOREM 5.19   Nonhomogeneous dynamical systems in which $f(n)$ is a polynomial   *Consider the dynamical system*

$$a(n+2) = b_1 a(n+1) + b_2 a(n) + f(n),$$

where the roots of the characteristic equation $x^2 = b_1 x + b_2$ are r and s, and $f(n)$ is an mth-degree polynomial, that is,

$$f(n) = b_{m+3} n^m + b_{m+2} n^{m-1} + \cdots + b_4 n + b_3.$$

The general solution is:

- $a(k) = c_1 r^k + c_2 s^k + c_3 + c_4 k + \cdots + c_{m+2} k^{m-1} + c_{m+3} k^m$ *if $r \neq 1$, $s \neq 1$, and $r \neq s$;*

- $a(k) = (c_1 + c_2 k) r^k + c_3 + c_4 k + \cdots + c_{m+2} k^{m-1} + c_{m+3} k^m$ *if $r = s \neq 1$;*

- $a(k) = c_1 r^k + c_2 + c_3 k + c_4 k^2 + \cdots + c_{m+2} k^m + c_{m+3} k^{m+1}$ *if $r \neq 1$ and $s = 1$; and*

- $a(k) = c_1 + c_2 k + c_3 k^2 + c_4 k^3 + \cdots + c_{m+2} k^{m+1} + c_{m+3} k^{m+2}$ *if $r = s = 1$.*

In all cases, c_1 and c_2 depend on the initial values, and the constants c_{m+3}, $c_{m+2}, \ldots, c_4$, and c_3 can be found by substituting into the dynamical system and then solving $m + 1$ equations for the $m + 1$ unknowns by setting the *coefficient of the n^m term, ..., the coefficient of the n term, and the constant term equal to zero.*

EXAMPLE 5.20   Consider the second-order nonhomogeneous dynamical system

$$a(n+2) = 4a(n+1) - 4a(n) + 7 - 2n. \qquad (5.20)$$

Since the characteristic equation for the corresponding linear dynamical system

$$a(n+2) = 4a(n+1) - 4a(n)$$

has $x = 2$ as a double root, it follows from theorem 5.19 part 2 that the general solution is of the form

$$a(k) = (c_1 + c_2 k)2^k + c_3 + c_4 k,$$

where c_1 and c_2 depend on the initial values, and c_3 and c_4 depend on the function $f(n) = 7 - 2n$. Omitting the term $(c_1 + c_2 k)2^k$ so as to simplify our computations, we substitute

$$a(n) = c_3 + c_4 n$$
$$a(n+1) = c_3 + c_4(n+1)$$
$$a(n+2) = c_3 + c_4(n+2)$$

into the dynamical system, giving

$$(c_3 + c_4 n + 2c_4) = 4(c_3 + c_4 + c_4 n) - 4(c_3 + c_4 n) + 7 - 2n,$$

or, after collecting terms on the left and simplifying,

$$(c_4 + 2)n + (c_3 - 2c_4 - 7) = 0.$$

Solving the two equations

$$c_4 + 2 = 0 \qquad \text{and} \qquad c_3 - 2c_4 - 7 = 0$$

gives $c_4 = -2$ and $c_3 = 3$.

Thus the general solution to the second-order nonhomogeneous dynamical system (5.20) is

$$a(k) = (c_1 + c_2 k)2^k + 3 - 2k,$$

where c_1 and c_2 depend on the initial values.

Suppose $a(0) = 2$ and $a(1) = 9$. To find the particular solution, we solve

$$2 = a(0) = (c_1 + 0)2^0 + 3 - 2(0) = c_1 + 3$$

and

$$9 = a(1) = (c_1 + c_2)2^1 + 3 - 2 = 2c_1 + 2c_2 + 3 - 2.$$

The first equation gives $c_1 = -1$. Substitution into the second equation gives $c_2 = 5$, so the particular solution is

$$a(k) = (-1 + 5k)2^k + 3 - 2k.$$ ■

EXAMPLE 5.21   Consider the dynamical system

$$a(n + 2) = 2a(n + 1) - a(n) + 6. \tag{5.21}$$

We see that $r = 1$ is a double root of the characteristic equation and $f(n) = 6$ is a zeroth-degree polynomial so that $m = 0$. Therefore, from theorem 5.19 part 4 we know that the general solution is of the form

$$a(k) = c_1 + c_2 k + c_3 k^2,$$

where c_1 and c_2 depend on the initial values, while c_3 depends on the nonhomogeneous term, 6. Substituting

$$a(n) = c_3 n^2, \quad a(n+1) = c_3(n+1)^2, \quad a(n+2) = c_3(n+2)^2$$

into the dynamical system gives

$$c_3 n^2 + 4c_3 n + 4c_3 = 2c_3 n^2 + 4c_3 n + 2c_3 - c_3 n^2 + 6.$$

Canceling the n and the n^2 terms gives $2c_3 = 6$, or

$$c_3 = 3.$$

The general solution to the dynamical system is then

$$a(k) = c_1 + c_2 k + 3k^2.$$

If $a(0) = 2$ and $a(1) = 1$, then solving

$$2 = c_1 \quad \text{and} \quad 1 = c_1 + c_2 + 3$$

gives the particular solution as

$$a(k) = 2 - 4k + 3k^2.$$

Observe that this solution is quadratic in that the points $(k, a(k))$ lie on a parabola.   ■

Let's now study dynamical systems in which the nonhomogeneous term is an exponential of the form $b_3 p^n$. The general solution is of the same form as in sections 4.1 and 4.2 except that there is an additional special case.

THEOREM 5.22 Nonhomogeneous dynamical systems in which $f(n)$ is an exponential   *Consider the second-order nonhomogeneous dynamical system*

$$a(n+2) = b_1 a(n+1) + b_2 a(n) + b_3 p^n,$$

where b_3 and p are given constants. Suppose the roots of the characteristic equation for the corresponding linear dynamical system are $x = r$ and $x = s$. The general solution is:

- $a(k) = c_1 r^k + c_2 s^k + c_3 p^k$ *if $r \neq p$, $s \neq p$, and $r \neq s$;*
- $a(k) = (c_1 + c_2 k) r^k + c_3 p^k$ *if $r = s \neq p$;*
- $a(k) = c_2 s^k + (c_1 + c_3 k) r^k$ *if $r = p \neq s$; and*
- $a(k) = (c_1 + c_2 k + c_3 k^2) r^k$ *if $r = s = p$, that is, if p is a double root of the characteristic equation.*

In each part, c_1 and c_2 depend on the initial values while c_3 can be obtained by substitution into the dynamical system.

EXAMPLE 5.23   Consider the nonhomogeneous equation

$$a(n+2) = 5a(n+1) - 6a(n) + 6 \times 4^n. \tag{5.22}$$

We know that the roots of the characteristic equation are 2 and 3. Thus we expect the general solution to be of the form

$$a(k) = c_1 2^k + c_2 3^k + c_3 4^k,$$

where c_1 and c_2 depend on the initial values, while c_3 depends on the nonhomogeneous term. Substituting $a(n) = c_3 4^n$ into the dynamical system gives

$$c_3 4^{n+2} = 5 c_3 4^{n+1} - 6 c_3 4^n + 6 \times 4^n.$$

Dividing by 4^n gives

$$16 c_3 = 20 c_3 - 6 c_3 + 6, \qquad \text{or} \qquad c_3 = 3.$$

Thus, the general solution to the dynamical system (5.22) is

$$a(k) = c_1 2^k + c_2 3^k + 3 \times 4^k,$$

where c_1 and c_2 depend on the initial values.   ■

EXAMPLE 5.24   Consider the dynamical system

$$a(n+2) = 4a(n+1) - 4a(n) + 8 \times 2^n. \qquad (5.23)$$

Here, $r = 2$ is a double root of the characteristic equation and is also involved in the nonhomogeneous term. From theorem 5.22 part 4, the general solution is of the form

$$a(k) = (c_1 + c_2 k + c_3 k^2)2^k,$$

where c_1 and c_2 depend on the initial values, while c_3 depends on the term $8 \times 2^n = 2^{n+3}$. Substituting

$$a(n) = c_3 n^2 2^n$$
$$a(n+1) = c_3(n^2 + 2n + 1)2^{n+1}$$
$$a(n+2) = c_3(n^2 + 4n + 4)2^{n+2}$$

into the dynamical system (5.23) and then dividing by 2^n gives

$$4c_3(n^2 + 4n + 4) = 8c_3(n^2 + 2n + 1) - 4c_3 n^2 + 8.$$

Since the n^2 terms and the n terms cancel, we obtain

$$16c_3 = 8c_3 + 8, \qquad \text{or} \qquad c_3 = 1.$$

Thus, the general solution to the dynamical system (5.23) is

$$a(k) = (c_1 + c_2 k + k^2)2^k. \qquad \blacksquare$$

We must make a comment concerning the proofs of theorems 5.19 and 5.22. The proof in each case is a matter of actually substituting the claimed form of the general solution into the dynamical system and solving for the constants $c_3, \ldots$. The algebra involved in the proof of each part of theorem 5.22 is relatively easy as can be seen below in the proof of part 4 below. The algebra involved in the proof of each part of theorem 5.19 is more difficult and will be omitted.

PROOF OF THEOREM 5.22 PART 4:   We are assuming that r is a double root of the characteristic equation, that is, the characteristic equation is

$$(x - r)^2 = 0, \qquad \text{or} \qquad x^2 - 2rx + r^2 = 0.$$

But this is the characteristic equation for the dynamical system

$$a(n+2) = 2ra(n+1) - r^2 a(n).$$

Thus, we are trying to find the general solution to the dynamical system

$$a(n+2) = 2ra(n+1) - r^2a(n) + b_3r^n.$$

Substituting $a(n) = (c_1 + c_2n + c_3n^2)r^n$ into the dynamical system gives

$$[c_1 + c_2(n+2) + c_3(n+2)^2]r^{n+2} =$$
$$2r[c_1 + c_2(n+1) + c_3(n+1)^2]r^{n+1} - r^2(c_1 + c_2n + c_3n^2)r^n + b_3r^n.$$

The terms involving c_1 and c_2 cancel out. Dividing by r^{n+2} gives

$$c_3(n+2)^2 = 2c_3(n+1)^2 - c_3n^2 + b_3r^{-2}.$$

The c_3n^2 and c_3n terms cancel leaving

$$4c_3 = 2c_3 + b_3r^{-2}, \qquad \text{or} \qquad c_3 = \frac{b_0}{2r^2}.$$

We see not only that the general solution to the equation

$$a(n+2) = 2ra(n+1) - r^2a(n) + b_3r^n$$

is of the form

$$a(k) = (c_1 + c_2k + c_3k^2)r^k$$

but that

$$c_3 = \frac{b_3}{2r^2}.$$

We could now solve for c_1 and c_2 in terms of $a(0)$ and $a(1)$ to complete the proof. ∎

PROBLEMS

1. Find the general solution for the following second-order dynamical systems.
    a. $a(n+2) = a(n+1) + 6a(n)$
    b. $a(n+2) = a(n+1) + 6a(n) - 14 \times 5^n$
    c. $a(n+2) = a(n+1) + 6a(n) - 12n + 8$
    d. $a(n+2) = a(n+1) + 6a(n) + 4 \times 2^n$
    e. $a(n+2) = a(n+1) + 6a(n) - 10 \times 3^{n+1}$

2. Find the general solution for the following second-order dynamical systems.
    a. $a(n+2) = -2a(n+1) - a(n)$
    b. $a(n+2) = -2a(n+1) - a(n) + 8$

     c. $a(n+2) = -2a(n+1) - a(n) + 9 \times 2^n$
     d. $a(n+2) = -2a(n+1) - a(n) + 4(-1)^n$

3. Find the general solution for the following second-order dynamical systems.
     a. $a(n+2) = -a(n+1) + 2a(n)$
     b. $a(n+2) = -a(n+1) + 2a(n) - 2^{n+2}$
     c. $a(n+2) = -a(n+1) + 2a(n) + 6(-2)^n$
     d. $a(n+2) = -a(n+1) + 2a(n) + 9$
     e. $a(n+2) = -a(n+1) + 2a(n) - 12n - 10$

4. Find the general solution for the following second-order dynamical systems.
     a. $a(n+2) = 2a(n+1) - a(n)$
     b. $a(n+2) = 2a(n+1) - a(n) - 6$
     c. $a(n+2) = 2a(n+1) - a(n) + 2^n$
     d. $a(n+2) = 2a(n+1) - a(n) + 6n + 4$

5. Find the particular solution to the dynamical system
     a. in problem 1 (b) given that $a(0) = 2$ and $a(1) = 4$
     b. in problem 2 (d) given that $a(0) = -3$ and $a(1) = -3$
     c. in problem 3 (d) given that $a(0) = 1$ and $a(1) = -11$
     d. in problem 4 (b) given that $a(0) = 7$ and $a(1) = 6$

6. See if you can combine theorems 5.19 and 5.22 to find the general solution to the dynamical systems
     a. $a(n+2) = a(n+1) + 6a(n) + (30n + 33)3^n$
     b. $a(n+2) = 2a(n+1) - a(n) - 6 + 2^n$

7. Suppose that the solution to a dynamical system is

$$a(k) = (c_1 + c_2 k + c_3 k^2) r^k.$$

Find c_1 and c_2 in terms of c_3, $a(0)$, and $a(1)$.

8. The roots of the characteristic equation for the dynamical system

$$a(n+2) = (r+s)a(n+1) - rsa(n) + b_3 r^n$$

are r and s. Show that the general solution is of the form

$$a(k) = (c_1 + c_3 k) r^k + c_2 s^k$$

by substituting this solution into the dynamical system and solving for c_3 in terms of b_3.

5.8 Gambler's ruin revisited

First, as in section 5.3, assume we are playing a game in which the probability of winning a dollar on any particular try is p, while the probability of losing a dollar on any particular try is $1 - p = q$. We quit the game when we have a total of N dollars or when we are broke. The question we now wish to answer is, "Given that we now have n dollars, on the average how many bets will we make?" Notice that we are not interested in whether we quit as a winner or a loser.

Suppose we now have n dollars. Let $e(n)$ represent the **expected** or average number of bets we will make before the game is over. Notice that $e(0) = 0$ and $e(N) = 0$ since in these two cases, we quit playing. These are our "initial" values.

Suppose $0 < n < N$. Then we have not yet quit and therefore expect to play the game at least one more time. So

$$e(n) = 1 + \text{how many more bets?}$$

We play the game and one of two things happens. Case 1 is that we win a dollar and continue playing if possible, and case 2 is that we lose a dollar and continue playing if possible. The trick is to find the expected number of bets we will make in each of these two cases. We then use a weighted average of these two cases by using the likelihood of each of these cases occurring.

If case 1 occurs, that is, if we win the next bet, we will have $n + 1$ dollars and we expect to make $e(n + 1)$ more bets. If case 2 occurs, we will have only $n - 1$ dollars and we expect to make $e(n - 1)$ more bets. Since p is the fraction of the time that case 1 occurs and q is the fraction of the time that case 2 occurs, the weighted average of the two cases is

$$pe(n + 1) + qe(n - 1).$$

For example, if you make 12 bets one-third of the time and you make 9 bets two-thirds of the time, then on the average you will make

$$\frac{1}{3}(12) + \frac{2}{3}(9) = 10$$

more bets, that is if you play three more times, you will on the average make 12 bets once and 9 bets twice for an average of 10 bets.

Thus, $e(n)$, the expected number of times to play before quitting, is 1 (since we now play the game once), plus the weighted average given above, that is,

$$e(n) = pe(n+1) + qe(n-1) + 1.$$

To solve this equation, we first substitute $n+1$ for n, giving

$$e(n+1) = pe(n+2) + qe(n) + 1.$$

Dividing by p and rearranging gives

$$e(n+2) = \frac{e(n+1)}{p} - \frac{qe(n)}{p} - \frac{1}{p}.$$

The two roots of the characteristic equation for the linear part of the equation are $x = 1$ and $x = q/p$, as in section 5.3.

If $p \neq q$, that is, if $p \neq 0.5$, then we have two distinct roots. But if $p = q = 0.5$, we have $r = 1$ as a double root.

First, let's assume that $p \neq q$, so that $q/p \neq 1$, and we have two distinct roots. Since $f(n) = -1/p$ is a 0th-degree polynomial, and $r = 1$ is one of the roots, by theorem 5.19, the solution to the nonhomogeneous equation is of the form

$$e(k) = c_1 + c_2 \left(\frac{q}{p}\right)^k + c_3 k,$$

where c_1 and c_2 depend on the initial values, while c_3 depends on the nonhomogeneous term, $-1/p$. Substituting into the second-order nonhomogeneous dynamical system gives

$$c_3(n+2) = \frac{c_3(n+1)}{p} - \frac{qc_3 n}{p} - \frac{1}{p}.$$

Multiplying both sides by p, substituting $1-p$ for q, and canceling the n terms gives

$$2c_3 p = c_3 - 1, \quad \text{or} \quad c_3 = \frac{-1}{2p-1} = \frac{1}{q-p}$$

using the fact that $2p - 1 = p - q$.

The general solution to the dynamical system is then

$$e(k) = c_1 + c_2 \left(\frac{q}{p}\right)^k + \frac{k}{q-p}.$$

Since

$$e(0) = 0 = c_1 + c_2 \quad \text{and} \quad e(N) = 0 = c_1 + c_2 \left(\frac{q}{p}\right)^N + \frac{N}{q-p},$$

we can solve these two equations for c_1 and c_2 to get

$$c_1 = \frac{N}{(q-p)[(q/p)^N - 1]} = -c_2,$$

and the particular solution to the nonhomogeneous equation is

$$e(k) = \frac{N[1 - (q/p)^k]}{(q-p)[(q/p)^N - 1]} + \frac{k}{q-p},$$

after collecting terms.

For example, suppose $p = 0.49$, $N = 20$, and $n = 1$, that is, we have 1 dollar left and will play until we are broke or have 20 dollars. In this case, $e(1) = 16.7$, that is, on the average, we will play 16.7 times before quitting. Notice this is an average, since about half the time we lose our dollar on the first try and quit. But there are times when we play for long periods, so that the average is 16.7.

Many gamblers have the best intentions about quitting, but once they reach their goal they are unable to quit. Under these conditions, they will eventually go broke. How long will they play before going broke? To rephrase the above more precisely, suppose a gambler has n dollars, the probability of winning any (1 dollar) bet is $p < 0.5$, and the gambler plays until broke. On the average, how many bets will the gambler make?

The solution is easy. Suppose the gambler makes an unrealistically high goal, say $N = 1\,000\,000$, but presently has $n = 50$. Then the gambler is (almost) certain to go broke and $e(50)$ would be a reasonably accurate answer to our question. To get the exact answer, let N go to infinity, that is, the expected number of bets would be

$$\lim_{N \to \infty} e(n) = \lim_{N \to \infty} \left(\frac{N[1 - (q/p)^n]}{(q-p)[(q/p)^N - 1]} + \frac{n}{q-p} \right).$$

It was shown in section 5.5 that if $s < 1$ then

$$\lim_{N \to \infty} Ns^N = 0.$$

Since $r > 1$, then $1/r < 1$ and

$$\lim_{N \to \infty} \frac{N}{r^N} = \lim_{N \to \infty} N(1/r)^N = 0.$$

This is because N grows linearly, but r^N grows exponentially. It easily follows that

$$\lim_{N \to \infty} \frac{N}{r^N - 1} = 0.$$

Since $p < 0.5$, then $q/p > 1$, and so

$$\lim_{N \to \infty} \frac{N[1 - (q/p)^n]}{(q-p)[(q/p)^N - 1]} = \lim_{N \to \infty} \frac{N}{[(q/p)^N - 1]} \frac{[1 - (q/p)^n]}{(q-p)} = 0$$

and

$$\lim_{N \to \infty} e(n) = \frac{n}{q-p}.$$

Thus if $p = 0.49$, we have $n = 1$ dollar, and we will play until we are broke, we expect to make $1/(0.51 - 0.49) = 50$ bets before going broke, as compared to the 16.7 bets when we stop at 20 dollars.

Now let's consider a fair game, that is, $p = q = 0.5$. Our dynamical system then becomes

$$e(n+2) = 2e(n+1) - e(n) - 2,$$

and $x = 1$ is a double root. By theorem 5.19, we know that the general solution is of the form

$$e(k) = c_1 + c_2 k + c_3 k^2,$$

where c_1 and c_2 depend on the initial values, while c_3 depends on the nonhomogeneous term, -2. Substituting into the dynamical system gives

$$c_3 n^2 + 4c_3 n + 4c_3 = 2c_3 n^2 + 4c_3 n + 2c_3 - c_3 n^2 - 2,$$

or

$$c_3 = -1.$$

Thus, the general solution is

$$e(k) = c_1 + c_2 k - k^2.$$

Since

$$e(0) = 0 = c_1 \qquad \text{and} \qquad e(N) = 0 = c_1 + c_2 N - N^2,$$

it follows that $c_1 = 0$ and $c_2 = N$, so that the particular solution is

$$e(k) = kN - k^2 = k(N - k).$$

In this case, if $n = 1$ and $N = 20$, then $e(n) = 19$ as compared with 16.7 when $p = 0.49$.

Notice that if we are in an honest game and decide to play until we go broke, then

$$\lim_{N \to \infty} e(n) = \lim_{N \to \infty} n(N - n) = \infty,$$

so we will be playing for a long time, on average.

PROBLEMS

1. Find the expected number of bets to be made if the probability of winning 1 dollar is 0.49, you start with 50 dollars, and you quit if you go broke or reach a total of 100 dollars.

2. Repeat problem 1 when
   a. $p = 0.49$, $n = 1$, and $N = 100$
   b. $p = 0.5$, $n = 50$, and $N = 100$
   c. $p = 0.5$, $n = 1$, and $N = 100$

3. Let $d(n)$ be the distance traveled by a falling object after n seconds. Then the average velocity during the nth second is given by $v(n) = d(n) - d(n-1)$. Observations indicate that the change in velocity during any second is constant and equals -32 feet per second squared, that is,

$$v(n+1) - v(n) = -32.$$

   a. Develop a second-order nonhomogeneous dynamical system for distance traveled, $d(n)$.
   b. Find the general solution to this dynamical system.
   c. Given that a falling object starts at $d(0) = 144$ feet and its average velocity for the first second is $v(1) = -16$ feet per second, find the particular solution for this falling object, and compute how long it will take to reach the ground.

5.9 A model of a national economy

In this section, we show how dynamical systems can be used in economics by presenting a simplified model of a nation's economy. To do this, we make four assumptions.

ASSUMPTION 1: The national income is composed of three elements: income from consumer expenditures, private investment, and government expenditures, denoted by c, p, and g, respectively. The total national income, denoted by t, is the sum of these three parts, that is,

$$t = c + p + g. \qquad \qquad \square$$

While consumption, investment, and government expenditures are being done continuously, they are only known at discrete periods of time, for example, quarterly reports issued by the government.

We shall assume that these components depend on a time period, n. Thus, the total national income during time period n is

$$t(n) = c(n) + p(n) + g(n). \qquad (5.22)$$

The above components of national income could be subdivided into more components and other components could be added, to give a more complex model of the economy.

ASSUMPTION 2:   We assume that individuals (companies, and the government) save a certain fixed proportion of their income, and therefore **spend** a fixed proportion of their incomes. This money is spent in the time period after it is earned, that is, consumption in time period $n + 1$ is proportional to the total income in time period n. Mathematically, this is

$$c(n + 1) = at(n). \qquad (5.23)$$

The constant a is known as the **marginal propensity to consume**, or MPC. Since what is not spent is saved, $1 - \text{MPC}$ is called the **marginal propensity to save**, or MPS.   □

Again, we realize this is a gross over-simplification. We could have assumed, for example, that there is a certain fixed consumption, and that a certain fixed proportion of the excess income is saved. This would give $c(n + 1) = at(n) + b$.

ASSUMPTION 3:   Private investment (in buying new machines or building new factories) depends on the **change** in consumption, that is, if consumption increases, more factories will be needed to produce the material being consumed, so more investment must be made. But if consumption decreases, then existing factories are sufficient to produce materials, so additional investment will be small or even negative from closing factories.   □

Let $p(n)$ be the additional investment made in time period n, where $p(n)$ may be negative. Thus, an investor, in time period $n+1$, compares the present consumption, $c(n + 1)$, to the past consumption, $c(n)$, and looks at the change in consumption, $c(n + 1) - c(n)$. If this is a large positive number, people are spending more, and thus the additional investment, $p(n+1)$, should be large. If $c(n+1) - c(n)$ is negative, then investment is reduced and $p(n + 1)$ is negative. In short, additional investment in time period $n + 1$ is proportional to the **change** in consumption from time period n to time period $n+1$, that is

$$p(n+1) = b(c(n+1) - c(n)).\tag{5.24}$$

The constant b in equation (5.24) is called the **constant of adjustment.**

One might argue that by the time the investors know $c(n+1)$ and $c(n)$, it is one time period later before they can act on this information. Thus, we would have $p(n+2) = b[c(n+1) - c(n)]$. But we could also argue that investors tend to watch current trends and have a good idea of what $c(n+1)$ will be ahead of its publication, and therefore we keep our original assumption. (Also, the equation given in this paragraph would lead to a third-order dynamical system which would lead to more difficult analysis.)

ASSUMPTION 4: Let's assume that government expenditures are constant. Whatever this amount, we let it represent one unit of money, that is,

$$g(n) = 1 \tag{5.25}$$

for all n.   □

While this last assumption seems unrealistic, it is more reasonable if we are assuming that all units of money are based on today's value of money, that is, the government's increases in expenditures equal the inflation rate. Thus, the **value** of the government's expenditures remains constant. While there are other equally valid assumptions about governmental expenditures, such as a proportion of the taxes, which is a proportion of total income, we will keep our relatively simple assumption 4.

Thus, our assumptions translate into the four equations (5.22), (5.23), (5.24), and (5.25).

We substitute 1 for $g(n)$, and then $n+2$ for n in equation (5.22), to get

$$t(n+2) = c(n+2) + p(n+2) + 1.$$

From the third assumption, equation (5.24), we make the substitution of $p(n+2) = b[c(n+2) - c(n+1)]$ into our last equation, giving

$$t(n+2) = c(n+2) + b[c(n+2) - c(n+1)] + 1$$
$$= (1+b)c(n+2) - bc(n+1) + 1.$$

From equation (5.23), we make the substitutions $c(n + 2) = at(n + 1)$, and $c(n+1) = at(n)$, giving the second-order nonhomogeneous dynamical system

$$t(n + 2) = (1 + b)at(n + 1) - bat(n) + 1 \qquad (5.26)$$

as our model for the national economy. This dynamical system is a variation of the Samuelson (1939) accelerator-multiplier model.

EXAMPLE 5.25   Consider the case in which $a = 0.9$ and $b = 1$. Dynamical system (5.26) becomes

$$t(n + 2) = 1.8t(n + 1) - 0.9t(n) + 1.$$

The roots of the characteristic equation are, by the quadratic formula, $r = 0.9 + 0.3i$ and $r = 0.9 - 0.3i$. Thus, the general solution is of the form

$$t(k) = c_1 (0.9 + 0.3i)^k + c_2 (0.9 - 0.3i)^k + c_3.$$

Substituting $t(n) = c_3$ into the nonhomogeneous equation and solving gives $c_3 = 10$. Thus, the general solution is

$$t(k) = c_1 (0.9 + 0.3i)^k + c_2 (0.9 - 0.3i)^k + 10.$$

Since

$$|0.9 \pm 0.3i| = \sqrt{(0.9)^2 + (0.3)^2} = \sqrt{0.9} = 0.9487 < 1,$$

the equilibrium value is stable. The complex roots cause the solution to oscillate to equilibrium. See figure 5.8, in which $t(0) = 4$ and $t(1) = 5$ for one such solution.   ∎

In this section, we will use the terminology "stable and unstable equilibrium" instead of "attracting and repelling fixed point". The reason is that we are describing an economy, so it makes sense to refer to an economy that is in equilibrium. We would then call the economy in example 5.25, a stable economy (not an attracting economy). As discussed before, the terminology you use depends on what you are doing.

EXAMPLE 5.26   Consider the case in which $a = 0.9$ and $b = 1.5$. Dynamical system (5.26) becomes

$$t(n + 2) = 2.25t(n + 1) - 1.35t(n) + 1$$

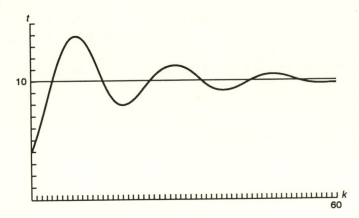

FIGURE 5.8.   One solution $(k, t(k))$ for the dynamical system $t(n+2) = 1.8t(n+1) - 0.9t(n) + 1$.

in this case. The roots of the corresponding characteristic equation are, from the quadratic formula, approximately $r = 1.125 \pm 0.29i$, and so the general solution is

$$t(k) = c_1 (1.125 + 0.29i)^k + c_2 (1.125 - 0.29i)^k + c_3.$$

By substituting $t(n) = c_3$ into the nonhomogeneous equation and solving, we again find that $c_3 = 10$. Thus, the general solution is

$$t(k) = c_1 (1.125 + 0.29i)^k + c_2 (1.125 - 0.29i)^k + 10.$$

Since

$$|1.125 \pm 0.29i| = \sqrt{1.35} = 1.16 > 1$$

the equilibrium value is unstable. The complex roots cause the solution to oscillate away from equilibrium. ∎

Let's now consider the general case of equation (5.26)

$$t(n+2) = a(1+b)t(n+1) - abt(n) + 1.$$

Let's first find the equilibrium value, which is also the constant term added to the general solution to the dynamical system. Substituting $t(n) = t$ into the equation gives

$$t = a(1+b)t - abt + 1, \qquad \text{or} \qquad (1 - a - ab + ab)t = 1.$$

Canceling the ab terms and then dividing by $(1 - a)$ gives

$$t = \frac{1}{1 - a} = \frac{1}{\text{MPS}}$$

as the equilibrium value. Notice that the economy's equilibrium value depends only on the value of a, the marginal propensity to consume.

We now know that the general solution to the dynamical system is

$$t(k) = c_1 (r_1)^k + c_2 (r_2)^k + \frac{1}{1-a},$$

where r_1 and r_2 are the roots of the characteristic equation

$$r^2 - a(1+b)r + ab = 0.$$

Using the quadratic formula, the roots are

$$r_1 = 0.5 \left(a(1+b) + \sqrt{a^2(1+b)^2 - 4ab} \right)$$

and

$$r_2 = 0.5 \left(a(1+b) - \sqrt{a^2(1+b)^2 - 4ab} \right).$$

We need to know if the roots are real or are complex. The roots are complex when the discriminant is negative, that is, when

$$a^2(1+b)^2 - 4ab < 0.$$

Since $a > 0$, we can divide both sides by a, giving

$$a(1+b)^2 < 4b, \qquad \text{or} \qquad a < \frac{4b}{(1+b)^2}.$$

Assume b is near one, that is, the investment is on the order of the increase in consumption (slightly more if we are optimistic, slightly less if we are pessimistic). This means that $4b/(1+b)^2 \approx 1$. Thus, we must have $a < 1$, which makes sense since we cannot continually spend more than we earn. Thus, it is reasonable to assume that the roots are complex for our model.

When the roots are complex, that is, $a^2(1+b)^2 - 4ab < 0$, we have that the roots are

$$r = \frac{a(1+b)}{2} \pm \frac{i}{2} \sqrt{4ab - a^2(1+b)^2}.$$

In this case, by theorem 5.14, we have that

$$|r| = \sqrt{ab}.$$

The equilibrium value is therefore stable when $\sqrt{ab} < 1$. Squaring both sides and dividing by b gives the condition that the equilibrium value is stable if

$$a < \frac{1}{b}.$$

Thus the economy is unstable when

$$a > \frac{1}{b},$$

that is, when the marginal propensity to spend is greater than the reciprocal of the constant of adjustment.

In figure 5.9 is a sketch of the regions of stability and instability for a and b. In particular, the region below the curve $a = 4b/(1 + b)^2$ corresponds to points (a, b) that give complex roots so the economy oscillates. The part of that region that is also below the curve $a = 1/b$ corresponds to points (a, b) that give a stable economy, so solutions oscillate to equilibrium. One such point in this region is $(a, b) = (0.9, 1)$ from example 5.25. Note that $a = 0.9 < 1/1 = 1/b$. The region between the curves $a = 4b/(1 + b)^2$ and $a = 1/b$ gives rise to unstable economies. In example 5.26, $(a, b) = (0.9, 1.5)$ and $a = 0.9 > 1/1.5 = 1/b$, so $t(k)$ was unstable.

The region above the curve $a = 4b/(1 + b)^2$ gives real roots to the characteristic equation, the largest of which is

$$r_1 = 0.5 \left(a(1 + b) + \sqrt{a^2(1 + b)^2 - 4ab} \right).$$

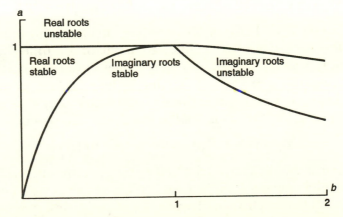

FIGURE 5.9. Regions of stability and instability for an economy satisfying $t(n+1) = a(1+b)t(n+1) - abt(n) + 1$. The curve $a = 4b/(1+b)^2$ starts at the origin and the curve $a = 1/b$ start at the point $(1,1)$.

It is a bit tricky to determine, but $\eta < 1$ when $a < 1$ and $b < 1$. The region of real roots corresponding to both stable and unstable economies is also given in figure 5.9.

Let's discuss this result. The constant b is a measure of how the investors react to good and bad news. Thus, the larger b, the more sensitive the investors are to consumption. Likewise, a is a measure of how consumers react to the economy, that is, the larger a, the more the consumers spend. Our result that we need $ab < 1$ to have a stable economy says that if one of these constants is large, the other must be small.

Suppose that the government has found that we are in an unstable economy. What can be done? They need to reduce either a or b. One possible method would be to increase taxes to reduce the amount available for spending, which would reduce a and make the economy more stable. Another method would be to encourage saving by not taxing certain savings accounts. But a reduction in a would mean an increase in $1 - a$, the marginal propensity to save. Recall that the equilibrium value was

$$t = 1/(1 - a),$$

so this policy would **reduce** the total national economy and possibly lead to a recession.

Another possible solution would be for the government to change from having constant expenditures to having variable expenditures. The original dynamical system was actually

$$t(n + 2) = (1 + b)at(n + 1) - bat(n) + g(n + 2),$$

with $g(n + 2) = 1$. Thus, as the economy oscillates with increasing amplitude, the government could have a one-or-two time change in expenditures to try to bring the $t(k)$ values closer to equilibrium. Thus, instead of $g(n + 2) = 1$, we might have let the government expenditures be a function of the size of the economy, that is, $g(n + 2) = f(t(n + 2))$. This would lead to the dynamical system

$$t(n + 2) = (1 + b)at(n + 1) - bat(n) + f(t(n + 2)).$$

Since there is usually a time delay before the government reacts, their reaction might use old data. So we might have $g(n+2) = f(t(n+1))$ or $g(n+2) = f(t(n))$. This would lead to a new dynamical system and its stability needs to be studied. One such analysis is in Baumol (1961). There should be some concern, because the latter two inputs might lead to instability. This will be seen in more detail in problem 7.

Economists often make models of a nation's economy, such as this one. They also use these models to help them make policy decisions to keep a prosperous economy. But just as we saw in the previous discussion, there is often no perfect solution, and they must try to choose the least of many evils.

PROBLEMS

1. Suppose the government spends 2 units in each time period instead of 1 unit. What is the equilibrium value?

2. Suppose that investment is made according to the formula $p(n+2) = b[c(n+1) - c(n)]$. With the other assumptions remaining the same, develop a third-order dynamical system that models this economy.

3. Suppose that, instead of the government having fixed expenditures, it increases its expenditures by 5 per cent each time period, that is, $g(n) = (1.05)^n$. Assuming that $a = 0.9$ and $b = 1$, find a second-order nonhomogeneous dynamical system to model the economy and then find the general solution to that equation.

4. Consider the model of the economy in which $a = 0.9$ and $b = 2$.
   a. Find the general solution to the equation. Observe that the equilibrium value 10 is unstable.
   b. Suppose we develop the strategy that if $t(n) < 9$, then the government expenditures are 2 units in the next time interval, that is, $g(n+1) = 2$ when $t(n) < 9$. Also, when $t(n) > 11$ then $g(n+1) = 0$. When $9 \leq t(n) \leq 11$, then $g(n+1) = 1$, as before. To see if this will stabilize the economy, use this strategy to compute $t(2)$ through $t(10)$ when $t(0) = 9.6$ and $t(1) = 9.4$.
   c. Repeat part (b), except that the government has a two time period delay, that is, when $t(n) < 9$ then $g(n+2) = 2$, and when $t(n) > 11$ then $g(n+2) = 0$.

5. Suppose that the dynamical system

$$t(n+2) = abt(n+1) - at(n) + 3$$

models our economy.
   a. What is the equilibrium value?
   b. What are the two roots of the characteristic equation corresponding to the linear second-order dynamical system?

     c. Assume that the values for a and b are such that the roots
     are complex. Find a condition that implies the equilibrium
     value is stable.

6. Suppose that the dynamical system

$$t(n+2) = abt(n+1) - ab^2 t(n) + 1$$

models our economy.

    a. What is the equilibrium value?

    b. What are the two roots of the characteristic equation?

    c. Assuming that the values for a and b are such that the roots
    are complex, find a condition that implies the equilibrium
    value is stable.

7. Suppose that the government, in an effort to stabilize the economy, decides to make its expenditures dependent on the size
of the economy. Consider our standard model, dynamical system (5.26). The government decides that when $t(n) > 10$,
they will reduce expenditures to bring it down. Similarly, when
$t(n) < 10$, they will increase expenditures to pull it up. One
possible government expenditure would be

$$g(n+2) = 1 + p[10 - t(m)]$$

for some time m.

    a. Consider the unstable economy of example 5.26, in which
    $a = 0.9$ and $b = 1.5$. Assume the government reaction is
    immediate, that is,

$$g(n+2) = 1 + p[10 - t(n+2)].$$

    The dynamical system that models this economy is then

$$t(n+2) = 2.25t(n+1) - 1.35t(n) + 1 + p[10 - t(n+2)].$$

    Find conditions on p so that the equilibrium value, still
    $t = 10$, is stable.

    b. Suppose that the government delays its reaction in part a)
    so that

$$g(n+2) = 1 + p[10 - t(n+1)],$$

    giving the dynamical system

$$t(n+2) = 2.25t(n+1) - 1.35t(n) + 1 + p[10 - t(n+1)].$$

    Show that the government's reaction has no effect on the
    stability of the equilibrium value $t = 10$ by comparing the

absolute value of the roots of the characteristic equation to the results of example 5.26.

c. Suppose we have the stable economy of example 5.25 in which $a = 0.9$ and $b = 1$. Suppose that the government delays its reaction so that

$$g(n+2) = 1 + p[10 - t(n)],$$

giving the dynamical system.

$$t(n+2) = 1.8t(n+1) - 0.9t(n) + 1 + p[10 - t(n)].$$

Find conditions on p that make the equilibrium value $t = 10$ unstable. Observe that this reaction makes the economy less stable or unstable for any positive choice of p.

5.10   Dynamical systems with order greater than two

We now know how to handle first and second-order linear dynamical systems. Higher-order (third, fourth, and so on) linear equations require nothing new.

EXAMPLE 5.27   Consider the third-order dynamical system

$$a(n+3) = 6a(n+2) - 11a(n+1) + 6a(n). \qquad (5.27)$$

The first step for finding the general solution is to substitute $a(n) = r^n$ into this equation, and then divide by r^n, to get the characteristic equation

$$r^3 = 6r^2 - 11r + 6.$$

The second step is to bring all the terms to one side, say the left, and factor, giving

$$(r-1)(r-2)(r-3) = 0.$$

Thus, the roots of the characteristic equation are $r = 1$, 2, and 3. Since there are no multiple roots, the general solution is

$$a(k) = c_1 + c_2 2^k + c_3 3^k,$$

since $1^k = 1$.

If we are given three initial values such as $a(0) = 2$, $a(1) = 3$, and $a(2) = 7$, we can then find the particular solution by solving the system of three equations

$$a(0) = c_1 + c_2 2^0 + c_3 3^0 = c_1 + c_2 + c_3 = 2,$$
$$a(1) = c_1 + c_2 2^1 + c_3 3^1 = c_1 + 2c_2 + 3c_3 = 3,$$
$$a(2) = c_1 + c_2 2^2 + c_3 3^2 = c_1 + 4c_2 + 9c_3 = 7.$$

It is easy to see that the solution to these equations is $c_1 = 2$, $c_2 = -1$, and $c_3 = 1$. The particular solution to dynamical system (5.27) is

$$a(k) = 2 - 2^k + 3^k.$$ ∎

As before, when r is a multiple root for the characteristic equation, the general solution contains terms of the form r^k, kr^k, $k^2 r^k$, and so forth, until we have the same number of terms as the multiplicity of r.

EXAMPLE 5.28   The fourth-order dynamical system

$$a(n+4) = 4a(n+3) - 16a(n+1) + 16a(n)$$

has characteristic equation

$$r^4 - 4r^3 + 16r - 16 = 0.$$

By a lucky guess, we see that $r = 2$ is a root of this equation. Recall that a polynomial has a root c if and only if the variable minus c, that is, $r - c$, is a factor of the polynomial. Dividing $(r - 2)$ into the characteristic equation gives

$$r^3 - 2r^2 - 4r + 8 = 0.$$

Another lucky guess of $r = 2$ tells us that 2 is (at least) a double root of the characteristic equation. Again, we divide the last equation by $(r - 2)$, giving

$$r^2 - 4 = 0.$$

Factoring this into $(r-2)(r+2)$, we see that the original characteristic equation factors into

$$r^4 - 4r^3 + 16r - 16 = (r - 2)^3 (r + 2) = 0.$$

Since 2 is a triple root of the equation, the general solution has terms of the form 2^k, $k2^k$, and $k^2 2^k$ (three terms), as well as the term $(-2)^k$ (since -2 is also a root). Thus, the general solution is, after factoring out a 2^k from the first three terms,

$$a(k) = 2^k(c_1 + c_2 k + c_3 k^2) + c_4(-2)^k,$$

where c_1, c_2, c_3, and c_4 are four constants which depend on the initial values $a(0)$, $a(1)$, $a(2)$, and $a(3)$. ∎

The main difficulty with higher-order equations is that it may be difficult or impossible to factor the characteristic equation. Often other techniques, such as Newton's method from calculus, must be used.

Let's consider an application of higher-order dynamical systems to the study of population growth. To make things easy, we will assume that growth rates are constant. Consider a species, say heffalumps, which can be broken into three equal age groups, 0–1 years, 1–2 years, and 2–3 years. Instead of years we could have said decades, etc. Let $a(n)$ represent the number of heffalumps in the 0–1 age group in year n. Likewise, let $b(n)$ and $c(n)$ represent the number of heffalumps in the 1–2 and 2–3 age groups, respectively, in year n.

There are two pieces of information we need: the survival rate for each age group, and the fertility rate for each age group.

First, consider the survival rate. Suppose half of the heffalumps born survive to their first birthday, that is, $b(n+1) = 0.5a(n)$. Also suppose two-thirds of one-year-old heffalumps survive to their second birthday, that is, $c(n+1) = (2/3)b(n)$. No heffalump lives to its third birthday.

Second, consider the fertility rate. Suppose that each year, on the average, each **pair** of heffalumps in the 0–1 age group has 1 offspring, each pair in the 1–2 age group has 10 offspring, and each pair in the 2–3 age group has 6 offspring. Since individuals in the 0–1 age group in year $n+1$ are offspring of some heffalump in year n, it follows that

$$a(n+1) = 0.5a(n) + 5b(n) + 3c(n).$$

(Note that each pair in the $a(n)$ group has an average of 1 offspring, so the total number of offspring from this age group is $1 \times a(n)/2$. Similarly for the other two age groups.)

We now need to derive a dynamical system involving only one age group. Since $b(n+1) = 0.5a(n)$, we make the substitutions $a(n) = 2b(n+1)$ and $a(n+1) = 2b(n+2)$ in the above equation, giving

$$2b(n+2) = b(n+1) + 5b(n) + 3c(n).$$

Since $c(n+1) = (2/3)b(n)$, we make the substitutions $b(n) = (3/2)c(n+1)$, $b(n+1) = (3/2)c(n+2)$, and $b(n+2) = (3/2)c(n+3)$

into the above equation, giving

$$2(3/2)c(n+3) = (3/2)c(n+2) + 5(3/2)c(n+1) + 3c(n).$$

After dividing both sides by 3 and simplifying, we get

$$c(n+3) = 0.5c(n+2) + 2.5c(n+1) + c(n).$$

The characteristic equation for this dynamical system is

$$r^3 - 0.5r^2 - 2.5r - 1 = 0,$$

which factors into

$$(r+0.5)(r+1)(r-2) = 0.$$

It then follows that the general solution for the 2–3 age group is

$$c(k) = c_1(-0.5)^k + c_2(-1)^k + c_3(2)^k.$$

Since $b(n) = (3/2)c(n+1)$, it follows that

$$
\begin{aligned}
b(k) &= 1.5c(k+1) \\
&= 1.5c_1(-0.5)^{k+1} + 1.5c_2(-1)^{k+1} + 1.5c_3(2)^{k+1} \\
&= -0.75c_1(-0.5)^k - 1.5c_2(-1)^k + 3c_3(2)^k.
\end{aligned}
$$

Similarly, we find that

$$
\begin{aligned}
a(k) &= 2b(k+1) = 2\left[-0.75c_1(-0.5)^{k+1} - 1.5c_2(-1)^{k+1} + 3c_3(2)^{k+1}\right] \\
&= -1.5c_1(-0.5)^{k+1} - 3c_2(-1)^{k+1} + 6c_3(2)^{k+1} \\
&= 0.75c_1(-0.5)^k + 3c_2(-1)^k + 12c_3(2)^k.
\end{aligned}
$$

Because of the $(2)^k$ term, the population grows exponentially. To get more information from our model, divide $a(k)$, $b(k)$, and $c(k)$ by $c_3(2)^k$. In this case,

$$\frac{a(k)}{c_3 2^k} = \frac{3c_1}{4c_3}(-0.25)^k + \frac{3c_2}{c_3}(-0.5)^k + 12,$$

$$\frac{b(k)}{c_3 2^k} = \frac{-3c_1}{4c_3}(-0.25)^k - \frac{3c_2}{2c_3}(-0.5)^k + 3,$$

$$\frac{c(k)}{c_3 2^k} = \frac{c_1}{c_3}(-0.25)^k + \frac{c_2}{c_3}(-0.5)^k + 1.$$

Since $(-0.25)^k$ and $(-0.5)^k$ go to zero, it follows that

$$\lim_{k \to \infty} \frac{a(k)}{c_3 2^k} = 12, \qquad \lim_{k \to \infty} \frac{b(k)}{c_3 2^k} = 3, \qquad \lim_{k \to \infty} \frac{c(k)}{c_3(2)^k} = 1.$$

In other words, for large k, the three age groups are in the ratio

$$(a(k) : b(k) : c(k)) = (12 : 3 : 1).$$

Thus, we have seen from this model that populations may grow exponentially, but that the ratios of the different age groups may become stable.

Some species reproduce in a continuous manner, so this would not be a good model. One species for which this model seems to work reasonably well is the blue whale. The blue whale mates between June 20 and August 20. The gestation period is 1 year and they nurse for 7 more months, so they only reproduce every 2 years.

When we break a population into m age groups, this approach leads to an mth-order linear dynamical system which means we must factor an mth-degree polynomial. If we were considering humans, we could break the (reproductive) age of the population into five age groups, 0–10, 10–20, ..., 40–50, or we could break the population into 50 age groups, 0–1, 1–2, ..., 49–50. The first grouping would be easier to work with; the second might give more information.

Suppose we have m roots to the characteristic equation, $r_1, r_2, \ldots, r_m$. Suppose that $r_1 > 0$ and $r_1 > |r_j|$ for $j = 2, 3, \ldots, m$. Then by dividing the size of each age group by $(r_1)^k$, as we divided $a(k)$, $b(k)$, and $c(k)$ by 2^k, we can find the relative age distribution of the population, (12:3:1) in the above discussion. There are mathematical techniques that enable us to find this value of r_1. They will be discussed in Chapter 8.

In some cases, the largest value of $|r_j|$ occurs when r_j is a complex number. As we remember, r^k, when r is complex, causes cycles. In this case we find that each age group has cyclic behavior, that is, the size of each age group rises and falls, causing what is known as population waves. See Hoppensteadt (1982) for further details.

Sometimes a species has a low reproductive rate until it reaches adulthood. For example, the American bison does not reproduce for the first 2 years. Once it reaches adulthood, the birth rate and death rate are approximately constant for the 2–3, 3–4, etc., age groups. We could develop a large number of equations that would reduce to one high-order dynamical system. But there is a trick that will keep the number of equations to a minimum.

Suppose that heffalumps can be broken into "three" age groups 0–1, 1–2, and adults. The number of each age group present at the beginning of the nth year is denoted $a(n)$, $b(n)$, and $c(n)$, respectively. Suppose, for example, that the fraction of each age group that survives to the next year is 0.8, 0.5, and 0.8, respectively. Since those in the 0–1 age group that survive a year will be in the 1–2 age group, and those in the 1–2 age group and in the adult age group that survive the year will then be in the adult age group, we can summarize these results with the two equations

$$b(n+1) = 0.8a(n) \quad \text{and} \quad c(n+1) = 0.5b(n) + 0.8c(n).$$

Suppose that the reproductive rates for the three age groups are 1.2, 10.2625, and 23.92, respectively. Since the offspring in each group during one time period will be in the 0–1 age group at the beginning of the next time period, we get the reproductive equation

$$a(n+1) = 1.2a(n) + 10.2625b(n) + 23.92c(n).$$

From the first two equations we get

$$a(n) = 1.25b(n+1) \quad \text{and} \quad b(n) = 2c(n+1) - 1.6c(n).$$

Using these two equations to substitute into the reproductive equation, we find (after some simplification) that

$$c(n+3) = 2c(n+2) + 7.25c(n+1) + 3c(n).$$

Factoring the characteristic equation gives

$$r^3 - 2r^2 - 7.25r - 3 = (r-4)(r+1.5)(r+0.5) = 0,$$

so the general solution is

$$c(k) = c_1 4^k + c_2(-1.5)^k + c_3(-0.5)^k.$$

Thus,

$$\lim_{k\to\infty} \frac{c(k)}{c_1 4^k} = \lim_{k\to\infty} 1 + \frac{c_2}{c_1}\left(-\frac{3}{8}\right)^k + \frac{c_3}{c_1}\left(-\frac{1}{8}\right)^k = 1.$$

As a note, by substituting $k+1$ for k, we also have

$$\lim_{k\to\infty} \frac{c(k+1)}{c_1 4^{k+1}} = 1.$$

The formulas $b(n) = 2c(n+1) - 1.6c(n)$ and $a(n) = 1.25b(n+1)$ give that

$$\lim_{k\to\infty} \frac{b(k)}{c_1 4^k} = \lim_{k\to\infty} \left(\frac{2c_3(k+1)}{c_1 4^k} - \frac{1.6c(k)}{c_1 4^k}\right)$$

$$= \lim_{k\to\infty} \left(8\frac{c(k+1)}{c_1 4^{k+1}} - 1.6\frac{c(k)}{c_1 4^k}\right)$$

$$= [8(1) - 1.6(1)] = 6.4$$

and

$$\lim_{k\to\infty} \frac{a(k)}{c_1 4^k} = \lim_{k\to\infty} \frac{1.25 b(k+1)}{c_1 4^k} = \lim_{k\to\infty} 5 \frac{b(k+1)}{c_1 4^{k+1}} = 5 \times 6.4 = 32.$$

From this, we find that eventually the relative ratios of the three age groups will be approximately $(A : B : C) = (32 : 6.4 : 1)$. This would mean that there would be 32 infants for every adult heffalump.

One species to which this model has been applied is the American bison. Bison break down into three groups: calves (0–1 age group), yearlings (1–2 age group), and adults. The calves have a survival rate of 0.6, the yearlings have a survival rate of 0.75, and the adults have a survival rate of 0.95. Only adults reproduce, with a reproduction rate of 0.42. Only females are counted in these rates. We will return to this in problem 5.

PROBLEMS

1. Suppose a species has two age groups. Suppose that the survival rate for the 0–1 age group is 0.8 and the reproductive rate is 2, while the reproductive rate for the 1–2 age group is 3.75. Find a second-order dynamical system to model this population and find the long term ratio of the 0–1 age group to the 1–2 age group.

2. Suppose a species has two age groups. Suppose that the survival rate for the 0–1 age group is 0.91 and the reproductive rate is 0.6, while the reproductive rate for the 1–2 age group is 1.0. Find a second-order dynamical system to model this population and find the long term ratio of the 0–1 age group to the 1–2 age group.

3. Suppose a species has two age groups: 0–1 and adults. Suppose that the survival rate for the 0–1 age group is 0.5 and the reproductive rate is 1.8, while the survival rate of the adults is 0.8 and the reproductive rate is 4. Find a second-order dynamical system to model this population and find the long term ratio of the 0–1 age group to the adults.

4. Consider the following dynamical system of three equations:

$$a(n+1) = 2b(n),$$
$$b(n+1) = 3c(n),$$
$$c(n+1) = a(n) - \frac{11}{3} b(n) + 6c(n).$$

    a. Rewrite these as a third-order equation in terms of $c(k)$.
    b. Find the general solution.
    c. Find the long term ratios $(a : b : c)$.

5. Find the long term ratios of calves to yearlings to adults among American bison. You will need to use Newton's method or a graphing program to approximate a root of the characteristic equation.

6. Suppose you enter a game in which the probability of losing 1 dollar on any bet is 3/4ths, the probability of winning 1 dollar on any bet is 3/16ths, and the probability of winning 2 dollars on any bet is 1/16th. Let $p(n)$ be the probability of eventually going broke, given that you presently have n dollars.
    a. Develop a dynamical system for $p(n+1)$ in terms of $p(n+3)$, $p(n+2)$, and $p(n)$, and rewrite it as a 3rd-order dynamical system.
    b. Find the general solution to the dynamical system in part a.
    c. Find the particular solution to the dynamical system, if you quit the game with 0, 3, or 4 dollars, that is, $p(0) = 1$, $p(3) = 0$, and $p(4) = 0$.
    d. Use the solution in part c to find $p(1)$ and $p(2)$.

7. Suppose you enter a game in which the probability of losing 1 dollar on any bet is 5/9ths, the probability of winning 1 dollar on any bet is 3/9ths, and the probability of winning 2 dollars on any bet is 1/9th. Let $p(n)$ be the probability of eventually going broke, given that you presently have n dollars.
    a. Develop a dynamical system for $p(n+1)$ in terms of $p(n+3)$, $p(n+2)$, and $p(n)$, and rewrite it as a 3rd-order dynamical system.
    b. Find the general solution to the dynamical system in part a.
    c. Find the particular solution to the dynamical system, if you quit the game with 0, 99, or 100 dollars, that is, $p(0) = 1$, $p(99) = 0$, and $p(100) = 0$. (This is messy.)
    d. Use the solution in part c to find $p(50)$.

5.11   Solutions involving trigonometric functions

You might wonder why solutions involving complex numbers tend to oscillate. The reason is that complex numbers are closely connected with trigonometric functions, specifically the sine and cosine functions. Recall that the point (a, b) in the plane corresponds to the complex number $a + bi$. But a point in the plane can also be written in polar coordinates, that is,

$$a + ib = r \cos \theta + i r \sin \theta,$$

where $r = \sqrt{a^2 + b^2}$ and θ is the angle that the line from the origin to the point (a, b) makes with the positive x-axis. Thus, by substitution, a solution involving complex numbers can be rewritten as

$$
\begin{aligned}
a(k) &= c_1 (a + bi)^k + c_2 (a - bi)^k \\
&= c_1 (r \cos \theta + i r \sin \theta)^k + c_2 (r \cos \theta - i r \sin \theta)^k \\
&= c_1 r^k (\cos \theta + i \sin \theta)^k + c_2 r^k (\cos \theta - i \sin \theta)^k.
\end{aligned}
$$

It happens that

$$(\cos \theta + i \sin \theta)^k = \cos k\theta + i \sin k\theta \qquad (5.28)$$

and

$$(\cos \theta - i \sin \theta)^k = \cos k\theta - i \sin k\theta. \qquad (5.29)$$

A proof of equation (5.28) is given at the end of this section. Substitution of equations (5.28) and (5.29) into the general solution and then simplification gives

$$a(k) = r^k (c_3 \cos k\theta + c_4 \sin k\theta),$$

where $c_3 = c_1 + c_2$ and $c_4 = i(c_1 - c_2)$.

Put another way, if the characteristic equation for a second-order linear dynamical system has $x = a \pm bi$ as its two roots, the general solution can be written as

$$a(k) = r^k (c_3 \cos k\theta + c_4 \sin k\theta),$$

where $r = \sqrt{a^2 + b^2}$ and θ is the angle between the line $y = bx/a$ and the x-axis. From this it can be seen that solutions involving complex numbers will oscillate. Let's state this formally as a theorem.

THEOREM 5.29 *Suppose the roots of the characteristic equation to the dynamical system*

$$a(n + 2) = b_1 a(n + 1) + b_2 a(n)$$

are

$$x = a \pm bi.$$

Then the general solution to this dynamical system can be given as

$$a(k) = c_1 (a + bi)^k + c_2 (a - bi)^k$$

or as

$$a(k) = r^k (c_3 \cos k\theta + c_4 \sin k\theta),$$

where

$$r^2 = a^2 + b^2 \quad and \quad \tan \theta = \frac{b}{a},$$

if $a \neq 0$. The constants c_1 and c_2, or the constants c_3 and c_4 depend on the initial values.

EXAMPLE 5.30   Let's reconsider dynamical system (5.8),

$$a(n+2) = -0.25 a(n),$$

of example 5.12. In figure 5.1, it was seen that the solution oscillates to zero, but each oscillation takes 4 units of time. Let's see why this happens.

The roots of the characteristic equation are $x = \pm i/2$. In this case, $r^2 = 0^2 + (1/2)^2 = 1/4$ so $r = 1/2$. The line from the origin to the point $(0, 1/2)$ makes an angle of $\theta = \pi/2$ with the x-axis, so the general solution can be written as

$$a(k) = (0.5)^k \left(c_3 \cos \frac{k\pi}{2} + c_4 \sin \frac{k\pi}{2} \right).$$

Now it is clear that $a(k)$ converges to 0 since $(0.5)^k$ goes to zero. It is also clear that the solution oscillates, each oscillation requiring 4 units of time since $\cos(0.5(k+4)\pi) = \cos(0.5k\pi)$.

Suppose we are given that $a(0) = 2$ and $a(1) = 3$. Then we have that

$$2 = a(0) = (0.5)^0 (c_3 \cos 0 + c_4 \sin 0) = c_3.$$

Also,

$$3 = a(1) = (0.5)^1 (c_3 \cos(\pi/2) + c_4 \sin(\pi/2)) = 0.5 c_4.$$

Thus, $c_3 = 2$ and $c_4 = 6$ and the particular solution is

$$a(k) = (0.5)^k \left(2 \cos \frac{k\pi}{2} + 6 \sin \frac{k\pi}{2} \right). \qquad \blacksquare$$

EXAMPLE 5.31   In the case of the dynamical system (5.9)

$$a(n+2) = 2a(n+1) - 2a(n)$$

of example 5.13, the roots of the characteristic equation are $x = 1 \pm i$, so that $r = \sqrt{2}$ and $\theta = \pi/4$. In this case, the general solution is

$$a(k) = (\sqrt{2})^k \left(c_3 \cos \frac{k\pi}{4} + c_4 \sin \frac{k\pi}{4} \right).$$

This clearly oscillates to infinity, each oscillation requiring 8 units of time since $\pi/4$ is one-eighth of 2π. A little more than one oscillation can be seen in figure 5.2. ∎

EXAMPLE 5.32   Consider the dynamical system (5.10),

$$a(n+2) = a(n+1) - a(n),$$

of example 5.15. The roots of the characteristic equation are $x = 0.5 \pm 0.5\sqrt{3}i$, so that $r = 1$ and $\theta = \pi/3$. In this case, the general solution is

$$a(k) = c_3 \cos \frac{k\pi}{3} + c_4 \sin \frac{k\pi}{3}.$$

This solution is periodic with period 6, which can be seen in figure 5.3. The reason is that $\cos(k+6)\theta = \cos k\theta$, that is, $\pi/3$ is one-sixth of 2π. ∎

EXAMPLE 5.33   Consider the dynamical system (5.11),

$$a(n+2) = 1.2a(n+1) - a(n),$$

of example 5.16. The roots of the characteristic equation are $x = 3/5 \pm 4i/5$, so that $r = 1$ and $\theta = \arctan 4/3$. The general solution is

$$a(k) = c_3 \cos k\theta + c_4 \sin k\theta.$$

If some integral multiple m of θ equals another of 2π, then the general solution is periodic with period m. But, in this example, θ is an irrational multiple of 2π, so the solution never repeats itself **exactly**, although it does comes very close to doing so. This can be seen in figure 5.4. ∎

PROOF OF EQUATION (5.28):   We know that the particular solution to the dynamical system

$$a(n+1) = (\cos \theta + i \sin \theta) a(n)$$

with $a(0) = 1$ is

$$a(k) = (\cos \theta + i \sin \theta)^k.$$

We will now show that the particular solution also is given by

$$a(k) = \cos k\theta + i \sin k\theta. \tag{5.30}$$

Since there is only one particular solution, these two solutions must be equal and we will have equation (5.28). From equation (5.30), we have by the formulas for addition of angles that

$$
\begin{aligned}
a(n+1) &= \cos(n+1)\theta + i\sin(n+1)\theta \\
&= \cos n\theta \cos\theta - \sin n\theta \sin\theta + i(\sin n\theta \cos\theta + \cos n\theta \sin\theta) \\
&= (\cos\theta + i\sin\theta)(\cos n\theta + i\sin n\theta) \\
&= (\cos\theta + i\sin\theta)a(n).
\end{aligned}
$$

Thus, equation (5.30) satisfies the dynamical system. Since it satisfies the initial condition, it is also the particular solution and our claim is proved.   ∎

PROBLEMS

1. Compute the polar coordinates θ and r corresponding to the following complex numbers.
   a. $2 - 2i$
   b. $-\sqrt{3} + i$
   c. $-15i$

2. Give the complex number that has the following polar form
   a. $r = 4$ and $\theta = \pi/6$.
   b. $r = \sqrt{3}$ and $\theta = 3\pi/4$.
   c. $r = 0.1$ and $\theta = 3\pi/2$.
   d. $r = 1$ and $\theta = 7\pi/6$.

3. Consider the dynamical system

$$a(n+2) = -a(n+1) - 0.5a(n).$$

   a. Find the general solution in terms of $\sin k\theta$ and $\cos k\theta$.
   b. Is zero an attracting or repelling fixed point?
   c. Given that $a(0) = 0$ and $a(1) = -1$, find the particular solution in terms of $\sin k\theta$ and $\cos k\theta$.

4. Consider the dynamical system

$$a(n+2) = (1/3)a(n+1) - (1/9)a(n) + 7.$$

   a. Find the general solution in terms of $\sin k\theta$ and $\cos k\theta$.
   b. Is $a = 9$ an attracting or repelling fixed point?

   c. Given that $a(0) = 11$ and $a(1) = 10$, find the particular
      solution in terms of $\sin k\theta$ and $\cos k\theta$.

5. Consider the dynamical system

$$a(n+2) = 6a(n+1) - 18a(n) + 13.$$

   a. Find the general solution in terms of $\sin k\theta$ and $\cos k\theta$.
   b. Is $a = 1$ an attracting or repelling fixed point?
   c. Given that $a(0) = 2$ and $a(1) = 7$, find the particular
      solution in terms of $\sin k\theta$ and $\cos k\theta$.

Introduction to nonlinear dynamical systems

6.1 A model of population growth

Consider a population of rabbits. For simplicity, assume that on the average each rabbit gives birth to two new rabbits in one unit of time and that no rabbit dies. The easiest way to model this problem is to consider the **change** in the size of the population of rabbits in one time period. Let $a(0)$ be the number of rabbits at time $t = 0$ (the time when we start our observations). Then $a(1) - a(0)$ is the change in the population in the first unit of time. But the change equals the number of new rabbits, that is, the number that were born during that time interval. Thus $a(1) - a(0) = 2a(0)$. Likewise the change from period 1 to period 2, $a(2) - a(1)$, equals the number born to the population of size $a(1)$. Thus

$$a(2) - a(1) = 2a(1).$$

Continuing in this manner we see that the change in population in a time period is twice the population at the beginning of the time period, that is

$$a(n+1) - a(n) = 2a(n), \qquad \text{or} \qquad a(n+1) = 3a(n).$$

The solution is $a(k) = 3^k a_0$.

We now wish to make this model more realistic. Now assume that the number of births in time period n is proportional to the size of the population in that time period, that is, the number of births equals $ba(n)$, where b is the birth rate. Likewise, the number of deaths in time period n is proportional to $a(n)$, that is, the number of deaths equals $da(n)$. The change in population in a time period is the number of births minus the number of deaths. Combining these assumptions gives

$$a(n+1) - a(n) = ba(n) - da(n), \qquad \text{or} \qquad a(n+1) = (1+r)a(n),$$

where $r = b - d$ is the (net) growth rate for the population. The solution, we know, is

$$a(k) = (1+r)^k a_0.$$

Let r be a moderate rate of growth, say $r = 0.2$, and let $a_0 = 100$. Then we obtain $a(10) = 619$, $a(20) = 3834$, $a(50) = 910\,044$, and $a(100) = 8\,281\,797\,452$. We see that while our model seems to make sense, the size of the population gets unrealistically large after a long period of time in that $a(k)$ goes to infinity exponentially. This is Malthus's theory that populations grow exponentially and he thus predicted a world-wide catastrophe that hasn't yet happened.

Does this mean our model is wrong? Yes and no. It depends on what information we want from it. For small values of time when the population size is relatively small, this model gives "good" estimates of population growth. For short periods of time, populations do appear to grow exponentially and growth rates are approximately constant. But over long periods of time, the growth rate r is not constant and changes as the size of the population changes. So we should replace the r in our dynamical system with $f(a(n))$, where f is a function of population size. In other words, our growth rate $f(a(n))$ changes as the size of the population changes.

The first assumption we make is that the environment of the population can only support a certain number, say ℓ, of the species. Thus, if $a(n) > \ell$, there will not be enough food or space available and more animals will die of starvation, etc., than are born. So, it follows that the growth rate is negative, $f(a(n)) < 0$, when $a(n) > \ell$.

Our second assumption is that if the population is less than ℓ, then there is extra food and space available, so the growth rate should be positive, that is, $f(a(n)) > 0$ when $a(n) < \ell$.

Our last assumption is that if the population is small relative to ℓ, that is, there is plenty of food and space for the existing population, then the growth rate should be close to the unrestricted growth rate r. But as the population increases, the growth rate should decrease and should, in fact, be zero when $a(n) = \ell$.

The simplest function that satisfies these conditions is the linear function

$$f(a(n)) = r\left(1 - \frac{a(n)}{\ell}\right).$$

Notice the following: (1) if $a(n)$ is small, $1 - a(n)/\ell$ is close to 1 and the growth rate, $f(a(n))$, is approximately r; (2) if $a(n) < \ell$, then $1 - a(n)/\ell > 0$ and the growth rate is positive; (3) if $a(n) = \ell$, the growth rate is zero; and (4) if $a(n) > \ell$, then $1 - a(n)/\ell < 0$ and the growth rate is negative. In fact, the larger $a(n)$, the more negative the growth rate.

The number r is called the **unrestricted growth rate** and the number ℓ is called the **carrying capacity** of the environment. The dynamical system that models population growth is then

$$a(n+1) - a(n) = r\left(1 - \frac{a(n)}{\ell}\right)a(n),$$

or, after simplification,

$$a(n+1) = (1+r)a(n) - ba^2(n), \qquad (6.1)$$

where $b = r/\ell$ and $a^2(n) = a(n)a(n)$. This is called the **logistic equation**. Mathematicians often call the term $-ba^2(n)$ a **damping term** because it dampens the growth of the population, that is, keeps it from going to infinity.

EXAMPLE 6.I   Suppose $r = 0.2$, $\ell = 8$ (where one unit represents 1000 of the species), and therefore $b = 0.2/8 = 0.025$. The logistic equation becomes

$$a(n+1) = 1.2a(n) - 0.025a^2(n). \qquad (6.2)$$

If $a_0 = 3$, then

$$a(1) = (1.2)3 - (0.025)9 = 3.375, \qquad a(2) = 3.765,$$

and so forth. Figure 6.1 gives graphs of solutions $(k, a(k))$ for several different initial values. The one with $a_0 = -0.1$ is unrealistic but mathematically interesting. The horizontal line, $\ell = 8$, is the limit to population size or the stable equilibrium value. Remember that only the points $(0, a(0))$, $(1, a(1))$, etc. are important, but these

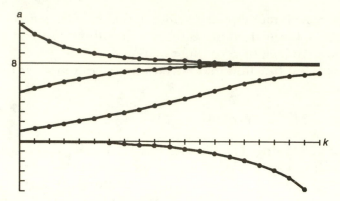

FIGURE 6.1. Points $(k, a(k))$ for the dynamical system $a(n + 1) = 1.2a(n) - 0.025a^2(n)$.

points have been connected with lines so that a pattern is easier to distinguish.   ■

Let's recall the definition of equilibrium value or fixed point.

DEFINITION 6.2   *Consider a first-order dynamical system*

$$a(n+1) = f(a(n)).$$

*A number a is called an **equilibrium value** or a **fixed point** for this dynamical system if a satisfies the equation*

$$a = f(a).$$

In this case,

$$a(k) = a \qquad for \quad k = 0, 1, \ldots$$

is a particular solution to the dynamical system.

Let's solve for the equilibrium values of the logistic equation (6.2) of example 6.1, by substituting $a = a(n + 1) = a(n)$. This gives

$$a = 1.2a - 0.025a^2.$$

Collecting terms on the left and factoring out an a gives

$$a(-0.2 + 0.025a) = 0.$$

The solutions are given by $a = 0$ and $-0.2 + 0.025a = 0$, that is,

$$a = 0 \qquad and \qquad a = 8.$$

This makes sense since a population of size 0 would remain there. Also, as discussed previously, if $a_0 = 8$, the carrying capacity, then the growth rate is zero and again the population remains constant.

Let's find the equilibrium value for the general logistic equation (6.1). Substituting a for $a(n+1)$ and $a(n)$ we have

$$a = (1 + r)a - \left(\frac{r}{\ell}\right)a^2,$$

or, after subtracting a from both sides then factoring out an a and an r,

$$0 = r\left(1 - \frac{a}{\ell}\right)a.$$

Dividing by r, which we know is not zero, we have the equilibrium values of

$$a = \ell \qquad \text{and} \qquad a = 0$$

as expected.

Let's recall the definitions of stable and unstable equilibrium values.

DEFINITION 6.3 *Suppose a first-order dynamical system has an equilibrium value a. This equilibrium value is said to be **stable** or **attracting** if there is a number ϵ such that, for every $a(0)$ satisfying*

$$|a_0 - a| < \epsilon,$$

we have that

$$\lim_{k \to \infty} a(k) = a.$$

*An equilibrium value is **unstable** or **repelling** if there is a number ϵ such that, for every $a(0)$ satisfying*

$$0 < |a_0 - a| < \epsilon,$$

there is some value of k for which

$$|a(k) - a| > \epsilon.$$

(This condition may not necessarily be met for all values of k).

Suppose that a_0 is "close" to a. Intuitively, the fixed point, a, is attracting if $a(k)$ goes towards a and is repelling if $a(k)$ goes away from a. There are fixed points that are neither stable nor unstable. For example, a could be semistable, that is, a attracts solutions that start on the right but repels solutions that start on the left (or vice versa).

Referring to figure 6.1, it appears that the fixed point $a = 8$ is attracting. In particular, it appears that if $1 < a_0 < 12$, then

$$\lim_{k \to \infty} a(k) = 8.$$

In fact, if you construct a cobweb graph for this function, you will see that if a_0 is between 0 and 48 (the roots of $1.2x - 0.025x^2$), then $a(k)$ will converge to 8. Note that the interval $(0, 48)$ is not symmetric about $a = 8$.

Also note that the fixed point $a = 0$ appears to be repelling.

EXAMPLE 6.4 Let $r = 1.4$ and $\ell = 10$ in the logistic equation (6.1). Then

$$a(n+1) = 2.4a(n) - 0.14a^2(n). \tag{6.3}$$

The equilibrium values are $a = 0$ and $a = 10$. A cobweb graph is given in figure 6.2 where $a_0 = 1$. Notice that in the linear model, if $r = 1.4$, $a(k)$ would increase rapidly toward positive infinity. But here, when a_0 is relatively "close" to the equilibrium value $a = 10$, then $a(k)$ goes to 10. Thus, $a = 10$ appears to be a stable equilibrium value. The cobweb graph also indicates that $a = 0$ is unstable.

By observing the graph of $g(x) = 2.4x - 0.14x^2$, it appears that if a_0 is in an interval in which $g(a_0) > 0$, then $a(k)$ goes to 10. Since the roots of $2.4x - 0.14x^2$ are $x = 0$ and $x = 17\frac{1}{7}$, it follows that if

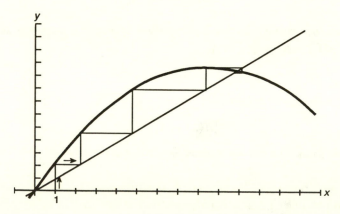

FIGURE 6.2. Cobweb for the dynamical system $a(n+1) = 2.4a(n) - 0.14a^2(n)$, with $a_0 = 1$.

$0 < a_0 < 17.14$, then $a(k) \rightarrow 10$. This interval is what we describe by the phrase "a_0 is close to a". ∎

There are several questions to be asked. What is the **general solution** to the logistic equation? For what values of r and ℓ is the equilibrium value $a = \ell$ stable? For what values of r and ℓ is the equilibrium value $a = 0$ stable? When an equilibrium value is stable, for what a_0 values does the solution $a(k)$ go to that equilibrium value?

To give a negative answer to the first question, it is **usually** impossible to find a solution to a nonlinear dynamical system. In section 2.4 we saw an exception to this rule. We can compute $a(0)$, $a(1)$, $\ldots$, $a(k)$ for large values of k using a spreadsheet or a graphing calculator with an $\boxed{\text{Ans}}$ button, so solutions exist. But we cannot find a "simple" algebraic expression for $a(k)$ in terms of k. This is very unsettling, for no matter how many values of $a(k)$ we compute, there is always the nagging feeling that something strange might happen on the next value we compute.

This reminds me of the story of two people having a discussion about prime numbers. The first argued "1's prime, 3's prime, 5's prime, 7's prime, so all odd integers are prime." "No," said the second, "you're reasoning is incomplete. 1, 3, 5, and 7 are prime. 9 is not prime, so let's chalk it up to experimental error. 11's prime and 13's prime. Now we can say that all odd integers are prime."

There are two morals to this story. First, no matter how many values of $a(k)$ we compute and no matter how convincing the pattern, we can still make no **certain** conclusion about the next value. Second, many people will find a pattern, but when one value disagrees with the pattern, they ignore that value instead of looking for a better pattern. But we digress.

Since we cannot write a solution to our equation, we must develop some other techniques to determine if the fixed points are attracting. Remember, knowing that a fixed point is attracting tells us a lot about the long term behavior of the solution to a dynamical system and thus about the real world it models.

Our definition of stability said that an equilibrium value a is **stable** if, when $a(0)$ is "close" to a, then the solution $a(k)$ gets "closer" to a.

Let's define $e(0) = a(0) - a$, where a is the equilibrium value that we are studying. The value $e(0)$ measures how close $a(0)$ is to the equilibrium value a, or is the **error** from equilibrium. Likewise we define $e(1) = a(1) - a$, $e(2) = a(2) - a$, $\ldots$, $e(n) = a(n) - a$,

$e(n+1) = a(n+1) - a$. Thus $a(k)$ gets closer to a if and only if $e(k)$ gets closer to 0. So it is enough to study $e(k)$. By adding a to both sides of the above, we have

$$a(n) = e(n) + a \qquad \text{and} \qquad a(n+1) = e(n+1) + a.$$

Now remember that $a(0)$ is "close" to a so that $e(0)$ is "close" to zero. If $e(0) = 0.1$, then $e^2(0) = 0.01$. If $e(0) = 0.01$, then $e^2(0) = 0.0001$. Observe that although $e(0)$ is small, $e^2(0)$ is much smaller. In terms of magnitude, $e^2(0)$ is small **compared** to $e(0)$. In the second case $e^2(0) = (1/100)\,e(0)$. We say that $e^2(0)$ is negligible compared to $e(0)$. When a number b is much smaller than a number c, we write $b \ll c$. So when $a(0)$ is close to equilibrium, $e^2(0) \ll e(0)$.

Let's go back to dynamical system (6.2)

$$a(n+1) = 1.2a(n) - 0.025a^2(n)$$

from example 6.1. Consider the equilibrium value $\ell = 8$. So from the above, we have

$$a(n) = 8 + e(n) \qquad \text{and} \qquad a(n+1) = 8 + e(n+1).$$

Substituting this into the above growth equation gives

$$[8 + e(n+1)] = 1.2[8 + e(n)] - 0.025[8 + e(n)]^2.$$

Rewriting this equation gives

$$8 + e(n+1) = 9.6 + 1.2e(n) - 0.025[64 + 16e(n) + e^2(n)].$$

or

$$e(n+1) = 0.8e(n) - 0.025e^2(n) \qquad (6.4)$$

after simplification.

For example, let $a(0)$ be close to equilibrium, say $a(0) = 8.5$ so that $e(0) = 0.5$. In table 6.1 we list $a(0)$ through $a(15)$ in the second column, and $e(0)$ through $e(15)$ in the third column. Notice that $a(k) - e(k) = 8$ in every case. This must be so, because $a(k) = e(k) + 8$. Also notice that $a(k)$ gets closer to 8 and $e(k)$ gets closer to 0.

How does this help? Dynamical system (6.4) for $e(n+1)$ is also nonlinear. The advantage studying this dynamical system is that $e(0)$ is **small** so that $0.025e^2(0)$ is "negligible." Thus we **drop** the term $0.025e^2(n)$ since it does not appear to affect the answer much. This gives us the equation

$$\tilde{e}(n+1) = 0.8\tilde{e}(n), \qquad (6.5)$$

TABLE 6.I. The $a(k)$ values for the dynamical system $a(n+1) = 1.2a(n) - 0.025a^2(n)$, along with the difference from equilibrium $e(k)$ and the approximations of the difference $\tilde{e}(k)$.

k	$a(k)$	$e(k)$	$\tilde{e}(k)$
0	8.5000	0.5000	0.5000
1	8.3937	0.3937	0.4000
2	8.3111	0.3111	0.3200
3	8.2464	0.2464	0.2560
4	8.1956	0.1956	0.2048
5	8.1555	0.1555	0.1638
6	8.1238	0.1238	0.1310
7	8.0987	0.0987	0.1048
8	8.0787	0.0787	0.0838
9	8.0628	0.0628	0.0671
10	8.0501	0.0501	0.0537
11	8.0400	0.0400	0.0429
12	8.0320	0.0320	0.0344
13	8.0255	0.0255	0.0275
14	8.0204	0.0204	0.0220
15	8.0163	0.0163	0.0176

where $\tilde{e}(0) = e(0)$. We are using $\tilde{e}$'s since they are only an approximation of the e's. Dynamical system (6.5) for $\tilde{e}(n)$ is a simple **linear dynamical system**. Its solution is

$$\tilde{e}(k) = (0.8)^k \tilde{e}(0).$$

In table 6.1, $\tilde{e}(0)$ through $\tilde{e}(15)$ are listed in the last column. Comparing the $\tilde{e}(k)$-values to the $e(k)$-values, you can see that it does give a good approximation.

Observe our solution for $\tilde{e}(k)$. We know that $\tilde{e}(k)$ decreases exponentially to zero since $0 < 0.8 < 1$. Since $\tilde{e}(k)$ is a good approximation for $e(k)$, we presume that $e(k)$ goes to 0 also, and so $a(k)$ goes to 8 and $\ell = 8$ is a stable equilibrium value.

Let's reconsider dynamical system (6.2)

$$a(n+1) = 1.2a(n) - 0.025a^2(n)$$

but this time, let's investigate the equilibrium value $a = 0$. Now we let

$$a(n) = 0 + e(n).$$

This is easy to substitute into the above equation. The result is

$$e(n+1) = 1.2e(n) - 0.025e^2(n).$$

Now we assume $a(0)$ is close to zero, so $e(0)$ is close to zero and so

$$0.025e^2(0) \ll e(0).$$

We therefore drop the term $0.025e^2(n)$ giving the approximating equation

$$\tilde{e}(n+1) = 1.2\tilde{e}(n).$$

The solution is

$$\tilde{e}(k) = 1.2^k \tilde{e}(0)$$

which goes to infinity. Our conclusion is that $\tilde{e}(k)$ is getting large and therefore it is getting further away from 0. The first few values of $\tilde{e}(k)$ approximate $e(k)$, so $e(k)$ is going away from 0. Since $a(k) = e(k)$, $a(k)$ is also going away from 0, so 0 is an unstable equilibrium.

An important point is this. Since $\tilde{e}(k)$ is becoming large, the term $0.025e^2(k)$ is also becoming large and at some point is no longer negligible. Thus, after the first few values, $\tilde{e}(k)$ is no longer a good approximation for $e(k)$. This can be seen in table 6.2 in which

TABLE 6.2. The $a(k) = e(k)$ values for the dynamical system $a(n+1) = 1.2a(n) - 0.025a^2(n)$, along with the approximations $\tilde{e}(k)$.

k	$a(k) = e(k)$	$\tilde{e}(k)$	k	$a(k) = e(k)$	$\tilde{e}(k)$
0	0.1000	0.1000	15	1.3371	1.5407
1	0.1197	0.1200	16	1.5598	1.8488
2	0.1433	0.1440	17	1.8109	2.2186
3	0.1715	0.1728	18	2.0911	2.6623
4	0.2051	0.2074	19	2.4001	3.1948
5	0.2450	0.2488	20	3.8337	3.8337
6	0.2925	0.2986	21	3.0961	4.6005
7	0.3489	0.3583	22	3.4757	5.5206
8	0.4156	0.4300	23	3.8688	6.6247
9	0.4944	0.5160	24	4.2684	7.9496
10	0.5872	0.6192	25	4.6666	9.5396
11	0.6960	0.7430	26	5.0555	11.4475
12	0.8231	0.8916	27	5.4276	13.7371
13	0.9708	1.0699	28	5.7767	16.4845
14	1.1414	1.2839	29	6.0978	19.7814

the $a(k) = e(k)$ values are given along with the approximating $\tilde{e}(k)$ values, with $a(0) = 0.1$ being close to zero.

Even though we cannot determine what eventually happens to $e(k)$, one thing is certain, $e(k)$, and consequently $a(k)$, does not remain close to 0 and so 0 is an unstable equilibrium.

This technique of dropping the squared term is one of the most important tools in nonlinear analysis. It is called **linearization**. The idea is to say as much as possible about a nonlinear equation using our knowledge of linear equations. The results are **local** in that we can only describe what is happening close to an equilibrium value.

PROBLEMS

1. For each of the following dynamical systems find the fixed points and use linearization to determine which are attracting and which are repelling. For those that are attracting, use cobwebs to determine an interval (c, d) for which, if a_0 is in that interval, then $a(k)$ goes to a.

   a. $a(n+1) = 1.7a(n) - 0.14a^2(n)$
   b. $a(n+1) = 0.8a(n) + 0.1a^2(n)$
   c. $a(n+1) = 3.4a(n) - 2.4a^2(n)$
   d. $a(n+1) = 0.2a(n) - 0.2a^3(n)$

2. Suppose that $r = 0.3$ and $\ell = 3$ in the logistic equation (6.1). In addition, assume that there is a constant immigration of 0.4 units of the species into the region each time period. Develop a dynamical system to model the size of the population, find the equilibrium values, and determine which equilibrium values are stable and the interval of stability.

6.2 Using linearization to study stability

In this section we will continue the study of stability for nonlinear dynamical systems. Let's start with a few examples,

EXAMPLE 6.5 Consider the dynamical system

$$a(n+1) = 0.1a^2(n) + 0.9a(n) - 0.2. \qquad (6.6)$$

To find the equilibrium values we solve the equation

$$a = 0.1a^2 + 0.9a - 0.2.$$

Multiplication by 10 and bringing all terms to the right gives

$$0 = a^2 - a - 2 \quad \text{or} \quad 0 = (a+1)(a-2),$$

so the equilibrium values are $a = -1$ and $a = 2$.

To study the stability for $a = -1$ we make the substitution

$$a(n) = e(n) - 1 \quad \text{and} \quad a(n+1) = e(n+1) - 1$$

into dynamical system (6.6) giving

$$e(n+1) - 1 = 0.1[e(n) - 1]^2 + 0.9[e(n) - 1] - 0.2.$$

After simplification, this reduces to

$$e(n+1) = 0.1e^2(n) + 0.7e(n).$$

If we pick $a(0)$ close to -1, then $e(0)$ is close to 0, and $e^2(0)$ is negligible compared to $e(0)$. Thus, the approximating dynamical system is

$$\tilde{e}(n+1) = 0.7\tilde{e}(n),$$

and its solution is $\tilde{e}(k) = 0.7^k \tilde{e}(0)$ which goes to zero. Therefore, $e(k)$ goes toward zero and $a(k)$ goes toward -1. Thus, $a = -1$ is a **stable** equilibrium value or **attracting fixed point**.

The equation $\tilde{e}(n+1) = 0.7\tilde{e}(n)$ implies that the "error" or distance from equilibrium, $e(k)$, for each time period is about 70 per cent of the previous error, that is we are about 30 per cent closer to equilibrium each time period. Another way of saying this is that

$$\frac{a(k+1) - a}{a(k) - a} \approx 0.7.$$

Note that if $a(0) = -0.9$, then

$$a(1) = -0.929 \quad \text{and} \quad a(2) = -0.949\,795\,9.$$

Then

$$\frac{a(1) - a}{a(0) - a} = \frac{-0.929 + 1}{-0.9 + 1} = 0.71 \quad \text{and} \quad \frac{a(2) - a}{a(1) - a} = 0.7071.$$

To study the stability for $a = 2$ we make the substitution

$$a(n) = e(n) + 2 \quad \text{and} \quad a(n+1) = e(n+1) + 2$$

into dynamical system (6.6) giving

$$e(n+1) + 2 = 0.1[e(n) + 2]^2 + 0.9[e(n) + 2] - 0.2.$$

After simplification, this reduces to

$$e(n+1) = 0.1e^2(n) + 1.3e(n).$$

If we pick $a(0)$ close to 2, then $e(0)$ is close to 0, and $e^2(0)$ is negligible compared to $e(0)$. Thus, the approximating dynamical system is

$$\tilde{e}(n+1) = 1.3\tilde{e}(n),$$

and its solution is $\tilde{e}(k) = 1.3^k\tilde{e}(0)$ which goes away from zero. Therefore, $e(k)$ goes away from zero and $a(k)$ goes away from 2. Thus, $a = 2$ is an **unstable** equilibrium value or **repelling fixed point**.

The equation $\tilde{e}(n+1) = 1.3\tilde{e}(n)$ implies that the "error" or distance from equilibrium, $e(k)$, for each time period is about 130 per cent of the previous error, that is we are about 30 per cent further from equilibrium each time period, at least initially. Another way of saying this is that

$$\frac{a(k+1) - a}{a(k) - a} \approx 1.3.$$

Note that if $a(0) = 1.99$, then $a(1) = 1.987\,01$ and $a(2) = 1.983\,129\,9$. Then

$$\frac{a(1) - 2}{a(0) - 2} = 1.299 \quad \text{and} \quad \frac{a(2) - a}{a(1) - a} = 1.298\,7. \qquad \blacksquare$$

Suppose we obtain an approximating equation

$$\tilde{e}(n+1) = r\tilde{e}(n)$$

for a dynamical system near an equilibrium value a, Then, if $|r| < 1$, we know that a is stable, and if $|r| > 1$ then a is unstable. But from the previous example we see that we know more. The size of r tells us how **fast** we go toward or away from equilibrium. The closer r is to 0, the faster we approach equilibrium. The larger the value of r (when greater than one), the faster we move away from equilibrium.

EXAMPLE 6.6 Consider the dynamical system

$$a(n+1) = 0.7a^2(n) + 0.3a(n) - 1.4. \qquad (6.7)$$

To find the equilibrium values we solve the equation

$$a = 0.7a^2 + 0.3a - 1.4.$$

Bringing all terms to the right and dividing by 0.7 gives

$$0 = a^2 - a - 2 = (a+1)(a-2),$$

so the equilibrium values are again $a = -1$ and $a = 2$.

To study the stability for $a = -1$ we make the substitution

$$a(n) = e(n) - 1 \quad \text{and} \quad a(n+1) = e(n+1) - 1$$

into dynamical system (6.7) giving

$$e(n+1) - 1 = 0.7[e(n) - 1]^2 + 0.3[e(n) - 1] - 1.4.$$

After simplification, this reduces to

$$e(n+1) = 0.7e^2(n) - 1.1e(n).$$

If we pick $a(0)$ close to -1, then $e(0)$ is close to 0, and $e^2(0)$ is negligible compared to $e(0)$. Thus, the approximating dynamical system is

$$\tilde{e}(n+1) = -1.1\tilde{e}(n),$$

and its solution is $\tilde{e}(k) = (-1.1)^k \tilde{e}(0)$ which oscillates away from zero. Therefore, $e(k)$ oscillates away from zero and $a(k)$ oscillates away from the **unstable** equilibrium $a = -1$. If $a(0) = -0.9$ then $a(1) = -1.103$, $a(2) = -0.879$, $a(3) = -1.123$, and $a(4) = -0.855$. Thus we see that the $a(k)$-values are getting further from -1, but that they are also oscillating in that $a(k)$ is greater than -1 when k is even and $a(k)$ is less than -1 is k is odd.

To study the stability for $a = 2$ we make the substitution

$$a(n) = e(n) + 2 \quad \text{and} \quad a(n+1) = e(n+1) + 2$$

into dynamical system (6.7) giving

$$e(n+1) + 2 = 0.7[e(n) + 2]^2 + 0.3[e(n) + 2] - 1.4.$$

After simplification, this reduces to

$$e(n+1) = 0.7e^2(n) + 3.1e(n).$$

If we pick $a(0)$ close to 2, then $e(0)$ is close to 0, and $e^2(0)$ is negligible compared to $e(0)$. Thus, the approximating dynamical system is

$$\tilde{e}(n+1) = 3.1\tilde{e}(n),$$

and its solution is $\tilde{e}(k) = 3.1^k \tilde{e}(0)$ which goes away from zero. Therefore, $e(k)$ goes away from zero and $a(k)$ goes away from 2. Thus, $a = 2$ is an **unstable** equilibrium value or **repelling fixed point**.

When one equilibrium value is stable, we know what may happen, that is, $a(k)$ goes toward that equilibrium value for nearby $a(0)$ values. But what happens when **all the equilibrium values are unstable**, as in this example? To answer this question, let's pick three $a(0)$ values and compute $a(1), a(2), \ldots$, until we can see a pattern. Our results are given in table 6.3. As you can see, some starting values reach a point in which the same two values keep repeating, that is, we have a **stable 2-cycle**. Other solutions go toward infinity, that is, they keep getting larger, without bound.   ■

When there are no stable equilibrium values, almost any type of behavior is possible. Solutions may go to stable m-cycles for some value of m, or they may wander around without a perceivable pattern. This later type of behavior is called **chaotic** behavior. The study of chaos is one of the newest and most exciting areas of mathematics. See Gleick (1987) for a good overview of chaos and the people who founded the area. For an elementary introduction to the mathematics of chaos, see Devaney (1989).

EXAMPLE 6.7   Consider the dynamical system

$$a(n+1) = a^2(n) + a(n) + 1. \tag{6.8}$$

To find the equilibrium values we solve the equation

$$a = a^2 + a + 1.$$

Canceling the a's gives

$$a^2 = -1,$$

so there are no real equilibrium values.

TABLE 6.3. Several partial solutions to the dynamical system $a(n+1) = 0.7a^2(n) + 0.3a(n) - 1.4$.

k	1st $a(k)$	2nd $a(k)$	3rd $a(k)$
0	−0.900	1.900	2.100
1	−1.103	1.697	2.317
2	−0.879	1.125	3.053
3	−1.123	−0.177	6.041
4	−0.855	−1.431	25.955
⋮	⋮	⋮	⋮
50	−0.471	−1.386	large
51	−1.386	−0.471	large
52	−0.471	−1.386	large
53	−1.386	−0.471	large

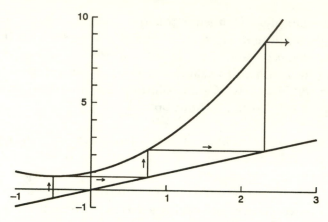

FIGURE 6.3. Cobweb diagram for the dynamical system $a(n+1) = a^2(n) + a(n) + 1$.

What happens when there are no equilibrium values? It should be clear from figure 6.3 that all solutions go toward infinity, that is, for any $a(0)$, we have

$$a(k) \to \infty. \qquad \blacksquare$$

Suppose we have a dynamical system

$$a(n+1) = ra^2(n) + ba(n) + c. \qquad (6.9)$$

such that there are no equilibrium values, that is, there are no real solutions to the equation

$$x = rx^2 + bx + c.$$

That means that the parabola

$$y = rx^2 + bx + c$$

and the line

$$y = x$$

do not intersect. If $r > 0$, then the parabola opens upward, as in figure 6.3. Since it doesn't intersect $y = x$, it must lie above $y = x$. Thus, as in figure 6.3, all solutions go to infinity.

If $r < 0$, the parabola opens downward, so the parabola must lie below $y = x$ for there to be no points of intersection. If you draw a figure similar to figure 6.3, but with the parabola opening downward, your cobwebs will show that all solutions go to negative infinity. We have thus proven the following theorem.

THEOREM 6.8   *Consider dynamical system (6.9)*

$$a(n+1) = ra^2(n) + ba(n) + c,$$

Suppose there are no (real) equilibrium values.

- *If $r > 0$ then $a(k) \to \infty$.*
- *If $r < 0$ then $a(k) \to -\infty$.*

EXAMPLE 6.9   Consider the dynamical system

$$a(n+1) = -0.2a^3(n) + 0.8a^2(n) + 0.4a(n). \tag{6.10}$$

To find the equilibrium values we solve the equation

$$a = -0.2a^3 + 0.8a^2 + 0.4a.$$

Subtracting a from both sides and factoring gives

$$0 = -0.2a^3 + 0.8a^2 - 0.6a = -0.2a(a^2 - 4a + 3) = -0.2a(a-3)(a-1),$$

so the equilibrium values are $a = 0$, $a = 1$, and $a = 3$.

To study the equilibrium value $a = 0$, we make the substitution $a(k) = e(k) + 0$ which easily gives us

$$e(n+1) = -0.2e^3(n) + 0.8e^2(n) + 0.4e(n).$$

If $a(0)$ is close to zero, then $e(0)$ is small and the terms $-0.2e^3(0)$ and $0.8e^2(0)$ are negligible by comparison and will be dropped, giving us the approximating equation

$$\bar{e}(n+1) = 0.4\bar{e}(n).$$

Since $\bar{e}(k) = 0.4^k \bar{e}(0) \to 0$, we have $e(k) \to 0$ and $a(k) \to 0$, so $a = 0$ is a stable equilibrium value.

To study the equilibrium value $a = 1$, we make the substitution $a(k) = e(k) + 1$ which gives us

$$e(n+1) + 1 = -0.2[e(n) + 1]^3 + 0.8[e(n) + 1]^2 + 0.4[e(n) + 1].$$

Multiplying out each of these terms and collecting like powers of $e(n)$ gives

$$e(n+1) = -0.2e^3(n) + 0.2e^2(n) + 1.4e(n).$$

Dropping the $e^3(n)$ and $e^2(n)$ terms for the same reason as before, gives the approximating equation

$$\bar{e}(n+1) = 1.4\bar{e}(n).$$

Since $\tilde{e}(k) = 1.4^k \tilde{e}(0) \to \infty$, we have $e(k)$ goes away from 0 and $a(k)$ goes away from $a = 1$. Therefore, $a = 1$ is an unstable equilibrium value.

Finally, we make the substitution $a(k) = e(k) + 3$ which gives us

$$e(n+1) + 3 = -0.2[e(n) + 3]^3 + 0.8[e(n) + 3]^2 + 0.4[e(n) + 3].$$

Multiplying out each of these terms and collecting like powers of $e(n)$ gives

$$e(n+1) = -0.2e^3(n) - e^2(n) - 0.2e(n).$$

Dropping the $e^3(n)$ and $e^2(n)$ gives the approximating equation

$$\tilde{e}(n+1) = -0.2\tilde{e}(n).$$

Since $\tilde{e}(k) = (-0.2)^k \tilde{e}(0) \to 0$, we have $e(k) \to 0$ and $a(k) \to 3$. Therefore, $a = 3$ is a stable equilibrium value.

A picture of the behavior of solutions is given in figure 6.4. ∎

In general, linearization requires a fair bit of algebra. Let's try to develop some tools, so that we can determine the stability of equilibrium values with less work. In particular, we want to develop results that will immediately tell us the stability of equilibrium values for the nonlinear dynamical system (6.9)

$$a(n+1) = ra^2(n) + ba(n) + c.$$

We already know the results if there are no equilibrium values. Let's now consider the simple case in which $c = 0$, that is, the dynamical

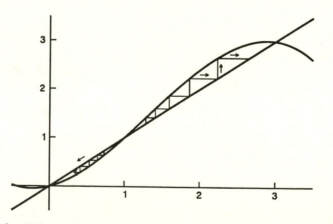

FIGURE 6.4.   Cobweb diagram for the dynamical system $a(n+1) = -0.2a^3(n) + 0.8a^2(n) + 0.4a(n)$. Notice that two of the points of intersection are stable equilibria, and the one in the middle is unstable.

system

$$a(n+1) = ra^2(n) + ba(n). \tag{6.11}$$

In this case, the fixed points are the solutions to $x = rx^2 + bx$. Subtracting x from both sides and factoring gives

$$0 = x(rx + b - 1).$$

Thus, dynamical system (6.11) has the two fixed points

$$x = 0 \quad \text{and} \quad x = \frac{1-b}{r}.$$

Making the substitution $a(k) = e(k) + 0$ in dynamical system (6.11) gives

$$e(n+1) = re^2(n) + be(n).$$

Assuming $a(0)$ is close to 0 implies that $e^2(0)$ is negligible compared to $e(0)$. Thus, we get the approximating equation

$$\tilde{e}(n+1) = b\tilde{e}(n).$$

The solution is $\tilde{e}(k) = b^k \tilde{e}(0)$ which goes to zero only if $|b| < 1$, that is, if $-1 < b < 1$. Thus, $a = 0$ is stable if $-1 < b < 1$, and $a = 0$ is unstable if $b > 1$ or if $b < -1$.

Let's now consider the equilibrium value $a = (1 - b)/r$. Making the substitution $a(k) = e(k) + (1 - b)/r$ gives

$$e(n+1) + \frac{1-b}{r} = r\left[e(n) + \frac{1-b}{r}\right]^2 + b\left[e(n) + \frac{1-b}{r}\right].$$

After simplifying (careful computation shows that all terms not involving an e cancel) and collecting like powers of $e(n)$, we get

$$e(n+1) = re^2(n) + (2-b)e(n).$$

Again, we drop the $e^2(n)$ term getting the approximating equation

$$\tilde{e}(n+1) = (2-b)\tilde{e}(n).$$

The solution is $\tilde{e}(k) = (2-b)^k \tilde{e}(0)$ which goes to zero only if

$$-1 < 2 - b < 1.$$

Subtracting 2 from both sides, dividing by -1, and reversing the inequalities gives that

$$\tilde{e}(k) \to 0 \quad \text{if} \quad 1 < b < 3.$$

Thus, $a = (1 - b)/r$ is stable if $1 < b < 3$.

We now summarize our results.

THEOREM 6.10   *Consider dynamical system (6.11)*

$$a(n+1) = ra^2(n) + ba(n).$$

The two fixed points are

$$a = 0 \qquad and \qquad a = \frac{1-b}{r}.$$

- *If $-1 < b < 1$, then $a = 0$ is stable and $a = (1 - b)/r$ is unstable.*
- *If $1 < b < 3$, then $a = (1 - b)/r$ is stable and $a = 0$ is unstable.*
- *If $b < -1$ or $b > 3$, then $a = 0$ and $a = (1 - b)/r$ are both unstable.*

It is interesting that r, the coefficient of the $a^2(n)$, plays no part in determining the stability of the equilibrium values for dynamical system (6.11).

EXAMPLE 6.11   Consider the dynamical system

$$a(n+1) = -0.5a^2(n) + 2.5a(n).$$

Applying theorem 6.10 with $r = -0.5$ and $b = 2.5$, we see that the equilibrium value $a = 0$ is unstable and the equilibrium value $a = (1 - b)/r = 3$ is stable.   ∎

We now apply theorem 6.10 to the logistic equation (6.1) for population growth

$$a(n+1) = (1+r)a(n) - ba^2(n) = -ba^2(n) + (1+r)a(n)$$

where $b = r/\ell$. Recall that r is the unrestricted growth rate and ℓ is the carrying capacity of the environment. The equilibrium values are $a = 0$ and $a = \ell$. By theorem 6.10, $a = \ell$ is stable if $1 < 1 + r < 3$, that is, if the growth rate is $0 < r < 2$. We thus see that if the growth rate is positive, $r > 0$, but not too large, $r < 2$, the population will stabilize at the carrying capacity, that is,

$$a(k) \to \ell.$$

Now let's see if we can develop results for the more general dynamical system (6.9)

$$a(n+1) = ra^2(n) + ba(n) + c.$$

Suppose it has an equilibrium value a. Then we make the substitution $a(k) = e(k) + a$, giving

$$e(n+1) + a = r[e(n) + a]^2 + b[e(n) + a] + c$$

or

$$e(n+1) + a = re^2(n) + (2ra + b)e(n) + ra^2 + ba + c.$$

Since a is an equilibrium value

$$a = ra^2 + ba + c,$$

so these terms must cancel giving us

$$e(n+1) = re^2(n) + (2ra + b)e(n).$$

Dropping the $e^2(n)$ terms, we get the approximating solution

$$\tilde{e}(n+1) = (2ra + b)\tilde{e}(n).$$

The solution is $\tilde{e}(k) = (2ra + b)^k \tilde{e}(0)$. Thus if

$$-1 < (2ra + b) < 1 \quad \text{then} \quad a \text{ is a stable equilibrium.}$$

We summarize our results with the following theorem.

THEOREM 6.12   *Consider dynamical system (6.9)*

$$a(n+1) = ra^2(n) + ba(n) + c.$$

If this equation has an equilibrium value a, then

- *a is stable if $-1 < 2ra + b < 1$ and*
- *a is unstable if $2ra + b < -1$ or $2ra + b > 1$.*

If there are no equilibrium values then

- *$a(k)$ goes to infinity if $r > 0$ and*
- *$a(k)$ goes to negative infinity if $r < 0$.*

It appears that c has no effect on the stability of the equilibrium values. This is not completely true, since the equilibrium value a depends on c, and a helps determine the stability.

EXAMPLE 6.13   Consider the dynamical system

$$a(n+1) = 0.1a^2(n) + 1.1a(n) - 0.6.$$

Simplifying the equation $x = 0.1x^2 + 1.1x - 0.6$ gives

$$0 = x^2 + x - 6 = (x - 2)(x + 3),$$

so the two fixed points are $a = 2$ and $a = -3$. Also, $b = 1.1$ and $r = 0.1$. Thus, using $a = 2$, we get that

$$2ra + b = 2 \times 0.1 \times 2 + 1.1 = 1.5$$

so $a = 2$ is unstable. Using $a = -3$, we get that

$$2ra + b = 2 \times 0.1 \times (-3) + 1.1 = 0.5$$

so $a = -3$ is stable.   ∎

PROBLEMS

1. The dynamical system

   $$a(n+1) = 3a(n) - a^2(n) + 3$$

   has the two equilibrium values $a = -1$ and $a = 3$.
   a. Show using linearization that $a = -1$ is unstable.
   b. Show using linearization that $a = 3$ is unstable.
   c. Check your results using theorem 6.12.

2. The dynamical system

   $$a(n+1) = 1.4a(n) - 0.2a^2(n) + 3$$

   has the two equilibrium values $a = -3$ and $a = 5$. Use linearization to find their stability. Check your answer using theorem 6.12.

3. The dynamical system

   $$a(n+1) = a^2(n) - a(n) + 1$$

   has one equilibrium value $a = 1$.
   a. Show that no conclusion about the stability of $a = 1$ can be made using linearization.
   b. Show that no conclusion about the stability of $a = 1$ can be made using theorem 6.12.
   c. Sketch a cobweb for this dynamical system and see that if $a(0) < 1$ (but close to 1), then $a(k)$ goes **toward** 1, but if $a(0) > 1$ (but close to 1), then $a(k)$ goes **away** from 1. Such an equilibrium value is called **semistable**.

4. The dynamical system

   $$a(n+1) = -2a^2(n) + 9a(n) - 8$$

has one equilibrium value $a = 2$.
  a. Show that no conclusion about the stability of $a = 2$ can be made using linearization.
  b. Show that no conclusion about the stability of $a = 2$ can be made using theorem 6.12.
  c. Sketch a cobweb for this dynamical system and see that if $a(0) > 2$ (but close to 2), then $a(k)$ goes **toward** 2, but if $a(0) < 2$ (but close to 2), then $a(k)$ goes **away** from 2. Such an equilibrium value is called **semistable**.

5. The dynamical system

$$a(n+1) = a^3(n) - a^2(n) + 1$$

has the two equilibrium values $a = -1$ and $a = 1$.
  a. Show using linearization that $a = -1$ is unstable.
  b. Show that no conclusion can be made for the fixed point $a = 1$ using linearization.
  c. Carefully draw a cobweb using the curve

$$f(x) = x^3 - x^2 + 1$$

and show that $a = 1$ is semistable.

6. Consider the dynamical system

$$a(n+1) = a^3(n) + (1 - r^2)a(n)$$

where r is some positive number.
  a. Check that the 3 equilibrium values are $a = 0$, $a = r$, and $a = -r$.
  b. For which values of r, if any, is the equilibrium value $a = 0$ stable?
  c. For which values of r, if any, is the equilibrium value $a = r$ stable?
  d. For which values of r, if any, is the equilibrium value $a = -r$ stable?

6.3 Harvesting strategies

Consider a population of some species of animal that is harvested or hunted, say deer. For convenience, let's assume that the carrying capacity is $\ell = 1$. For example, if the carrying capacity is really 10 000, then 10 000 of the species is considered to be one unit. Thus, $a(k) = 0.5$ means that at time k, there are 5000 of the species.

Remember that the growth rate for our species is given as $r[1 - a(n)/\ell] = r[1 - a(n)]$, where r is the unrestricted growth rate, and the size of the units are chosen so that one unit equals the carrying capacity of the environment. For ease of explanation, let's assume that one unit equals $10\,000$ deer. Let's also assume we have determined that the unrestricted growth rate is $r = 0.8$. Since the change in population, $a(n+1) - a(n)$, equals the growth rate times the size of the population, our dynamical system becomes

$$a(n+1) - a(n) = 0.8(1 - a(n))a(n),$$

or, after simplification,

$$a(n+1) = 1.8a(n) - 0.8a^2(n).$$

In this section, we will use the theory of nonlinear dynamics to study two different harvesting or hunting strategies for this population.

6.3.1   Fixed harvest

Suppose that in our forest we allow hunters to kill h units of deer per season, where h may be a fraction. Then the dynamical system that models growth becomes

$$a(n+1) = 1.8a(n) - 0.8a^2(n) - h,$$

since $a(n+1)$ is reduced by the h deer that were killed. We are assuming that all the deer are killed at the end of the time period, for otherwise, the killing of the deer would affect the number of births and deaths, and consequently, the growth rate. But for simplicity, we will ignore this problem.

EXAMPLE 6.14 Let $h = 0.072$, that is, 720 deer are killed each year when one unit equals $10\,000$ deer. Then our dynamical system becomes

$$a(n+1) = 1.8a(n) - 0.8a^2(n) - 0.072. \tag{6.12}$$

The fixed points are the solutions to the equation

$$a = 1.8a - 0.8a^2 - 0.072,$$

that is, the fixed points are $a = 0.9$ and $a = 0.1$.

To check the stability of the fixed points, we apply theorem 6.12. Recall that a fixed point a for the dynamical system (6.9)

$$a(n+1) = ra^2(n) + ba(n) + c$$

is stable if $|2ra + b| < 1$. For dynamical system (6.12), $r = -0.8$ and $b = 1.8$. For the equilibrium value $a = 0.9$, we find that $2ra + b = 0.36 < 1$ so $a = 0.9$ is **stable**. Likewise, for $a = 0.1$, we find that $2ra + b = 1.64 > 1$ so this equilibrium value is **unstable**.

In figure 6.5, cobwebs are drawn for this dynamical system using two different values of a_0. Note that when the a_0 values are close to 0.9, the $a(k)$ values go to 0.9, while if a_0 is close to 0.1, the $a(k)$ values go away from 0.1. In fact they go towards extinction if $a_0 < 0.1$ and towards 0.9 if $a_0 > 0.1$. What we have shown is that when we kill 720 deer each year, $a = 0.9$ is a stable equilibrium value, that is, the population of deer will stabilize at 9000. The increased growth rate at this level (there is more food available since there are fewer deer) makes up for the deer being killed. Being stable says that even if a minor catastrophe happened and the population of deer dropped to some lower level, say 2000, or $a_0 = 0.2$, eventually the population would grow back to 9000. The unstable equilibrium value $a = 0.1$ tells how much of a disaster we can deal with, that is, if the disaster caused the population to go below 1000, the deer would die out instead of going back to 9000. ∎

EXAMPLE 6.15 We now let $h = 0.24$, that is, the hunters kill 2400 deer each year. Thus the growth equation is

$$a(n+1) = 1.8a(n) - 0.8a^2(n) - 0.24.$$

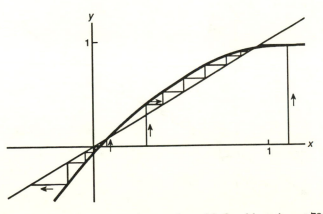

FIGURE 6.5. Cobweb for the logistic equation with fixed hunting at 720 deer per year.

The fixed points are the solutions to $a = 1.8a - 0.8a^2 - 0.24$, and after solving using the quadratic formula, we obtain

$$a = 0.5 \pm \sqrt{-0.05}.$$

These roots are complex, and thus there are no **real** fixed points. By theorem 6.8, the numbers $a(k)$ go to negative infinity for all initial values, since -0.8, the coefficient of $a^2(n)$, is negative. This can be seen in the cobweb of figure 6.6. This implies that we are killing too many deer. Either the deer will be exterminated or we will have to reduce the amount of hunting. ■

What we have seen is that, for two different hunting schemes, that is, for $h = 0.072$ and $h = 0.24$, we had a different number of fixed points: 2 and 0, respectively.

We would like to predict what happens for **any** fixed harvesting plan, that is, for any value of h. The equation to be considered is the dynamical system

$$a(n+1) = 1.8a(n) - 0.8a^2(n) - h. \tag{6.13}$$

We now assume that h is some fixed constant. The fixed points are the solutions to the equation

$$a = 1.8a - 0.8a^2 - h,$$

where h is a constant and a is the as yet unknown fixed point. After simplifying then dividing by 0.8, the equation becomes

$$a^2 - a + 1.25h = 0.$$

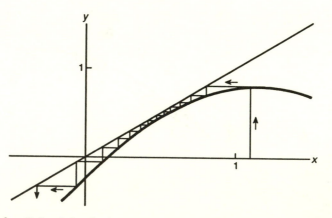

FIGURE 6.6. Cobweb for the logistic equation with fixed hunting at 2400 deer per year.

Using the quadratic formula, the solutions are

$$a = \frac{1 \pm \sqrt{1 - 5h}}{2}.$$

We have three cases. The first case is when $1 - 5h < 0$, that is, $h > 0.2$. When $h > 0.2$, there are **no equilibrium values**. Since the coefficient of the $a^2(n)$ term is negative in dynamical system (6.13), then theorem 6.8 implies that $a(k)$ goes to negative infinity.

The second case is when $1 - 5h > 0$, that is, when $h < 0.2$. In this case, we have the two equilibrium values

$$a = \frac{1 \pm \sqrt{1 - 5h}}{2}.$$

Let's examine the stability of these using theorem 6.12. To apply this theorem, we need to compute $2ra + b$, where $r = -0.8$ and $b = 1.8$ in dynamical system (6.10).

For the equilibrium value $a = [1 - \sqrt{1 - 5h}]/2$, we get that

$$2ra + b = -0.8[1 - \sqrt{1 - 5h}] + 1.8 = 1 + 0.8\sqrt{1 - 5h} > 1$$

when $h < 0.2$. Thus, this equilibrium value is unstable.

For the equilibrium value $a = [1 + \sqrt{1 - 5h}]/2$, we get that

$$2ra + b = -0.8[1 + \sqrt{1 - 5h}] + 1.8 = 1 - 0.8\sqrt{1 - 5h} < 1$$

when $h < 0.2$. Thus, this equilibrium value is stable for h-values in which

$$1 - 0.8\sqrt{1 - 5h} > -1.$$

To solve this inequality, subtract 1 from both sides, multiply both sides by $-5/4$ (making sure you reverse the inequality sign), and square both sides, giving

$$25/4 > 1 - 5h, \quad \text{or} \quad h > -1.05.$$

In conclusion, if $-1.05 < h < 0.2$, then the fixed point $a = 0.5(1 + \sqrt{1 - 5h})$ is attracting. Thus, for hunting strategies in which less than 2000 deer are removed each year, the population will stabilize. Any $h < 0.2$ is a **sustainable harvesting strategy**.

Note that $h < 0$ means we are adding more of the species to the environment instead of taking some of the species out of the environment. This is equivalent to stocking lakes or immigration. For $h < -1.05$ many different types of behaviors can occur. For h slightly below -1.05, say $h = -1.2$, solutions either go to negative infinity or to the stable 2-cycle consisting of the two values 2.183 and 1.317. The particular behavior that occurs depends on $a(0)$. For $h < -1.2$ other types of behavior can occur, such as stable 4-cycles, 8-cycles, ..., and even what mathematicians call **chaos**. For more discussion on this topic see Gleick (1987) and Devaney (1989). You should experiment with several different h values.

The last case is when $h = 0.2$. In this case the dynamical system which models population growth becomes

$$a(n+1) = 1.8a(n) - 0.8a^2(n) - 0.2.$$

The equilibrium values satisfy

$$a = 1.8a - 0.8a^2 - 0.2, \qquad \text{or} \qquad a^2 - a + 0.25 = 0.$$

This factors into $(a-0.5)^2 = 0$, and there is only one (multiple) root,

$$a = 0.5.$$

In this case, $2ra + b = 1$ so theorem 6.12 does not apply. In particular, if we make the substitution $a(k) = e(k) + 0.5$ we get the dynamical system

$$e(n+1) = e(n) - 0.8e^2(n).$$

Dropping the $e^2(n)$ term gives the approximating equation

$$\tilde{e}(n+1) = \tilde{e}(n).$$

This says that the error term remains approximately the same or doesn't change by much. But it is still changing and the approximating equation does not tell us if it is slowly getting smaller or slowly getting bigger. In other words, **linearization fails to work in this case**.

In figure 6.7, it can be seen that $a = 0.5$ is semistable in the sense that if $a_0 > 0.5$ then $a(k)$ goes to 0.5, but if $a_0 < 0.5$ then $a(k)$ goes to negative infinity.

This implies that when we kill 2000 deer a year, the population of deer will drop to 5000. But this population is very fragile and if a minor catastrophe struck, dropping the population below 5000, say 4999, then the population would die out or we would have to adjust our hunting habits until the population recovered.

What we have shown is that the **maximum sustainable harvest** is $h = 0.2$. For a harvest above that, the population dies out. This is a very tenuous position, because at $h = 0.2$, the equilibrium value $a = 0.5$ is unstable from below. To be safe, we might restrict h to some number slightly below 0.2. Even this would be risky, since the stable

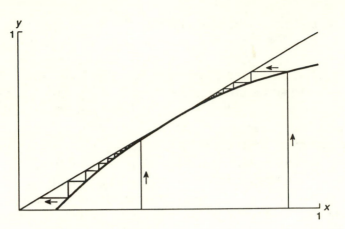

FIGURE 6.7. Cobweb for the logistic equation with fixed hunting at 2000 deer per year.

and unstable equilibrium values will be close together and a small catastrophe could cause the population to drop below the unstable equilibrium, causing the population to start to die out.

6.3.2 Proportional harvest

Because of the unsatisfactory solution to our fixed harvesting problem, we might try an alternative strategy, that is, to hunt or harvest, not a fixed number of the population, but **a fixed proportion of the population**. Let p represent the proportion of the population that will be removed. Then the total number removed, or harvested, will be $pa(n)$ and the dynamical system becomes

$$a(n+1) = 1.8a(n) - 0.8a^2(n) - pa(n),$$

or, after simplification,

$$a(n+1) = (1.8 - p)a(n) - 0.8a^2(n). \qquad (6.14)$$

The equilibrium values, that is, the solutions to

$$a = (1.8 - p)a - 0.8a^2,$$

are $a = 0$ and $a = 1 - 1.25p$.

Since there is no constant term, dynamical system (6.14) satisfies the conditions of theorem 6.10. Thus $a = 0$ is stable if, $1.8 - p$, the coefficient of the $a(n)$ term is between -1 and 1, and $a = 1 - 1.25p$ is stable if $1 < 1.8 - p < 3$.

Since $-1 < 1.8 - p < 1$ implies that $2.8 > p > 0.8$, then $a = 0$ **is stable if** $0.8 < p < 2.8$ **and is unstable if** $p < 0.8$ **or** $p > 2.8$. If you draw a cobweb, you will see that $a = 0$ is semistable when $p = 0.8$.

Solving $1 < 1.8 - p < 3$, we find that $a = 1 - 1.25p$ **is stable if** $-1.2 < p < 0.8$ **and is unstable if** $p < -1.2$ or $p > 0.8$. When $p = 0.8$, then we have $a = 1 - 1.25(0.8) = 0$, which is semistable as discussed in the previous paragraph.

Note that there are two equilibrium values when $p \neq 0.8$ and there is one semistable equilibrium value when $p = 0.8$.

Since p is the proportion of the species harvested, we will restrict our attention to $0 < p < 1$. We are interested in harvesting strategies that have a positive stable equilibrium value for the species. We have found that p values in the interval $0 < p < 0.8$ give us the positive stable equilibrium value of $a = 1 - 1.25p$. Our harvest h depends on our strategy p. In particular, the **harvest** is $pa(k)$, that is,

$$h = pa(k).$$

Since $a = 1 - 1.25p$ is a stable equilibrium value, $a(k)$ goes to a. This means that if we harvest the same proportion p of the species every year, the species population will stabilize at $a = 1 - 1.25p$ and our harvest will stabilize at

$$h = pa = p(1 - 1.25p) = p - 1.25p^2.$$

What value of p gives us our maximum harvest? In figure 6.8 is the graph of the harvest versus the proportion harvested. This is a parabola. The maximum harvest occurs at the vertex of this parabola which is the point

$$(p, h) = (0.4, 0.2).$$

Thus the maximum harvest occurs when we harvest $p = 0.4$ or 40 per cent of the species each time period, and that maximum harvest is $h = 0.2$ or 2000 deer. Notice that this harvest is stable since $p = 0.4 < 0.8$, so that $a = 1 - 1.25 \times 0.4 = 0.5$ is stable.

6.3.3 Comparison of harvesting strategies

We have two harvesting strategies. The first is to harvest a fixed amount of the species in each time period. The second is to harvest a fixed proportion of the species in each time period. Notice that the maximum sustainable harvest of 2000 of the species is the same for

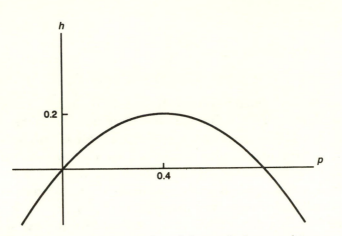

FIGURE 6.8. A graph of the proportion of the species harvested p versus the total amount of harvest h, that is, the graph $h = p - 1.25p^2$.

both policies. But this harvest is stable using a proportional harvest while it is semistable using a fixed harvest.

What is happening in the second strategy is that, when p is slightly larger than 0.4 or when the population drops because of outside influences, the actual harvest $h = pa(k)$ also drops, which allows the population to recover.

It should be clear that the fixed proportion strategy is the superior harvesting strategy, but will be harder to implement. Normally, we can't count the number of the species in order to compute our kill $pa(k)$. There are statistical methods for estimating the number $a(k)$.

An alternative might be to use a constant **effort** in our harvesting. For example, if we hunt for three days each week, it might be argued that our kill will be proportional to $a(k)$. Notice that if we harvest a fixed number of the species, as the population drops, we will have to work harder to find the same number as before.

Another observation is that **the maximum harvest occurs when we harvest 40 per cent of the species.** In real life, people might try increasing their hunting efforts in order to increase their catch, say in the fishing industry. But we have shown that, over the long term, they will reduce the population, and their catch will drop. We are faced with the environmental reality that **if we increase our effort in harvesting, we will decrease our harvest.**

PROBLEMS

1. Consider the equation

$$a(n+1) = 5a(n) - a^2(n) - h.$$

   a. Find the fixed points for this equation in terms of h. Notice that (1) if $h < 4$ there are two fixed points, (2) if $h = 4$ there is one fixed point, and (3) if $h > 4$ there are no fixed points.
   b. Determine for what values of h each of the two equilibrium values are attracting and repelling given that $h < 4$.
   c. Use theorem 6.8 to determine whether $a(k)$ goes to positive infinity or negative infinity when $h > 4$.
   d. Sketch a cobweb when $h = 4$ to determine the stability of the one equilibrium value, $a = 2$.

2. Consider the equation

$$a(n+1) = -pa(n) + a^3(n).$$

   The fixed points for this dynamical system are $a = 0$, $a = \sqrt{1+p}$, and $a = -\sqrt{1+p}$.
   a. Using linearization, find conditions on p which imply stability for $a = 0$.
   b. Sketch a cobweb for this equation to determine the stability of $a = 0$ when $p = -1$.
   c. For what p values, with $p > -1$, are $a = \pm\sqrt{1+p}$ stable? Use linearization.

3. Consider the dynamical system

$$a(n+1) - a(n) = 0.5(1 - 0.5a(n))a(n) - pa(n),$$

   which corresponds to a fixed proportion harvesting policy for a population in which the unrestricted growth rate is $r = 0.5$ and the carrying capacity is $\ell = 2$.
   a. Find the nonzero fixed point for this equation, and determine for what values of p it is stable.
   b. Find the maximum sustainable harvest by graphing an appropriate parabola $h = f(p)$ for this equation and note that it is stable.

CHAPTER 7

Vectors and matrices

7.1   Introduction to vectors and matrices

In section 3.4 we had bags containing differently colored marbles. A marble was drawn from one bag and its color determined the bag in which the next marble was drawn. For example, suppose we have a red bag and a blue bag. The red bag contains 6 red and 4 blue marbles while the blue bag contains 7 blue and 3 red marbles. A marble is drawn from a bag, its color is recorded and the marble is then returned to the bag from which it was drawn. The next marble is drawn from the bag that is the same color as the last marble drawn. We let $r(n)$ and $b(n)$ be the probabilities that the nth marble drawn is red and blue, respectively.

This was an example of a Markov chain. We computed the probabilities using the multiplication principle. To compute $r(n+1)$, we noted that there were 2 ways that the $n+1$th marble could be red and that was for the nth and $n+1$th to be red, or for the nth to be blue and the $n+1$th to be red. By the multiplication principle, the probability that both are red is the probability that the nth is red, $r(n)$, times the probability the $n+1$th is red **given** that the nth is red, 0.6 in this case. Similarly, the probability the nth is blue and then the $n+1$th is red is $0.3b(n)$. Adding these probabilities gives that $r(n+1) = 0.6r(n) + 0.3b(n)$. Similarly, $b(n+1) = 0.4r(n) + 0.7b(n)$.

This gives us the dynamical system of 2 equations

$$r(n+1) = 0.6r(n) + 0.3b(n) \tag{7.1}$$

$$b(n+1) = 0.4r(n) + 0.7b(n) \tag{7.2}$$

The purpose of this chapter is to develop a shorthand notation for displaying systems of equations such as dynamical system (7.1) and (7.2). In particular, we rewrite (7.1) and (7.2) as

$$\begin{pmatrix} r(n+1) \\ b(n+1) \end{pmatrix} = \begin{pmatrix} 0.6 & 0.3 \\ 0.4 & 0.7 \end{pmatrix} \begin{pmatrix} r(n) \\ b(n) \end{pmatrix}.$$

We then denote

$$P(n) = \begin{pmatrix} r(n) \\ b(n) \end{pmatrix} \quad \text{and} \quad R = \begin{pmatrix} 0.6 & 0.3 \\ 0.4 & 0.7 \end{pmatrix},$$

and rewrite dynamical system (7.1) and (7.2) as

$$P(n+1) = RP(n). \tag{7.3}$$

In our notation, $P(n)$ is called a **vector** and R is called a **matrix**.

If we are given $r(0)$ and $b(0)$, that is, we are given $P(0)$, then we can formally compute $P(1) = RP(0)$. Since the order of multiplication is important when dealing with matrices, we carefully compute that

$$P(2) = RP(1) = R[RP(0)] = (RR)P(0) = R^2P(0),$$

and we see that for dynamical systems of several equations we have the same result as we did for first-order linear dynamical systems of a single equation, that is, the solution is

$$P(k) = R^k P(0).$$

The problem is, what do we mean by $RP(n)$ when R is an array of numbers and $P(n)$ is a column of numbers. Also, what do we mean by $R \times R$ when R is an array of numbers.

A **matrix** is an array of ℓ rows of numbers, containing m numbers in each row. Such an array is called an ℓ-by-m matrix. The above R is a 2-by-2 matrix. As another example,

$$R = \begin{pmatrix} 1 & 2 & 3 & 4 \\ 5 & 6 & 7 & 8 \end{pmatrix}$$

is a 2-by-4 matrix. The number ℓ, 2 in this case, is the number of **rows** in the matrix, while the number m, 4 in this case, is the number of **columns** in the matrix. We can identify each number in the matrix by first giving the row the number is in and then giving the column the number is in. Thus, the number in row 2, column 3 is the number 7. We give a shorthand notation for this as $r_{23} = 7$. Thus, r_{jk} represents the number in row j and column k. As another example, $r_{13} = 3$.

Using this notation, we represent an arbitrary ℓ-by-m matrix as

$$
R = \begin{pmatrix}
r_{11} & r_{12} & \cdots & r_{1m} \\
r_{21} & r_{22} & \cdots & r_{2m} \\
\vdots & \vdots & \ddots & \vdots \\
r_{\ell 1} & r_{\ell 2} & \cdots & r_{\ell m}
\end{pmatrix}
$$

Matrices that have exactly one column are used quite frequently and, as we will see, have a specialized place in applications. For this reason, we give such matrices a special name. A matrix with one column is called a **vector**. For example,

$$
A = \begin{pmatrix}
1 \\
3 \\
5
\end{pmatrix}
$$

is a vector (or a 3-by-1 matrix). Since a vector has only one column, we can identify a number in a vector just by specifying the row in which the number is found. For example, the number in the second row of the vector A is $a_2 = 3$. Thus, a_j represents the number in the jth row of a vector. If a vector contains exactly ℓ rows, we say that this is a vector in ℓ-space or an ℓ-vector or an ℓ-dimensional vector. The vector A above is a 3-vector while the vector $P(n)$ in dynamical system (7.3) is a 2-vector.

Matrices with exactly one row are called row vectors. In this case, we would have to differentiate between row vectors and column vectors. To avoid this problem, we will use only column vectors.

To simplify our notation, we tend to let the letters a, b, c, r, s, x, y, and so forth, represent real (or, rarely, imaginary) numbers; the capital letters A, B, C, P, Q, X, and Y represent vectors; and the capital letters R, S, and M represent matrices (with more then one column). If numbers depend on n, we will denote them as $a(n)$, $b(n)$, etc.; and if vectors depend on n, we will denote them as $A(n)$, $B(n)$, etc.

We now define our new **matrix algebra** rules. We start with the rule for adding vectors. Similar to the idea that you cannot add apples and oranges, two vectors can be added only if they have the same number of rows. In that case, you add the numbers in the corresponding rows. For example, if

$$A = \begin{pmatrix} 2 \\ 3 \\ 2 \end{pmatrix} \qquad \text{and} \qquad B = \begin{pmatrix} 4 \\ 5 \\ 0 \end{pmatrix}$$

then

$$A + B = \begin{pmatrix} 2 \\ 3 \\ 2 \end{pmatrix} + \begin{pmatrix} 4 \\ 5 \\ 0 \end{pmatrix} = \begin{pmatrix} 2+4 \\ 3+5 \\ 2+0 \end{pmatrix} = \begin{pmatrix} 6 \\ 8 \\ 2 \end{pmatrix}.$$

We define the multiplication of a vector by a number or scalar as the product of each row of the vector by the number. For example, using the vectors A and B given above, we have

$$3A - 2B = 3 \begin{pmatrix} 2 \\ 3 \\ 2 \end{pmatrix} - 2 \begin{pmatrix} 4 \\ 5 \\ 0 \end{pmatrix}$$

$$= \begin{pmatrix} 3 \times 2 \\ 3 \times 3 \\ 3 \times 2 \end{pmatrix} + \begin{pmatrix} -2 \times 4 \\ -2 \times 5 \\ -2 \times 0 \end{pmatrix} = \begin{pmatrix} -2 \\ -1 \\ 6 \end{pmatrix}.$$

Notice that the sum of two ℓ-dimensional vectors results in a new ℓ-dimensional vector.

In a similar manner, we define the addition of two ℓ-by-m matrices as the sum of the corresponding terms, as in

$$\begin{pmatrix} 2 & 3 & 4 \\ 4 & 1 & 9 \end{pmatrix} + \begin{pmatrix} 1 & 5 & 2 \\ 3 & 3 & 0 \end{pmatrix} = \begin{pmatrix} 2+1 & 3+5 & 4+2 \\ 4+3 & 1+3 & 9+0 \end{pmatrix}$$

$$= \begin{pmatrix} 3 & 8 & 6 \\ 7 & 4 & 9 \end{pmatrix}.$$

Likewise, the product of a matrix and a scalar is the product of each term of the matrix by the scalar, as in

$$4 \begin{pmatrix} 2 & 3 & 4 \\ 3 & 0 & 8 \end{pmatrix} = \begin{pmatrix} 4 \times 2 & 4 \times 3 & 4 \times 4 \\ 4 \times 3 & 4 \times 0 & 4 \times 8 \end{pmatrix} = \begin{pmatrix} 8 & 12 & 16 \\ 12 & 0 & 32 \end{pmatrix}.$$

We now define the multiplication of a matrix times a vector. Looking back at dynamical system (7.3), we see that we use the first row of the matrix and the vector $P(n)$ to get the first number in the new vector $P(n+1)$. Similarly, we use the second row of the matrix and the vector to get the second number in the new vector.

Let's consider this idea more formally. To multiply a matrix and a vector, the matrix must have the same number of columns as the vector has rows; that is, to multiply an ℓ-by-m matrix M and a k-dimensional vector A we must have that $k = m$. Consider the 2-by-3 matrix M and the 3-dimensional vector A defined by

$$M = \begin{pmatrix} -2 & 3 & 4 \\ 1 & 7 & -5 \end{pmatrix} \quad \text{and} \quad A = \begin{pmatrix} 11 \\ -4 \\ 3 \end{pmatrix}.$$

The product MA will result in a 2-dimensional vector B, that is, the product of an ℓ-by-m matrix and an m-dimensional vector is an ℓ-dimensional vector. To get b_1, the first number of the vector B, we multiply the numbers in the first row of M by the corresponding numbers of the vector A, and then add the result; that is,

$$b_1 = r_{11} a_1 + r_{12} a_2 + r_{13} a_3 = -2 \times 11 + 3 \times (-4) + 4 \times 3 = -22.$$

Likewise,

$$b_2 = r_{21} a_1 + r_{22} a_2 + r_{23} a_3 = 1 \times 11 + 7 \times (-4) - 5 \times 3 = -32.$$

Thus,

$$MA = \begin{pmatrix} -2 & 3 & 4 \\ 1 & 7 & -5 \end{pmatrix} \begin{pmatrix} 11 \\ -4 \\ 3 \end{pmatrix}$$

$$= \begin{pmatrix} -2 \times 11 + 3 \times (-4) + 4 \times 3 \\ 1 \times 11 + 7 \times (-4) - 5 \times 3 \end{pmatrix}$$

$$= \begin{pmatrix} -22 \\ -32 \end{pmatrix}.$$

One purpose for this method of multiplication is that we can now write systems of equations in a shorthand form. Consider the equations

$$2x + 3y - z = 9$$
$$x - 2y + 2z = -5$$
$$x + y + 3z = 0.$$

By letting

$$M = \begin{pmatrix} 2 & 3 & -1 \\ 1 & -2 & 2 \\ 1 & 1 & 3 \end{pmatrix}, \quad X = \begin{pmatrix} x \\ y \\ z \end{pmatrix}, \quad \text{and} \quad A = \begin{pmatrix} 9 \\ -5 \\ 0 \end{pmatrix},$$

we can write the above three equations as

$$MX = A.$$

We can also give a method for multiplying two matrices together. In order to multiply a matrix times a vector, we need that the number of columns of the matrix equals the number of rows of the vector. Similarly, when multiplying two matrices, we need that the number of columns of the first matrix must be the same as the number of rows of the second matrix.

Suppose we wish to compute the product of two 3-by-3 matrices, R and S, say

$$RS = \begin{pmatrix} 2 & -4 & 3 \\ -3 & 0 & 1 \\ 7 & 2 & 10 \end{pmatrix} \begin{pmatrix} 5 & 6 & -2 \\ -5 & 2 & 1 \\ 1 & 0 & 2 \end{pmatrix}.$$

We will get a new matrix that has the same number of rows as R and the same number of columns as S. To get the **first** column of the new matrix, multiply the matrix R times the **first** column of S, as if it were a vector. Thus the first column of the new matrix is

$$\begin{pmatrix} 2 & -4 & 3 \\ -3 & 0 & 1 \\ 7 & 2 & 10 \end{pmatrix} \begin{pmatrix} 5 \\ -5 \\ 1 \end{pmatrix} = \begin{pmatrix} 2 \times 5 - 4 \times (-5) + 3 \times 1 \\ -3 \times 5 + 0 \times (-5) + 1 \times 1 \\ 7 \times 5 + 2 \times (-5) + 10 \times 1 \end{pmatrix}$$

$$= \begin{pmatrix} 33 \\ -14 \\ 35 \end{pmatrix}.$$

To get the second column of the new matrix, multiply R times the second column of S, as if it were a vector, giving

$$\begin{pmatrix} 2 & -4 & 3 \\ -3 & 0 & 1 \\ 7 & 2 & 10 \end{pmatrix} \begin{pmatrix} 6 \\ 2 \\ 0 \end{pmatrix} = \begin{pmatrix} 4 \\ -18 \\ 46 \end{pmatrix}.$$

Likewise, the third column of the new matrix is

$$\begin{pmatrix} 2 & -4 & 3 \\ -3 & 0 & 1 \\ 7 & 2 & 10 \end{pmatrix} \begin{pmatrix} -2 \\ 1 \\ 2 \end{pmatrix} = \begin{pmatrix} -2 \\ 8 \\ 8 \end{pmatrix}.$$

Thus we get that

$$RS = \begin{pmatrix} 2 & -4 & 3 \\ -3 & 0 & 1 \\ 7 & 2 & 10 \end{pmatrix} \begin{pmatrix} 5 & 6 & -2 \\ -5 & 2 & 1 \\ 1 & 0 & 2 \end{pmatrix} = \begin{pmatrix} 33 & 4 & -2 \\ -14 & -18 & 8 \\ 35 & 46 & 8 \end{pmatrix}.$$

Let us consider an example in which R is a 2-by-3 matrix and S is a 3-by-2 matrix. The product RS is a matrix with 2 rows (the number of rows of R) and 2 columns (the number of columns of S). Similarly, SR is a 3-by-3 matrix. For example, let

$$R = \begin{pmatrix} 2 & 6 & -2 \\ 4 & -1 & 0 \end{pmatrix} \quad \text{and} \quad S = \begin{pmatrix} 2 & -2 \\ 3 & 6 \\ 0 & 1 \end{pmatrix}.$$

Then

$$RS = \begin{pmatrix} 2 & 6 & -2 \\ 4 & -1 & 0 \end{pmatrix} \begin{pmatrix} 2 & -2 \\ 3 & 6 \\ 0 & 1 \end{pmatrix}$$

$$= \begin{pmatrix} 2 \times 2 + 6 \times 3 - 2 \times 0 & 2 \times (-2) + 6 \times 6 - 2 \times 1 \\ 4 \times 2 - 1 \times 3 + 0 \times 0 & 4 \times (-2) - 1 \times 6 + 0 \times 1 \end{pmatrix}$$

$$= \begin{pmatrix} 22 & 30 \\ 5 & -14 \end{pmatrix}.$$

Likewise,

$$SR = \begin{pmatrix} 2 & -2 \\ 3 & 6 \\ 0 & 1 \end{pmatrix} \begin{pmatrix} 2 & 6 & -2 \\ 4 & -1 & 0 \end{pmatrix} = \begin{pmatrix} -4 & 14 & -4 \\ 30 & 12 & -6 \\ 4 & -1 & 0 \end{pmatrix}.$$

One important observation is that **the order in which the matrices are multiplied is important**, that is, in general $RS \neq SR$, so be careful.

Let's get an idea why the order of multiplication is important by considering a simple example.

EXAMPLE 7.1 Suppose a florist makes two kinds of flower arrangements. The deluxe arrangement uses 10 roses and 8 daisies while the standard arrangement uses 5 roses and 6 daisies. Roses cost 80 cents each and daisies cost 10 cents each. How much does each arrangement cost in materials?

We make a 2-by-2 matrix M in which the first row is the number of roses and daisies, respectively, that the deluxe arrangement uses. The second row is the number of roses and daisies, respectively, that the standard arrangement uses. We make a 2-vector A that is the cost per rose and cost per daisy, respectively. These are

$$M = \begin{pmatrix} 10 & 8 \\ 5 & 6 \end{pmatrix}, \quad A = \begin{pmatrix} 80 \\ 10 \end{pmatrix}.$$

Thus, the vector

$$MA = \begin{pmatrix} 880 \\ 460 \end{pmatrix}$$

gives the cost for materials for the deluxe and standard arrangements, respectively.

It should be clear that AM has no meaning.

Consider the 5-by-2 matrix R which gives the number of deluxe and standard arrangements made on each of the days, Monday through Friday.

$$R = \begin{matrix} & \\ \text{Monday} \\ \text{Tuesday} \\ \text{Wednesday} \\ \text{Thursday} \\ \text{Friday} \end{matrix} \begin{pmatrix} \text{deluxe} & \text{standard} \\ 20 & 30 \\ 15 & 40 \\ 10 & 30 \\ 30 & 20 \\ 20 & 15 \end{pmatrix}.$$

Then RMA is a 5-vector that gives the costs each day for the arrangements made. You should compute this vector. ∎

In similar, but more complex situations, matrix multiplication gives an organized manner for displaying data and an easy method for doing computations using a computer.

In summary, let R be an ℓ-by-m matrix and let S be an m-by-k matrix. Let r_{ij} be the number in the ith row and jth column of R and let s_{ij} be the number in the ith row and jth column of S. The matrix $T = RS$ has ℓ rows and k columns, and t_{ij}, the number in its ith row and jth column, is

$$t_{ij} = r_{i1} s_{1j} + r_{i2} s_{2j} + \cdots + r_{im} s_{mj}.$$

If R is an m-by-m matrix, that is, it has the same number of rows as columns, then R is a **square matrix**. In this case we can multiply R times itself, getting $RR = R^2$. This is particularly important in dynamical systems of the form $A(n+1) = RA(n)$. Specifically, $A(1) = RA(0)$, $A(2) = RA(1) = R^2 A(0)$, and more generally,

$$A(k) = R^k A(0).$$

Thus, you can see the need for computing powers of matrices. As an example, if

$$R = \begin{pmatrix} 2 & -1 \\ 0 & 3 \end{pmatrix},$$

then

$$R^2 = \begin{pmatrix} 2 & -1 \\ 0 & 3 \end{pmatrix} \begin{pmatrix} 2 & -1 \\ 0 & 3 \end{pmatrix} = \begin{pmatrix} 4 & -5 \\ 0 & 9 \end{pmatrix}.$$

We can compute higher powers of R recursively, such as,

$$R^3 = RR^2 = \begin{pmatrix} 2 & -1 \\ 0 & 3 \end{pmatrix} \begin{pmatrix} 4 & -5 \\ 0 & 9 \end{pmatrix} = \begin{pmatrix} 8 & -19 \\ 0 & 27 \end{pmatrix}.$$

Notice that this is one case in which the order of multiplication is not important; that is $R^3 = RR^2 = R^2R$. To check this,

$$R^2R = \begin{pmatrix} 4 & -5 \\ 0 & 9 \end{pmatrix} \begin{pmatrix} 2 & -1 \\ 0 & 3 \end{pmatrix} = \begin{pmatrix} 8 & -19 \\ 0 & 27 \end{pmatrix}.$$

Likewise, R^4 can be computed as R^3R, RR^3, or $(R^2)^2 = R^2R^2$.

PROBLEM

1. Consider the vectors

$$A = \begin{pmatrix} 2 \\ 0 \\ -3 \end{pmatrix}, \quad B = \begin{pmatrix} 4 \\ -1 \\ 5 \end{pmatrix}, \quad \text{and} \quad C = \begin{pmatrix} 5 \\ 4 \end{pmatrix}$$

and the matrices

$$R = \begin{pmatrix} 3 & 0 & 8 \\ 2 & -1 & 0 \\ 5 & 7 & -2 \end{pmatrix}, \quad S = \begin{pmatrix} 1 & -2 \\ 3 & 5 \\ 0 & 4 \end{pmatrix}, \quad M = \begin{pmatrix} -3 & 5 & 0 \\ 1 & 0 & 9 \end{pmatrix}.$$

Compute each of the following, if possible.

a. *RA*,	**b.** *RS*,	**c.** *SR*,	**d.** *MR*,
e. *MB*,	**f.** *SB*,	**g.** *SC*,	**h.** R^2,
i. S^2,	**j.** *RSC*,	**k.** *SM*,	**l.** *MS*,
m. *MSC*,	**n.** *SMB*,	**o.** $2A - 3B$,	**p.** $3R$,
q. $2R - SM$.			

7.2 Rules of linear algebra

The study of the manipulations of vectors and matrices is called **linear algebra**. Most of the rules are similar to the rules for manipulating real numbers. For example, assuming that A, B, and C are m-vectors; M, R, and S are m-by-m matrices; and a is a real number, we have

$$A + B = B + A$$

$$(A + B) + C = A + (B + C) = A + B + C$$

$$aA = Aa$$

$$M + R = R + M$$

$$(M + R) + S = M + (R + S) = M + R + S$$

$$(aM)A = a(MA) = M(aA) = aMA$$

$$(MR)A = M(RA) = MRA$$

$$(MR)S = M(RS) = MRS$$

$$M(R + S) = MR + MS$$

and so forth.

The proofs of these rules are easy, but tedious. For example, letting $m = 3$, we have that

$$A+B = \begin{pmatrix} a_1 \\ a_2 \\ a_3 \end{pmatrix} + \begin{pmatrix} b_1 \\ b_2 \\ b_3 \end{pmatrix} = \begin{pmatrix} a_1 + b_1 \\ a_2 + b_2 \\ a_3 + b_3 \end{pmatrix} = \begin{pmatrix} b_1 \\ b_2 \\ b_3 \end{pmatrix} + \begin{pmatrix} a_1 \\ a_2 \\ a_3 \end{pmatrix} = B+A.$$

The rule that $(MR)A = M(RA)$ is somewhat long and tedious, but not difficult. It consists of writing M, R, and A symbolically, as we wrote A and B above. We compute the new m-by-m matrix MR, and then multiply the vector A by this matrix, giving the vector $(MR)A$. Next we compute the m-vector RA, and then multiply this vector by M, giving the vector $M(RA)$. It can then be seen that the two resulting vectors are the same, by carefully comparing each of the components of these vectors. A detailed proof is not illuminating. It should be clear that the reader could prove this for 2-by-2 or 3-by-3 matrices, given enough time and energy.

In multiplying matrices, we saw in the last section that the order of multiplication is important in that, in general,

$$RS \neq SR.$$

There are cases in which $RS = SR$. For example, let $S = R^2$. Then $RS = RR^2 = RRR = R^2R = SR$. But these are the exceptions, not the rule.

Most of the matrices we deal with will be square matrices. To make our discussion easier, we will make a few definitions. If a matrix is square, then the diagonal from upper left to lower right, which consists of the numbers $a_{11}, a_{22}, \ldots, a_{mm}$, is called the **main diagonal**. The square matrix with all ones in the main diagonal, that is $a_{11} =$

$a_{22} = \cdots = a_{mm} = 1$, and all the other components zero, is called the
identity matrix or I. For example, the 3-by-3 identity matrix is

$$I = \begin{pmatrix} 1 & 0 & 0 \\ 0 & 1 & 0 \\ 0 & 0 & 1 \end{pmatrix}.$$

Usually, the dimension m is implied by the particular discussion.
Technically, we should identify the 2-by-2 identity matrix as I_2, the
3-by-3 identity matrix as I_3, and so forth, but for simplicity, we will
not do this unless it is necessary.

The identity matrix is given this name, because it acts much like
the scalar multiplicative identity, 1; that is,

$$IM = MI = M$$

for all matrices M. To see this for the 3-by-3 case, perform the
multiplication

$$\begin{pmatrix} 1 & 0 & 0 \\ 0 & 1 & 0 \\ 0 & 0 & 1 \end{pmatrix} \begin{pmatrix} a_{11} & a_{12} & a_{13} \\ a_{21} & a_{22} & a_{23} \\ a_{31} & a_{32} & a_{33} \end{pmatrix}.$$

You should get that the result is

$$\begin{pmatrix} a_{11} & a_{12} & a_{13} \\ a_{21} & a_{22} & a_{23} \\ a_{31} & a_{32} & a_{33} \end{pmatrix}.$$

Similarly, if you multiply a vector A by the identity matrix, you get
the same vector back; that is,

$$IA = A.$$

Consider the matrices

$$M = \begin{pmatrix} 1 & 1 \\ 2 & 4 \end{pmatrix} \quad \text{and} \quad R = \begin{pmatrix} 2 & -0.5 \\ -1 & 0.5 \end{pmatrix}.$$

Note that $MR = I$ and $RM = I$. The matrix R is called the **inverse** of
matrix M and is denoted $R = M^{-1}$ (or you could call M the inverse
of R and denote it by R^{-1}). When a matrix has an inverse it is called
nonsingular or **invertible**. If a matrix does not have an inverse, it is
called **singular**. The matrix

$$S = \begin{pmatrix} 2 & 3 \\ 0 & 0 \end{pmatrix}$$

is singular. If you multiply S by any 2-by-2 matrix, you will get zeros in the second row, so S times any matrix cannot equal the identity matrix.

In section 7.5 we will discuss the problem of determining when a matrix is singular or nonsingular and give a method for finding its inverse when it is nonsingular.

PROBLEM

1. Consider the matrices and vector

$$M = \begin{pmatrix} 2 & 3 \\ 4 & 5 \end{pmatrix}, \qquad R = \begin{pmatrix} 2 & 2 \\ 3 & 1 \end{pmatrix},$$

$$S = \begin{pmatrix} 3 & 2 \\ 2 & 0 \end{pmatrix}, \qquad A = \begin{pmatrix} 3 \\ 4 \end{pmatrix}.$$

a. Compute matrices MR and RS, and vector RA.
b. Show that the matrix $(MR)S$ equals matrix $M(RS)$ by computing each of these.
c. Show that the vector $(MR)A$ equals vector $M(RA)$ by computing each of these.
d. Show that matrix MR does not equal matrix RM.
e. Show that matrix IR equals matrix R.
f. Show that vector IA equals vector A.

7.3   Gauss–Jordan elimination

In section 5.10, we found solutions to higher-order dynamical systems. The general solutions were of the form

$$a(k) = a_1 r_1^k + a_2 r_2^k + a_3 r_3^k.$$

We were then given initial values $a(0)$, $a(1)$, and $a(2)$ giving the 3 equations

$$a_1 + a_2 + a_3 = a(0)$$

$$a_1 r_1 + a_2 r_2 + a_3 r_3 = a(1)$$

$$a_1 r_1^2 + a_2 r_2^2 + a_3 r_3^2 = a(2)$$

which we needed to solve for the 3 unknowns, a_1, a_2, and a_3. In that section, we used ad hoc methods to solve these equations. In this section, we develop systematic methods to solve several equations for some of the unknowns.

For example, it is easy to solve the system of linear equations

$$2x + 3y = 4 \quad \text{and} \quad x - y = 2.$$

We could multiply the second equation by 3, giving the equations

$$2x + 3y = 4 \quad \text{and} \quad 3x - 3y = 6.$$

Adding these two equations yields

$$5x = 10 \quad \text{or} \quad x = 2.$$

Substituting 2 into either of the first 2 equations, say the first, gives $2(2) + 3y = 4$ or $y = 0$.

The object of this section is to give a **systematic method** for solving systems of linear equations or for determining when the equations do not have a solution. This method is called **Gauss–Jordan elimination**. Quite often, it is easier to solve the equations directly than it is to use this method. On the other hand, once we have a systematic method for solving linear equations, we can write a computer program that will solve all of our equations for us.

Our first step is to develop a shorthand for writing our system of equations. A system of linear equations can be written in the form

$$MX = B$$

for some matrix M and some vectors X and B. In the case of the two equations given above, we would have

$$M = \begin{pmatrix} 2 & 3 \\ 1 & -1 \end{pmatrix}, \quad X = \begin{pmatrix} x \\ y \end{pmatrix}, \quad \text{and} \quad B = \begin{pmatrix} 4 \\ 2 \end{pmatrix}.$$

We now form what is called the **augmented matrix** M'. This matrix is formed by placing the vector B as an additional column of M. Thus,

$$M' = \begin{pmatrix} 2 & 3 & 4 \\ 1 & -1 & 2 \end{pmatrix}.$$

EXAMPLE 7.2 As another example, the augmented matrix

$$M' = \begin{pmatrix} 1 & -1 & 1 & -2 \\ 2 & 5 & 8 & 4 \\ 0 & 2 & 1 & 3 \end{pmatrix}$$

would correspond to the system of equations

$$x - y + z = -2$$
$$2x + 5y + 8z = 4$$

$$2y + z = 3.$$

The augmented matrix gives us all the information that the system of equations does, but it is more concise.

We can now think of each row of M' as being an equation with the variables x, y, and z being given implicitly. We now review operations we can perform when solving equations.

- We can multiply both sides of an equation by a non-zero constant.

- We can add or subtract a multiple of one equation to or from another equation.

- We can rearrange the order in which we write our equations; that is, we can exchange the position of two of the equations.

To save space, we can perform these procedures on the augmented matrix instead of the equations. To see this, let's solve the system of three equations (7.4). On the left of the page the equations will be manipulated, while on the right of the page the augmented matrix will be manipulated.

$$
\begin{array}{rcr}
x - y + z &=& -2 \\
2x + 5y + 8z &=& 4 \\
2y + z &=& 3
\end{array}
\qquad
\begin{pmatrix}
1 & -1 & 1 & -2 \\
2 & 5 & 8 & 4 \\
0 & 2 & 1 & 3
\end{pmatrix}.
$$

We will not change the first and third equation (row), but will subtract twice the first equation (row) from the second. Therefore the second equation (row) becomes

$$
\begin{array}{r}
2x + 5y + 8z = 4 \\
-2[\; x - y + z = -2] \\
\hline
7y + 6z = 8
\end{array}
\qquad
\begin{array}{r}
2\;\;\;5\;\;\;8\;\;\;4 \\
-2[\,1\;\;-1\;\;1\;\;-2] \\
\hline
0\;\;\;7\;\;\;6\;\;\;8
\end{array}
$$

Thus, we now get the system of equations (augmented matrix)

$$
\begin{array}{rcr}
x - y + z &=& -2 \\
7y + 6z &=& 8 \\
2y + z &=& 3
\end{array}
\qquad
\begin{pmatrix}
1 & -1 & 1 & -2 \\
0 & 7 & 6 & 8 \\
0 & 2 & 1 & 3
\end{pmatrix}.
$$

Now divide the second equation (row) by 7, giving

$$
\begin{array}{rcr}
x - y + z &=& -2 \\
y + (6/7)z &=& 8/7 \\
2y + z &=& 3
\end{array}
\qquad
\begin{pmatrix}
1 & -1 & 1 & -2 \\
0 & 1 & 6/7 & 8/7 \\
0 & 2 & 1 & 3
\end{pmatrix}.
$$

Now leave the first two equations (rows) unchanged, but subtract two times the second equation from the third equation, giving

$$
\begin{aligned}
2y + \quad z \; &= \; 3 \\
-2[\; y \; + (6/7)z &= 8/7] \\
\hline
- (5/7)z &= 5/7
\end{aligned}
\qquad
\begin{aligned}
0 \; 2 \quad 1 \qquad 3 \\
-2[\; 0 \; 1 \; 6/7 \; 8/7] \\
\hline
0 \; 0 \; -5/7 \; 5/7
\end{aligned}
$$

as the third equation (row). Our equations (rows) become

$$
\begin{aligned}
x - y + \qquad z &= -2 \\
y + (6/7)z &= 8/7 \\
- (5/7)z &= 5/7
\end{aligned}
\qquad
\begin{pmatrix}
1 & -1 & 1 & -2 \\
0 & 1 & 6/7 & 8/7 \\
0 & 0 & -5/7 & 5/7
\end{pmatrix}.
$$

Multiply the third equation by $-7/5$, giving

$$
\begin{aligned}
x - y + \qquad z &= -2 \\
y + (6/7)z &= 8/7 \\
z &= -1
\end{aligned}
\qquad
\begin{pmatrix}
1 & -1 & 1 & -2 \\
0 & 1 & 6/7 & 8/7 \\
0 & 0 & 1 & -1
\end{pmatrix}.
$$

From the third equation (row), we see that $z = -1$. We now go back and eliminate the z variable from the first 2 equations. To do this, we subtract $(6/7)$ths of the third row from the second row to give

$$
\begin{aligned}
x - y + z &= -2 \\
y &= 2 \\
z &= -1
\end{aligned}
\qquad
\begin{pmatrix}
1 & -1 & 1 & -2 \\
0 & 1 & 0 & 2 \\
0 & 0 & 1 & -1
\end{pmatrix}.
$$

We now subtract the third row from the first row to give

$$
\begin{aligned}
x - y &= -1 \\
y &= 2 \\
z &= -1
\end{aligned}
\qquad
\begin{pmatrix}
1 & -1 & 0 & -1 \\
0 & 1 & 0 & 2 \\
0 & 0 & 1 & -1
\end{pmatrix}.
$$

Finally we eliminate y from the first equation by adding the second row to the first row giving

$$
\begin{aligned}
x &= 1 \\
y &= 2 \\
z &= -1
\end{aligned}
\qquad
\begin{pmatrix}
1 & 0 & 0 & 1 \\
0 & 1 & 0 & 2 \\
0 & 0 & 1 & -1
\end{pmatrix}.
$$

We can now read from either the equations or from the augmented matrix that

$$
x = 1, \qquad y = 2, \qquad \text{and} \qquad z = -1. \qquad \blacksquare
$$

We had two goals to accomplish in our manipulation of the equations or rows in example 7.2. The first goal is to manipulate the equations or rows using our three operations given above, so that we obtain all 1's in the main diagonal (of the unaugmented matrix). In other words, we wish to get

$$a_{jj} = 1$$

for every value j. Notice that $a_{11} = a_{22} = a_{33} = 1$ in the last augmented matrix given above.

The second goal is to get all 0's everywhere but the main diagonal and the last column. In this case, there is exactly one solution to the set of equations, which can be read from the last column. When the system of equations has exactly one solution the system is called **consistent**.

EXAMPLE 7.3 Let's consider another example. Find the solution to the system of equations

$$\begin{aligned} x + 2y - 3z &= 2 \\ -x - y + 2z &= 1. \\ x + 3y - 4z &= 3 \end{aligned}$$

We first form the augmented matrix,

$$M' = \begin{pmatrix} 1 & 2 & -3 & 2 \\ -1 & -1 & 2 & 1 \\ 1 & 3 & -4 & 3 \end{pmatrix}.$$

Now add the first row to the second row, giving

$$\begin{pmatrix} 1 & 2 & -3 & 2 \\ 0 & 1 & -1 & 3 \\ 1 & 3 & -4 & 3 \end{pmatrix}.$$

Now subtract the first row from the third row, giving

$$\begin{pmatrix} 1 & 2 & -3 & 2 \\ 0 & 1 & -1 & 3 \\ 0 & 1 & -1 & 1 \end{pmatrix}.$$

Now subtract the second row from the third row, giving

$$\begin{pmatrix} 1 & 2 & -3 & 2 \\ 0 & 1 & -1 & 3 \\ 0 & 0 & 0 & -2 \end{pmatrix}.$$

Solving the original three equations is the same as solving the above three equations. But the third row of the last augmented matrix corresponds to the equation

$$0 = -2.$$

Since this last expression is always false, there are no solutions to the last three equations, and therefore there are **no solutions** to the original set of three equations. A system of equations that has no solution is called **inconsistent**. ∎

EXAMPLE 7.4 Consider the system of equations

$$\begin{aligned} x - 3y + z &= 2 \\ -2x + y + 2z &= 1. \\ x + 2y - 3z &= -3 \end{aligned}$$

The augmented matrix is

$$\begin{pmatrix} 1 & -3 & 1 & 2 \\ -2 & 1 & 2 & 1 \\ 1 & 2 & -3 & -3 \end{pmatrix}.$$

Add twice the first row to the second row, giving

$$\begin{pmatrix} 1 & -3 & 1 & 2 \\ 0 & -5 & 4 & -5 \\ 1 & 2 & -3 & -3 \end{pmatrix}.$$

Divide the second row by -5, giving

$$\begin{pmatrix} 1 & -3 & 1 & 2 \\ 0 & 1 & -4/5 & -1 \\ 1 & 2 & -3 & -3 \end{pmatrix}.$$

Subtract the first row from the third row, giving

$$\begin{pmatrix} 1 & -3 & 1 & 2 \\ 0 & 1 & -4/5 & -1 \\ 0 & 5 & -4 & -5 \end{pmatrix}.$$

Subtract five times the second row from the third row, giving

$$\begin{pmatrix} 1 & -3 & 1 & 2 \\ 0 & 1 & -4/5 & -1 \\ 0 & 0 & 0 & 0 \end{pmatrix}.$$

Add three times the second row to the first row, giving

$$\begin{pmatrix} 1 & 0 & -7/5 & -1 \\ 0 & 1 & -4/5 & -1 \\ 0 & 0 & 0 & 0 \end{pmatrix}.$$

This is the same as the equations

$$\begin{aligned} x \quad\quad - 1.4z &= -1 \\ y - 0.8z &= -1. \\ 0 &= 0 \end{aligned}$$

Since we have, in reality, only two equations with three unknowns, we can give any value we want for z, and then we can determine x and y. If we let $z = 0$, then $y = -1$ and $x = -1$. If we let $z = 5$, then $y = 3$ and $x = 6$. If we let $z = 5a$ to avoid fractions, then $y = 4a - 1$, and $x = 7a - 1$. Notice that there are an infinite number of solutions, one for each different value of z. In vector notation, we give the solution as

$$\begin{pmatrix} x \\ y \\ z \end{pmatrix} = a \begin{pmatrix} 7 \\ 4 \\ 5 \end{pmatrix} + \begin{pmatrix} -1 \\ -1 \\ 0 \end{pmatrix}.$$

where a can be any real number. A system of equations with at least one solution is also called **consistent**. ■

One observation is that we have exactly one solution to a system of m equations with m unknowns when we can obtain 1's along the main diagonal of the unaugmented matrix with all zeros above and below the main diagonal. Otherwise, we have either zero or infinitely many solutions.

Observe that by the above procedure, we are able to produce an augmented matrix of the following form.

- If a row does not consist of all 0's, then the first number in the row is a one.
- The first 1 in any row is to the right of the first 1 in the row above it.
- Find the first 1 in a row. There are only 0's above and below that 1, that is, that column has one 1 and the rest 0's.
- Rows that are all 0's are at the bottom of the augmented matrix.

Such a matrix is said to be in **reduced row echelon form**. Given a system of equations, the goal is to rewrite the corresponding augmented matrix in reduced row echelon form. The solutions to the system of equations, if any, can then be read off easily from the matrix. The method for reducing the matrix to this form can also be easily converted to a computer program.

EXAMPLE 7.5   As our last example, let's consider the system of equations

$$\begin{aligned}
y - z + 2w &= 2 \\
x + y + z + 3w &= 7 \\
x + 2y + 5w &= 9 \\
x - y + 3z - w &= 3
\end{aligned}$$

The augmented matrix is

$$\begin{pmatrix}
0 & 1 & -1 & 2 & 2 \\
1 & 1 & 1 & 3 & 7 \\
1 & 2 & 0 & 5 & 9 \\
1 & -1 & 3 & -1 & 3
\end{pmatrix}.$$

Switch the first and second rows, giving

$$\begin{pmatrix}
1 & 1 & 1 & 3 & 7 \\
0 & 1 & -1 & 2 & 2 \\
1 & 2 & 0 & 5 & 9 \\
1 & -1 & 3 & -1 & 3
\end{pmatrix}.$$

Subtract the first row from the third row, giving

$$\begin{pmatrix}
1 & 1 & 1 & 3 & 7 \\
0 & 1 & -1 & 2 & 2 \\
0 & 1 & -1 & 2 & 2 \\
1 & -1 & 3 & -1 & 3
\end{pmatrix}.$$

Subtract the first row from the fourth row giving

$$\begin{pmatrix}
1 & 1 & 1 & 3 & 7 \\
0 & 1 & -1 & 2 & 2 \\
0 & 1 & -1 & 2 & 2 \\
0 & -2 & 2 & -4 & -4
\end{pmatrix}.$$

Let's combine three steps by (1) subtracting the second row from the first row, (2) subtracting the second row from the third row, and

(3) adding twice the second row to the fourth row, giving

$$\begin{pmatrix} 1 & 0 & 2 & 1 & 5 \\ 0 & 1 & -1 & 2 & 2 \\ 0 & 0 & 0 & 0 & 0 \\ 0 & 0 & 0 & 0 & 0 \end{pmatrix}.$$

We now have the matrix in reduced row echelon form. It corresponds to the two equations

$$x + 2z + w = 5$$

$$y - z + 2w = 2$$

Letting $z = a$ and $w = b$ be any numbers, we can then solve for x and y. We can write this in vector form by taking the z and w terms to the right side of the equations and adding the two equations $z = a$ and $w = b$ giving

$$x = -2a - b + 5$$

$$y = a - 2b + 2$$

$$z = a$$

$$w = b$$

or in vector notation

$$\begin{pmatrix} x \\ y \\ z \\ w \end{pmatrix} = a \begin{pmatrix} -2 \\ 1 \\ 1 \\ 0 \end{pmatrix} + b \begin{pmatrix} -1 \\ -2 \\ 0 \\ 1 \end{pmatrix} + \begin{pmatrix} 5 \\ 2 \\ 0 \\ 0 \end{pmatrix}.$$ ∎

PROBLEM

1. Find all solutions for the following systems of equations, if any, by reducing the corresponding augmented matrix to reduced row echelon form. When there are an infinite number of solutions, give them as the sums of a vector plus arbitrary multiples of other vectors.

a. $2x + 3y = 5$
    $-3x + 4y = 1$

b. $x - 5y + 2z = 10$
    $2x - y + 5z = -2$
    $3x + 2y - z = -16$

c. $\begin{aligned} 2x - y + 3z + 2w &= 10 \\ -x + 2y \quad\quad + 5w &= 10 \\ 5y + z + w &= -7 \\ x + y + z + 3w &= 8 \end{aligned}$

d. $\begin{aligned} x - 2y + 3z &= 5 \\ 2x + 2y - z &= 6 \\ -x - 10y + 11z &= 3 \end{aligned}$

e. $\begin{aligned} x - 2y + 3z &= 5 \\ 2x + 2y - z &= 6 \\ -x - 10y + 11z &= 4 \end{aligned}$

f. $\begin{aligned} x + y \quad\quad - w &= 4 \\ x + y + z + w &= 5 \\ x + y + 3z + 5w &= 7 \\ z + 2w &= 1 \end{aligned}$

7.4 Determinants

We wish to develop a method for determining when a system of equations has exactly one solution. To do this, let's consider an arbitrary system of two equations and two unknowns, say

$$ax + by = c_1$$
$$rx + sy = c_2$$

where a, b, r, s, c_1, and c_2 are given constants. The augmented matrix is

$$\begin{pmatrix} a & b & c_1 \\ r & s & c_2 \end{pmatrix}.$$

Assume that one of a, b, r, or s does not equal zero. (If they all equal zero, then there are no solutions if either c_1 or c_2 is nonzero, and there are infinitely many solutions if $c_1 = c_2 = 0$.) We will assume, without loss of generality, that $a \neq 0$.

Divide the first row by a, giving

$$\begin{pmatrix} 1 & b/a & c_1/a \\ r & s & c_2 \end{pmatrix}.$$

Subtract r times the first row from the second row, giving

$$\begin{pmatrix} 1 & b/a & c_1/a \\ 0 & s - rb/a & c_2 - rc_1/a \end{pmatrix}.$$

To simplify matters, multiply the second row by a, giving

$$\begin{pmatrix} 1 & b/a & c_1/a \\ 0 & as - rb & ac_2 - rc_1 \end{pmatrix}.$$

If $as - rb = 0$, then (1) there are no solutions if $ac_2 - rc_1 \neq 0$; and (2) there are infinitely many solutions if $ac_2 - rc_1 = 0$. If

$$as - rb \neq 0,$$

then we can proceed to write the matrix in reduced row echelon form and read off the unique solution.

Let's review. Consider the system of equations

$$MX = C$$

where

$$M = \begin{pmatrix} a & b \\ r & s \end{pmatrix}, \qquad X = \begin{pmatrix} x \\ y \end{pmatrix}, \qquad C = \begin{pmatrix} c_1 \\ c_2 \end{pmatrix}.$$

This system has exactly one solution if and only if $as - rb \neq 0$. We now make a definition.

DEFINITION 7.6   *We define the **determinant** of the 2-by-2 matrix M as*

$$|M| = \begin{vmatrix} a & b \\ r & s \end{vmatrix} = as - rb.$$

We can now state that a system of 2 equations and 2 unknowns has exactly one solution if and only if $|M| \neq 0$.

Suppose we have three equations and three unknowns, say $MX = C$, or

$$\begin{pmatrix} a & b & c \\ r & s & t \\ u & v & w \end{pmatrix} \begin{pmatrix} x \\ y \\ z \end{pmatrix} = \begin{pmatrix} c_1 \\ c_2 \\ c_3 \end{pmatrix}.$$

The augmented matrix is

$$M' = \begin{pmatrix} a & b & c & c_1 \\ r & s & t & c_2 \\ u & v & w & c_3 \end{pmatrix}.$$

Again, we assume that $a \neq 0$. To avoid fractions, multiply the second and third rows by a, giving

$$M' = \begin{pmatrix} a & b & c & c_1 \\ ar & as & at & ac_2 \\ au & av & aw & ac_3 \end{pmatrix}.$$

Now subtract r times the first row from the second row and, in order to save a step, subtract u times the first row from the third row, giving

$$M' = \begin{pmatrix} a & b & c & c_1 \\ 0 & as - br & at - cr & ac_2 - rc_1 \\ 0 & av - bu & aw - cu & ac_3 - uc_1 \end{pmatrix}.$$

Now, if we can find a solution, y and z, to the second and third equations (rows), we can substitute these values into the first equation (row) to find the value of x, that is, we wish to solve the system of equations

$$\begin{aligned} (as - br)y + (at - cr)z &= ac_2 - rc_1 \\ (av - bu)y + (aw - cu)z &= ac_3 - uc_1 \end{aligned}.$$

This system of equations has exactly one solution if and only if the determinant of the matrix

$$\begin{pmatrix} as - br & at - cr \\ av - bu & aw - cu \end{pmatrix}$$

does not equal zero; that is, if

$$(as - br)(aw - cu) - (at - cr)(av - bu) \neq 0.$$

Multiplying these terms together, canceling, and factoring out an a, gives

$$a(asw + btu + crv - atv - brw - csu) \neq 0.$$

Since $a \neq 0$, we must have the term in parentheses not equal to zero in order to have exactly one solution to the system of three equations given above.

We therefore make the definition

DEFINITION 7.7   *The determinant of the 3-by-3 matrix*

$$M = \begin{pmatrix} a & b & c \\ r & s & t \\ u & v & w \end{pmatrix}$$

is

$$|M| = asw + btu + crv - atv - brw - csu,$$

To summarize, the system of equations

$$MX = B,$$

where M is either a 2-by-2 or 3-by-3 matrix has exactly one solution if and only if $|M|$, the determinant of M, does not equal zero.

There are similar definitions of determinant for m-by-m matrices, where $m \geq 4$, but we will not go into that. Interested readers are referred to books on linear algebra.

Consider the system of equations

$$2x + 5y = -7$$
$$-4x + 2y = 15 \, \dot{}$$

The determinant of the matrix

$$M = \begin{pmatrix} 2 & 5 \\ -4 & 2 \end{pmatrix}$$

is $2 \times 2 - 5 \times (-4) = 24 \neq 0$, so this system of equations has exactly one solution. This solution is $x = -89/24$ and $y = 1/12$, which can be found by reducing the augmented matrix to reduced row echelon form.

Consider the system of equations

$$2x + y + 3z = c_1$$
$$-x + 2y + z = c_2 \, .$$
$$x + 8y + 9z = c_3$$

The determinant of the matrix

$$M = \begin{pmatrix} 2 & 1 & 3 \\ -1 & 2 & 1 \\ 1 & 8 & 9 \end{pmatrix}$$

is

$$|M| = (2 \times 2 \times 9) + (1 \times 1 \times 1) + (3 \times (-1) \times 8)$$
$$- (2 \times 1 \times 8) - (1 \times (-1) \times 9) - (3 \times 2 \times 1)$$
$$= 36 + 1 - 24 - 16 + 9 - 6 = 0,$$

thus, the above system of equations has either zero or infinitely many solutions, depending on the choices for c_1, c_2, and c_3. For example, if $c_1 = 1$, $c_2 = 2$, and $c_3 = 3$, there are no solutions to the system of equations. On the other hand, if $c_1 = 1$, $c_2 = 2$, and $c_3 = 8$, there are infinitely many solutions to this system of equations. You should check this out.

PROBLEM

1. Find the determinants of the following matrices.

a. $\begin{pmatrix} 2 & 4 \\ 0 & 3 \end{pmatrix}$

b. $\begin{pmatrix} 5 & -2 \\ 2 & 12 \end{pmatrix}$

c. $\begin{pmatrix} 1 & 2 & -3 \\ 3 & -2 & 5 \\ 4 & 1 & 4 \end{pmatrix}$

d. $\begin{pmatrix} 1 & 5 & 0 \\ 2 & -3 & 1 \\ 8 & 1 & 3 \end{pmatrix}$

7.5 Inverse matrices

The **inverse** of an m-by-m matrix M is an m-by-m matrix M^{-1} such that

$$MM^{-1} = I \quad \text{and} \quad M^{-1}M = I.$$

For example, the matrix

$$M = \begin{pmatrix} 1 & -4 \\ -1 & 5 \end{pmatrix}$$

has as its inverse

$$M^{-1} = \begin{pmatrix} 5 & 4 \\ 1 & 1 \end{pmatrix}.$$

You should check to see that

$$\begin{pmatrix} 1 & -4 \\ -1 & 5 \end{pmatrix}\begin{pmatrix} 5 & 4 \\ 1 & 1 \end{pmatrix} = \begin{pmatrix} 1 & 0 \\ 0 & 1 \end{pmatrix}$$

and

$$\begin{pmatrix} 5 & 4 \\ 1 & 1 \end{pmatrix}\begin{pmatrix} 1 & -4 \\ -1 & 5 \end{pmatrix} = \begin{pmatrix} 1 & 0 \\ 0 & 1 \end{pmatrix}.$$

Some matrices have an inverse while others do not. The purpose of this section is to determine when a matrix has an inverse, to find that inverse when it does, and to see how inverse matrices can be used to easily solve systems of equations.

The first step is to observe that a matrix can have at most one inverse. Suppose that matrix M has two inverses, M^{-1} and m^{-1}. Then

$$m^{-1} = m^{-1}I = m^{-1}(MM^{-1}) = (m^{-1}M)M^{-1} = IM^{-1} = M^{-1}$$

and we see that $M^{-1} = m^{-1}$, so there is only one inverse.

Now, suppose we wish to solve the system of m equations with m unknowns

$$MX = B. \tag{7.4}$$

If M has an inverse, then we can multiply both sides of equation (7.4) on the left by M^{-1} giving

$$M^{-1}MX = M^{-1}B \quad \text{or} \quad X = M^{-1}B.$$

For example, consider the system of equations

$$x - 4y = 3$$
$$-x + 5y = 2,$$

which can be rewritten as

$$MX = B$$

where

$$M = \begin{pmatrix} 1 & -4 \\ -1 & 5 \end{pmatrix}, \quad X = \begin{pmatrix} x \\ y \end{pmatrix}, \quad \text{and} \quad B = \begin{pmatrix} 3 \\ 2 \end{pmatrix}.$$

Then the solution is

$$X = M^{-1}B = \begin{pmatrix} 5 & 4 \\ 1 & 1 \end{pmatrix} \begin{pmatrix} 3 \\ 2 \end{pmatrix} = \begin{pmatrix} 23 \\ 5 \end{pmatrix}.$$

On the other hand, consider the system of equations

$$x + y = 3$$
$$5x + 5y = 2,$$

which can be rewritten as

$$RX = B$$

where

$$R = \begin{pmatrix} 1 & 1 \\ 5 & 5 \end{pmatrix}$$

and X and B are as before. If R has an inverse, then a solution to this system of equations is $R^{-1}B$. Let's try to solve this system of equations putting

$$R' = \begin{pmatrix} 1 & 1 & 3 \\ 5 & 5 & 2 \end{pmatrix}$$

into reduced row echelon form. Subtracting 5 times the first row from the second gives

$$\begin{pmatrix} 1 & 1 & 3 \\ 0 & 0 & -13 \end{pmatrix}.$$

The last row states that $0 = -13$. Thus, this system of equations has no solution and so this matrix R has no inverse.

Let's further analyze why R has no inverse by attempting to find R^{-1} and seeing what goes wrong. To find R^{-1} we need to find x, y, z, and w such that

$$RR^{-1} = \begin{pmatrix} 1 & 1 \\ 5 & 5 \end{pmatrix} \begin{pmatrix} x & z \\ y & w \end{pmatrix} = \begin{pmatrix} 1 & 0 \\ 0 & 1 \end{pmatrix} = I.$$

Multiplying the matrices gives

$$\begin{pmatrix} x+y & z+w \\ 5x+5y & 5z+5w \end{pmatrix} = \begin{pmatrix} 1 & 0 \\ 0 & 1 \end{pmatrix}.$$

Equating corresponding components of these matrices gives us four equations and four unknowns. Two of the equations are

$$x + y = 1$$
$$5x + 5y = 0.$$

To solve these equations you must put

$$R' = \begin{pmatrix} 1 & 1 & 1 \\ 5 & 5 & 0 \end{pmatrix}$$

into reduced row echelon form. But as before, subtracting 5 times the first row from the second gives

$$\begin{pmatrix} 1 & 1 & 1 \\ 0 & 0 & -5 \end{pmatrix}$$

and it is seen that these equations have no solution.

Similarly, when using Gauss–Jordan elimination to solve the equations

$$z + w = 0$$

$$5z + 5w = 1.$$

we again see that there is no solution. The solution to these four equations would give the inverse matrix R^{-1}. Since they have no solution, there is no inverse.

This should indicate that a system of equations, $MX = B$ has a unique solution if and only if the matrix M has an inverse, and that the solution is $X = M^{-1}B$.

Even though we already know the inverse of

$$M = \begin{pmatrix} 1 & -4 \\ -1 & 5 \end{pmatrix},$$

let's pretend we don't, and try to find it. To find M^{-1} we need to find x, y, z, and w such that

$$MM^{-1} = \begin{pmatrix} 1 & -4 \\ -1 & 5 \end{pmatrix} \begin{pmatrix} x & z \\ y & w \end{pmatrix} = \begin{pmatrix} 1 & 0 \\ 0 & 1 \end{pmatrix} = I.$$

Multiplying the matrices gives

$$\begin{pmatrix} x - 4y & z - 4w \\ -x + 5y & -z + 5w \end{pmatrix} = \begin{pmatrix} 1 & 0 \\ 0 & 1 \end{pmatrix}.$$

Equating corresponding components of these matrices gives us four equations and four unknowns. Two of the equations are

$$x - 4y = 1$$
$$-x + 5y = 0.$$

To solve these equations you must put

$$R' = \begin{pmatrix} 1 & -4 & 1 \\ -1 & 5 & 0 \end{pmatrix}$$

into reduced row echelon form. Adding the first row to the second gives

$$\begin{pmatrix} 1 & -4 & 1 \\ 0 & 1 & 1 \end{pmatrix}.$$

Adding 4 times the second row to the first gives

$$\begin{pmatrix} 1 & 0 & 5 \\ 0 & 1 & 1 \end{pmatrix},$$

and we get that $x = 5$ and $y = 1$.

The other two equations are

$$z - 4w = 0$$
$$-z + 5w = 1.$$

To solve these equations you must put

$$M' = \begin{pmatrix} 1 & -4 & 0 \\ -1 & 5 & 1 \end{pmatrix}$$

into reduced row echelon form. Adding the first row to the second gives

$$\begin{pmatrix} 1 & -4 & 0 \\ 0 & 1 & 1 \end{pmatrix}.$$

Adding 4 times the second row to the first gives

$$\begin{pmatrix} 1 & 0 & 4 \\ 0 & 1 & 1 \end{pmatrix},$$

and we get that $z = 4$ and $w = 1$ and that

$$M^{-1} = \begin{pmatrix} 5 & 4 \\ 1 & 1 \end{pmatrix},$$

which we already knew.

In solving for M^{-1}, we first solved the system of equations $MX = A$, and then we solved the system of equations $MZ = C$, where A was the first column of the identity matrix I and C was the second column of I. To save time, we can solve both sets of equations simultaneously by constructing an augmented matrix $M \mid I$ by attaching matrix I to matrix M, and then putting it into reduced row echelon form. To be specific, we construct

$$M \mid I = \begin{pmatrix} 1 & -4 & 1 & 0 \\ -1 & 5 & 0 & 1 \end{pmatrix}.$$

We then add the first row to the second, giving

$$\begin{pmatrix} 1 & -4 & 1 & 0 \\ 0 & 1 & 1 & 1 \end{pmatrix}.$$

Then we add four times the second row to the first, giving

$$\begin{pmatrix} 1 & 0 & 5 & 4 \\ 0 & 1 & 1 & 1 \end{pmatrix}.$$

We now have the matrix $I \mid M^{-1}$, that is, the last two columns are M^{-1}.

In order to find the inverse of an m-by-m matrix M, we need to find a unique solution to m systems of equations, that is, we must solve: $MX = B_1$ where B_1 is the first column of the identity matrix I; $MX = B_2$ where B_2 is the second column of I; and so forth. Thus, if matrix M has an inverse, its determinant must be nonzero, $|M| \neq 0$. On the other hand, if M does not have an inverse, we will not be able to solve $MX = B$ uniquely for B equal to some column of I, so $|M| = 0$.

THEOREM 7.8   *An m-by-m matrix M has an inverse matrix M^{-1} if and only if the determinant of M is nonzero, that is, $|M| \neq 0$. Suppose that M has an inverse. Form the m-by-$2m$ augmented matrix $M \mid I$ by affixing the identity matrix to M. Apply Gauss–Jordan elimination to get this matrix into reduced row echelon form. The resulting augmented matrix will be in the form $I \mid M^{-1}$, that is, the last m columns will be the inverse matrix.*

EXAMPLE 7.9   Let's find the inverse of the matrix

$$M = \begin{pmatrix} 1 & -1 & 2 \\ -1 & 2 & 1 \\ -1 & 4 & 7 \end{pmatrix}$$

if it exists. Since

$$|M| = 1(2)(7) + (-1)(1)(-1) + 2(-1)(4)$$
$$- [1(1)(4) + (-1)(-1)(7) + 2(2)(-1)]$$
$$= 14 + 1 - 8 - (4 + 7 - 4) = 0,$$

we know that M does not have an inverse.   ■

EXAMPLE 7.10   Let's find the inverse of the matrix

$$M = \begin{pmatrix} 1 & -1 & 2 \\ -1 & 2 & 1 \\ 1 & -2 & 0 \end{pmatrix}$$

if it exists. Since

$$|M| = 1(2)(0) + (-1)(1)(1) + 2(-1)(-2)$$
$$- [1(1)(-2) + (-1)(-1)(0) + 2(2)(1)]$$
$$= 0 - 1 + 4 - (-2 + 0 + 4) = 1 \neq 0,$$

we know that M has an inverse.

Let's find that inverse by forming

$$M \mid I = \begin{pmatrix} 1 & -1 & 2 & 1 & 0 & 0 \\ -1 & 2 & 1 & 0 & 1 & 0 \\ 1 & -2 & 0 & 0 & 0 & 1 \end{pmatrix}.$$

Add the first row to the second, giving

$$\begin{pmatrix} 1 & -1 & 2 & 1 & 0 & 0 \\ 0 & 1 & 3 & 1 & 1 & 0 \\ 1 & -2 & 0 & 0 & 0 & 1 \end{pmatrix}.$$

Subtract the first row from the third, giving

$$\begin{pmatrix} 1 & -1 & 2 & 1 & 0 & 0 \\ 0 & 1 & 3 & 1 & 1 & 0 \\ 0 & -1 & -2 & -1 & 0 & 1 \end{pmatrix}.$$

Add the second row to the third, giving

$$\begin{pmatrix} 1 & -1 & 2 & 1 & 0 & 0 \\ 0 & 1 & 3 & 1 & 1 & 0 \\ 0 & 0 & 1 & 0 & 1 & 1 \end{pmatrix}.$$

Add the second row to the first, giving

$$\begin{pmatrix} 1 & 0 & 5 & 2 & 1 & 0 \\ 0 & 1 & 3 & 1 & 1 & 0 \\ 0 & 0 & 1 & 0 & 1 & 1 \end{pmatrix}.$$

Subtract 5 times the third row from the first, giving

$$\begin{pmatrix} 1 & 0 & 0 & 2 & -4 & -5 \\ 0 & 1 & 3 & 1 & 1 & 0 \\ 0 & 0 & 1 & 0 & 1 & 1 \end{pmatrix}.$$

Subtract 3 times the third row from the second, giving

$$\begin{pmatrix} 1 & 0 & 0 & 2 & -4 & -5 \\ 0 & 1 & 0 & 1 & -2 & -3 \\ 0 & 0 & 1 & 0 & 1 & 1 \end{pmatrix}.$$

We now have that the inverse matrix is the last 3 columns, that is,

$$M^{-1} = \begin{pmatrix} 2 & -4 & -5 \\ 1 & -2 & -3 \\ 0 & 1 & 1 \end{pmatrix}.$$

You should check this by computing MM^{-1} to see that it equals I.

Let's use the inverse matrix to find the solution to the system of equations

$$\begin{aligned} x - y + 2z &= 2 \\ -x + 2y + z &= -1. \\ x - 2y &= 3 \end{aligned}$$

The solution to $MX = B$ is

$$X = M^{-1}B = \begin{pmatrix} 2 & -4 & -5 \\ 1 & -2 & -3 \\ 0 & 1 & 1 \end{pmatrix} \begin{pmatrix} 2 \\ -1 \\ 3 \end{pmatrix} = \begin{pmatrix} -7 \\ -5 \\ 2 \end{pmatrix}.$$

You should substitute these values in for x, y, and z to see that they satisfy the equations. ∎

PROBLEMS

1. Find the determinants of the following matrices.   If the determinant is nonzero, find the inverse matrix.

   a. $\begin{pmatrix} 2 & 4 \\ 0 & 1 \end{pmatrix}$

   b. $\begin{pmatrix} 4 & 2 \\ 2 & 2 \end{pmatrix}$

   c. $\begin{pmatrix} 1 & 2 & -3 \\ 3 & -2 & 5 \\ 5 & -6 & 13 \end{pmatrix}$

   d. $\begin{pmatrix} 5 & -2 & -2 \\ -2 & 1 & 1 \\ -2 & 0 & 1 \end{pmatrix}$

2. Find the solution to the system of equations

   $$\begin{aligned} 4x + 2y &= 7 \\ 2x + 2y &= -3 \end{aligned}$$

   using your answer to problem 1 part (b).

3. Find the solution to the system of equations

   $$\begin{aligned} 5x - 2y - 2z &= 4 \\ -2x + y + z &= 0 \\ -2x + z &= 1 \end{aligned}$$

   using your answer to problem 1 part (d).

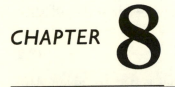

Dynamical systems of several equations

8.1 Introduction to dynamical systems of several equations

In section 5.10, we developed a model for population growth in which the population was divided into three age groups. With $a(n)$, $b(n)$, and $c(n)$ representing the first, second, and third age groups, respectively, in time period n, we derived the following dynamical system of three equations:

$$a(n+1) = 0.5a(n) + 5b(n) + 3c(n)$$
$$b(n+1) = 0.5a(n)$$
$$c(n+1) = (2/3)b(n).$$

In this chapter we will learn how to use matrix algebra to study dynamical systems of several equations such as this one.

To understand how the population modeled by the above three equations behaves, we need to compute solutions for $a(k)$, $b(k)$, and $c(k)$. For shorthand, we construct the 3-vector $A(n)$ and the matrix R as

$$A(n) = \begin{pmatrix} a(n) \\ b(n) \\ c(n) \end{pmatrix} \quad \text{and} \quad R = \begin{pmatrix} 0.5 & 5 & 3 \\ 0.5 & 0 & 0 \\ 0 & \frac{2}{3} & 0 \end{pmatrix}. \quad (8.1)$$

333

Using the notation (8.1), the previous dynamical system of three equations can be rewritten in the form

$$A(n+1) = RA(n). \tag{8.2}$$

If we are given $a(0)$, $b(0)$, and $c(0)$, that is, we are given $A(0)$, then we can formally compute $A(1) = RA(0)$. Since the order of multiplication is important when dealing with matrices, we carefully compute that

$$A(2) = RA(1) = R[RA(0)] = (RR)A(0) = R^2 A(0),$$

and we see that for dynamical systems of several equations we have the same result as we did for first-order linear dynamical systems of a single equation, that is, the solution is

$$A(k) = R^k A(0).$$

Note that this argument gives a unique solution to dynamical system (8.1) for any m-by-m matrix R and initial m-vector $A(0)$.

REMARK: With many graphing calculators, it is relatively easy to compute $A(1)$ through $A(k)$ for reasonably small k. For example, on the TI-81 calculator, store matrix R in $\boxed{[A]}$ and $A(0)$ in $\boxed{[B]}$. Then type $\boxed{[B]}$ $\boxed{\text{ENTER}}$. Next type $\boxed{[A]}$ $\boxed{*}$ $\boxed{\text{ANS}}$ (which is really $R * A(0)$), $\boxed{\text{ENTER}}$, giving $A(1)$. The next time you press $\boxed{\text{ENTER}}$, the calculator gives you $A(2)$, by computing $\boxed{[A]}$ $\boxed{*}$ $\boxed{\text{ANS}}$ $= RA(1)$. Each time you press $\boxed{\text{ENTER}}$, you get the next $A(k)$-vector. In the case of our population growth model, let

$$A(0) = \begin{pmatrix} 20 \\ 20 \\ 20 \end{pmatrix}$$

meaning that there are 20 of the species in each age group. Using the calculator, we find that

$$A(1) = \begin{pmatrix} 170 \\ 10 \\ 13.3 \end{pmatrix}, \qquad A(2) = \begin{pmatrix} 175 \\ 85 \\ 6.7 \end{pmatrix}, \qquad A(3) = \begin{pmatrix} 532.5 \\ 87.5 \\ 56.7 \end{pmatrix}.$$

Similarly, we can compute R^k on the calculator for reasonably small k. Store R in $\boxed{[A]}$, press $\boxed{[A]}$ and $\boxed{\text{ENTER}}$, then press $\boxed{[A]}$ $\boxed{*}$ $\boxed{\text{ANS}}$ and $\boxed{\text{ENTER}}$. The answer given is R^2. Pressing $\boxed{\text{ENTER}}$ again gives R^3. Each time you press $\boxed{\text{ENTER}}$, you get the next power of R. In the population model, we find that

$$R^4 = \begin{pmatrix} 9.2 & 23.6 & 10.9 \\ 1.8 & 7.4 & 4.1 \\ 0.9 & 1.5 & 0.5 \end{pmatrix}$$

where each component has been rounded to one decimal place.   □

The problem is that matrix multiplication is time consuming and not very illuminating, as is seen from looking at $A(3)$ and R^4 for the population model. Thus, for this population model, knowing that $A(100) = R^{100}A(0)$ will not help much in learning about the long term behavior of the population, which we discovered in section 5.10. The goal of this chapter is to develop methods (using matrix algebra) for studying the solutions to dynamical systems of several equations.

PROBLEM

1. In section 5.4 we developed the dynamical system of two equations

$$p(n+1) = 0.5p(n) + 0.5a(n) \qquad \text{and} \qquad a(n+1) = p(n)$$

to model a sex-linked gene.
   a. Rewrite this equation in matrix notation, carefully specifying the vector $P(n)$ and the matrix R.
   b. Compute R^3.
   c. If you have access to appropriate technology, compute R^{20} and guess at the 2-by-2 matrix M such that

$$\lim_{k \to \infty} R^k = M.$$

   d. Given that $p(0) = 0.8$ and $a(0) = 0.3$, find the vector $P(4)$.

8.2 Characteristic values

By using vectors and matrices, we can rewrite a first-order dynamical system of several equations as one first-order linear dynamical system relating a vector at time $n+1$ to the vector at time n. For example, the dynamical system $a(n+1) = 0.5a(n) + 5b(n) + 3c(n)$, $b(n+1) = 0.5a(n)$, and $c(n+1) = (2/3)b(n)$, which we used to model population growth, was rewritten as

$$A(n+1) = RA(n)$$

where

$$A(n) = \begin{pmatrix} a(n) \\ b(n) \\ c(n) \end{pmatrix} \quad \text{and} \quad R = \begin{pmatrix} 0.5 & 5 & 3 \\ 0.5 & 0 & 0 \\ 0 & \frac{2}{3} & 0 \end{pmatrix}.$$

Since the solution to this equation is

$$A(k) = R^k A(0),$$

in order to compute $A(10)$, for instance, we would need to compute the tenth power of the matrix R. It is clear that this would require quite a bit of tedious work. Our goal in this section is to reduce this work considerably.

Let's consider two examples. Specifically, let's compute $R^3 A$ and $R^3 B$, where

$$R = \begin{pmatrix} 1 & -1 \\ 2 & 4 \end{pmatrix}, \quad A = \begin{pmatrix} 2 \\ 1 \end{pmatrix}, \quad B = \begin{pmatrix} 1 \\ -1 \end{pmatrix}.$$

In this case,

$$RA = \begin{pmatrix} 1 & -1 \\ 2 & 4 \end{pmatrix} \begin{pmatrix} 2 \\ 1 \end{pmatrix} = \begin{pmatrix} 1(2) - 1(1) \\ 2(2) + 4(1) \end{pmatrix} = \begin{pmatrix} 1 \\ 8 \end{pmatrix}.$$

Therefore,

$$R^2 A = R(RA) = \begin{pmatrix} 1 & -1 \\ 2 & 4 \end{pmatrix} \begin{pmatrix} 1 \\ 8 \end{pmatrix} = \begin{pmatrix} -7 \\ 34 \end{pmatrix},$$

and

$$R^3 A = R(R^2 A) = \begin{pmatrix} 1 & -1 \\ 2 & 4 \end{pmatrix} \begin{pmatrix} -7 \\ 34 \end{pmatrix} = \begin{pmatrix} -41 \\ 122 \end{pmatrix}.$$

As you can see, it would take some time to compute $R^{20} A$, manually.

Now, let's compute $R^3 B$.

$$RB = \begin{pmatrix} 1 & -1 \\ 2 & 4 \end{pmatrix} \begin{pmatrix} 1 \\ -1 \end{pmatrix} = \begin{pmatrix} 1(1) - 1(-1) \\ 2(1) + 4(-1) \end{pmatrix}$$

$$= \begin{pmatrix} 2 \\ -2 \end{pmatrix} = 2 \begin{pmatrix} 1 \\ -1 \end{pmatrix}.$$

Notice that $RB = 2B$ in this case. Therefore,

$$R^2 B = R(RB) = R(2B) = 2(RB) = 2(2B) = 4B = 4 \begin{pmatrix} 1 \\ -1 \end{pmatrix}.$$

Similarly,

$$R^3 B = R(R^2 B) = R(2^2 B) = 2^2(RB) = 2^3 B.$$

In this case, it is easy to see that

$$R^{20} B = 2^{20} B = 2^{20} \begin{pmatrix} 1 \\ -1 \end{pmatrix}.$$

Try computing $R^3 C$, where

$$C = \begin{pmatrix} 1 \\ -2 \end{pmatrix}.$$

You should find that $RC = 3C$ and therefore $R^3 C = 3^3 C$.

We see from the above that, when R is an m-by-m matrix, there may be an m-vector A and a constant r such that

$$RA = rA.$$

Such a number r is called a **characteristic value** for the matrix R, and the vector A is called a **characteristic vector** corresponding to the characteristic value r. Specifically, the matrix

$$R = \begin{pmatrix} 1 & -1 \\ 2 & 4 \end{pmatrix}$$

has characteristic values $r = 2$ and $r = 3$. The vector B given above is a characteristic vector corresponding to $r = 2$, and the vector C given above is a characteristic vector corresponding to $r = 3$.

Any multiple of B will also be a characteristic vector corresponding to $r = 2$, that is, for a fixed constant b,

$$R(bB) = b(RB) = b(2B) = 2(bB),$$

and so bB is a characteristic vector corresponding to $r = 2$ also.

We have seen that in many cases, R^k acts like r^k for some numbers r. Often the largest of these numbers r is the most important. We therefore make the following definition.

DEFINITION 8.1   *Consider an m-by-m matrix R with characteristic values, $r_1, \ldots, r_m$. We define the norm of R to be*

$$\|R\| = \max\{|r_1|, |r_2|, \ldots, |r_m|\}.$$

REMARK: An m-by-m matrix has at most m characteristic values. The reason for this will be made clear later in this section.   □

Since

$$R = \begin{pmatrix} 1 & -1 \\ 2 & 4 \end{pmatrix}$$

has characteristic values $r = 2$ and $r = 3$, then

$$\|R\| = 3.$$

The question is, how do you find the characteristic values and corresponding characteristic vectors for a given matrix R?

Consider the matrix

$$R = \begin{pmatrix} 3 & 1 \\ 4 & 3 \end{pmatrix}.$$

We wish to find a constant r and a vector

$$A = \begin{pmatrix} x \\ y \end{pmatrix}$$

(where x and y are not both zero) such that

$$RA = rA,$$

that is,

$$\begin{pmatrix} 3 & 1 \\ 4 & 3 \end{pmatrix} \begin{pmatrix} x \\ y \end{pmatrix} = r \begin{pmatrix} x \\ y \end{pmatrix},$$

or

$$3x + y = rx$$
$$4x + 3y = ry.$$

Collecting terms on the left yields

$$(3 - r)x + y = 0,$$
$$4x + (3 - r)y = 0,$$

or

$$\begin{pmatrix} 3-r & 1 \\ 4 & 3-r \end{pmatrix} \begin{pmatrix} x \\ y \end{pmatrix} = \begin{pmatrix} 0 \\ 0 \end{pmatrix}.$$

One solution to this system of equations is obviously $x = 0$ and $y = 0$. We are looking for a solution such that x and y are not both zero. But, in that case, there is **more than one solution** to this system of equations. Thus, the determinant of the matrix must be zero, that is,

$$\begin{vmatrix} 3-r & 1 \\ 4 & 3-r \end{vmatrix} = (3-r)(3-r) - (1)(4) = 0.$$

This simplifies to

$$r^2 - 6r + 5 = 0.$$

Factoring this equation gives $r = 1$ and $r = 5$, so the characteristic values for the matrix

$$R = \begin{pmatrix} 3 & 1 \\ 4 & 3 \end{pmatrix}$$

are $r = 1$ and $r = 5$. This also tells us that

$$\|R\| = 5.$$

Our next problem is to find a characteristic vector corresponding to the value of $r = 1$, that is, find values for x and y that satisfy the equations

$$(3-r)x + y = 0 \quad \text{and} \quad 4x + (3-r)y = 0$$

when $r = 1$. Substituting 1 for r in these equations gives us

$$2x + y = 0 \quad \text{and} \quad 4x + 2y = 0.$$

Since the second equation is a multiple of the first, any solution to the equation

$$2x + y = 0$$

will work. We can give any value we wish for y, and then solve for x. Thus, letting $y = 2$ (to avoid fractions) we get $x = -1$ and a characteristic vector is

$$A = \begin{pmatrix} -1 \\ 2 \end{pmatrix}.$$

We could also have used Gauss–Jordan elimination to find the vector A.

We get other characteristic vectors by picking different values for y. Letting $y = 2a$, we get $x = -a$, and all characteristic vectors corresponding to $r = 1$ are of the form

$$a \begin{pmatrix} -1 \\ 2 \end{pmatrix}.$$

You should check the computations to see that

$$RA = A.$$

To find a characteristic vector corresponding to $r = 5$, we substitute 5 for r in the above equations, giving

$$-2x + y = 0 \qquad \text{and} \qquad 4x - 2y = 0.$$

Again the second equation is a multiple of the first, so we only need a solution to the equation

$$-2x + y = 0.$$

Thus we again see that we can pick any value for y, say $y = a$. Therefore, $x = y/2 = a/2$, and

$$a \begin{pmatrix} 0.5 \\ 1 \end{pmatrix}$$

is a characteristic vector for $r = 5$, for every value of a. In particular, picking $a = 2$ (to eliminate decimals), we get

$$B = \begin{pmatrix} 1 \\ 2 \end{pmatrix}.$$

Again you should compute RB to see that it equals $5B$.

Let's develop a systematic method for finding the characteristic values for an m-by-m matrix R. We are looking for a number r and a vector A (with at least one nonzero entry), such that

$$RA = rA.$$

Since $A = IA$, where I is the identity matrix, we have

$$RA = rIA, \qquad \text{or} \qquad (R - rI)A = 0.$$

We see that we are looking for a number r and a vector A such that $(R - rI)A = 0$, where A is not all zeros. Since the vector containing all zeros is also a solution to this equation, we must have more than one solution (actually an infinite number) to this equation. Thus, the determinant of the m-by-m matrix $(R - rI)$, must be zero. This leads to the following conclusion.

THEOREM 8.2   *The characteristic values of an m-by-m matrix R are the values r such that the determinant of the matrix $(R - rI)$ is zero, that is, the solutions r to the equation*

$$|R - rI| = 0.$$

EXAMPLE 8.3   Find the characteristic values for the matrix

$$R = \begin{pmatrix} 1 & 2 & 1 \\ -1 & 1 & 0 \\ 3 & -6 & -1 \end{pmatrix}.$$

First we construct the matrix

$$R - rI = \begin{pmatrix} 1 & 2 & 1 \\ -1 & 1 & 0 \\ 3 & -6 & -1 \end{pmatrix} - r \begin{pmatrix} 1 & 0 & 0 \\ 0 & 1 & 0 \\ 0 & 0 & 1 \end{pmatrix}$$

$$= \begin{pmatrix} 1 - r & 2 & 1 \\ -1 & 1 - r & 0 \\ 3 & -6 & -1 - r \end{pmatrix}.$$

The determinant is

$$|R - rI| = (1 - r)(1 - r)(-1 - r) + (2)(0)(3) + (1)(-1)(-6)$$
$$- (1 - r)(0)(-6) - (2)(-1)(-1 - r) - (1)(1 - r)(3)$$
$$= -r^3 + r^2 + 2r \qquad (8.3)$$

after simplification. To solve the equation

$$|R - rI| = -r^3 + r^2 + 2r = 0,$$

we only need to factor the polynomial, giving

$$r(-r + 2)(r + 1) = 0.$$

Thus, the characteristic values are

$$r = 0, \qquad r = 2, \qquad r = -1.$$

and the norm of R is

$$\|R\| = 2. \qquad \blacksquare$$

REMARK: The determinant of the matrix $|R - r|$ is an mth degree polynomial in r and can thus have at most m distinct roots. Thus R can have at most m distinct characteristic values, the roots of the polynomial $|R - r|$. The idea behind finding characteristic values is easy, but in practice it may be difficult to accomplish since we would need to factor an mth degree polynomial.

Since the characteristic values are the roots of the polynomial $|R - rI|$ there will be times that the characteristic values are **complex numbers.** If r_1 is a double root of the above polynomial, that is, $(r - r_1)^2$ is a factor of the polynomial, we say that r_1 is a characteristic value of order 2. If $(r - r_1)^j$ is a factor of the polynomial $|R - rI|$, then r_1 is called a characteristic value of order j.

Once the characteristic values are found, we need to find corresponding characteristic vectors. To do this, we pick a characteristic value, say r_1, construct the matrix

$$R - r_1 I,$$

and then solve the system of equations

$$(R - r_1 I)A = 0$$

for the unknowns $a_1, a_2, \ldots, a_m$, where

$$A = \begin{pmatrix} a_1 \\ \vdots \\ a_m \end{pmatrix},$$

possibly using Gauss–Jordan elimination.   □

EXAMPLE 8.4   To find a characteristic vector corresponding to the characteristic value $r = 2$ for the matrix R of example 8.3, we construct the matrix

$$R - 2I = \begin{pmatrix} 1 & 2 & 1 \\ -1 & 1 & 0 \\ 3 & -6 & -1 \end{pmatrix} - 2\begin{pmatrix} 1 & 0 & 0 \\ 0 & 1 & 0 \\ 0 & 0 & 1 \end{pmatrix}$$

$$= \begin{pmatrix} -1 & 2 & 1 \\ -1 & -1 & 0 \\ 3 & -6 & -3 \end{pmatrix}.$$

We then solve the equations

$$(R - 2I)A = \begin{pmatrix} -1 & 2 & 1 \\ -1 & -1 & 0 \\ 3 & -6 & -3 \end{pmatrix}\begin{pmatrix} x \\ y \\ z \end{pmatrix} = \begin{pmatrix} 0 \\ 0 \\ 0 \end{pmatrix},$$

or

$$-x + 2y + z = 0, \qquad -x - y = 0, \qquad 3x - 6y - 3z = 0.$$

We construct the augmented matrix

$$M' = \begin{pmatrix} -1 & 2 & 1 & 0 \\ -1 & -1 & 0 & 0 \\ 3 & -6 & -3 & 0 \end{pmatrix}.$$

Multiplying the first row by -1 gives

$$M' = \begin{pmatrix} 1 & -2 & -1 & 0 \\ -1 & -1 & 0 & 0 \\ 3 & -6 & -3 & 0 \end{pmatrix}.$$

Adding the first row to the second gives

$$M' = \begin{pmatrix} 1 & -2 & -1 & 0 \\ 0 & -3 & -1 & 0 \\ 3 & -6 & -3 & 0 \end{pmatrix}.$$

Subtracting 3 times the first row from the third gives

$$M' = \begin{pmatrix} 1 & -2 & -1 & 0 \\ 0 & -3 & -1 & 0 \\ 0 & 0 & 0 & 0 \end{pmatrix}.$$

Dividing the second row by -3 gives

$$M' = \begin{pmatrix} 1 & -2 & -1 & 0 \\ 0 & 1 & 1/3 & 0 \\ 0 & 0 & 0 & 0 \end{pmatrix}.$$

Adding twice the second row to the first gives

$$M' = \begin{pmatrix} 1 & 0 & -1/3 & 0 \\ 0 & 1 & 1/3 & 0 \\ 0 & 0 & 0 & 0 \end{pmatrix}.$$

This gives that $x = (1/3)z$ and $y = -(1/3)z$. Letting $z = 3$, we get
that $x = 1$, $y = -1$, and a characteristic vector is

$$A = \begin{pmatrix} 1 \\ -1 \\ 3 \end{pmatrix}.$$

Any multiple of this vector will also work.

To find a characteristic vector B for the value $r = 0$ we need to
solve the system of equations

$$(R - 0I)B = RB = \begin{pmatrix} 1 & 2 & 1 \\ -1 & 1 & 0 \\ 3 & -6 & -1 \end{pmatrix} \begin{pmatrix} x \\ y \\ z \end{pmatrix} = \begin{pmatrix} 0 \\ 0 \\ 0 \end{pmatrix},$$

or

$$x + 2y + z = 0, \qquad -x + y = 0, \qquad 3x - 6y - z = 0.$$

You should find that a solution (or characteristic vector) is

$$B = \begin{pmatrix} 1 \\ 1 \\ -3 \end{pmatrix}$$

or any multiple of this vector.

Likewise, to find a characteristic vector C for $r = -1$ you need to solve the system of equations

$$(R + I)C = \begin{pmatrix} 2 & 2 & 1 \\ -1 & 2 & 0 \\ 3 & -6 & 0 \end{pmatrix} \begin{pmatrix} x \\ y \\ z \end{pmatrix} = \begin{pmatrix} 0 \\ 0 \\ 0 \end{pmatrix},$$

or

$$2x + 2y + z = 0, \qquad -x + 2y = 0, \qquad 3x - 6y = 0.$$

You should find that a characteristic vector is

$$C = \begin{pmatrix} 2 \\ 1 \\ -6 \end{pmatrix}$$

or any multiple of this vector.   ∎

If r is a characteristic value of order 2, we would expect to find two characteristic vectors, A and B, where B is not a multiple of A. We will not deal with this case. Refer to any book on linear algebra, if you are interested.

To review the results of examples 8.3 and 8.4, the matrix

$$R = \begin{pmatrix} 1 & 2 & 1 \\ -1 & 1 & 0 \\ 3 & -6 & -1 \end{pmatrix}$$

has the characteristic values $r = 2$, 0, and -1, and corresponding characteristic vectors

$$A = \begin{pmatrix} 1 \\ -1 \\ 3 \end{pmatrix}, \qquad B = \begin{pmatrix} 1 \\ 1 \\ -3 \end{pmatrix}, \qquad C = \begin{pmatrix} 2 \\ 1 \\ -6 \end{pmatrix}.$$

Thus, $RA = 2A$, $RB = 0B = 0$, and $RC = -C$.

We now wish to compute RX, where X is some vector other than one of the characteristic vectors. The idea is to write X as a sum of multiples of the characteristic vectors, say

$$X = c_1 A + c_2 B + c_3 C$$

for some constants c_1, c_2, and c_3. If we can do this, then

$$RX = R(c_1 A + c_2 B + c_3 C) = c_1 RA + c_2 RB + c_3 RC$$
$$= c_1(2)A + c_2(0)B + c_3(-1)C = 2c_1 A - c_3 C.$$

Then,

$$R^k X = c_1 R^k A + c_2 R^k B + c_3 R^k C = c_1 2^k A + c_3(-1)^k C.$$

Suppose that

$$X = \begin{pmatrix} 3 \\ -4 \\ 3 \end{pmatrix}.$$

We now wish to solve the system of equations

$$X = c_1 A + c_2 B + c_3 C$$

for the constants c_1, c_2, and c_3, that is, solve

$$c_1 \begin{pmatrix} 1 \\ -1 \\ 3 \end{pmatrix} + c_2 \begin{pmatrix} 1 \\ 1 \\ -3 \end{pmatrix} + c_3 \begin{pmatrix} 2 \\ 1 \\ -6 \end{pmatrix} = \begin{pmatrix} 3 \\ -4 \\ 3 \end{pmatrix}$$

or

$$c_1 + c_2 + 2c_3 = 3, \qquad -c_1 + c_2 + c_3 = -4, \qquad 3c_1 - 3c_2 - 6c_3 = 3.$$

You should find it relatively simple to solve these three equations. If you do you will find that

$$c_1 = 2, \qquad c_2 = -5, \qquad c_3 = 3.$$

Thus,

$$X = 2A - 5B + 3C$$

and

$$R^k X = R^k(2A - 5B + 3C) = 2(2^k)A + 3(-1)^k C$$
$$= 2^{k+1} \begin{pmatrix} 1 \\ -1 \\ 3 \end{pmatrix} + 3(-1)^k \begin{pmatrix} 2 \\ 1 \\ -6 \end{pmatrix}$$

(since $R^k B = 0B$).

To review, suppose we want to find the solution to the dynamical system of m equations given by $A(n+1) = RA(n)$ with $A(0) = X$. Here we have an m-by-m matrix R, an m-vector X. The solution, which we want to compute, is $A(k) = R^k X$. To accomplish this computation, we follow these four steps.

1. **Compute the characteristic values:** Set the determinant of the matrix $R - rI$ equal to zero and solve, that is, find solutions r to the equation

$$|R - rI| = 0.$$

   This means finding the m roots, $r_1, r_2, \ldots, r_m$, of an mth-degree polynomial. These roots are the m characteristic values of R. We will assume that these are all distinct.

2. **Compute the characteristic vectors:** To do this, solve the system of m equations

$$(R - r_1 I) A_1 = 0,$$

   where the m unknowns are the m components of the vector A_1. Note that the solution to this system of equations will not be unique, that is, there will be one independent component, and the other components will be given as multiples of this component. Repeat this process for $r_2, r_3, \ldots, r_m$, finding corresponding characteristic vectors $A_2, \ldots, A_m$. We now have m characteristic values and their corresponding characteristic vectors.

3. **Write X as a sum of characteristic vectors:** Remember that X and $A_1, \ldots, A_m$ are all known at this point. Now solve the system of equations

$$c_1 A_1 + c_2 A_2 + \cdots + c_m A_m = X$$

   for the unknowns, $c_1, \ldots, c_m$. This solution will be unique.

4. **Compute our answer, that is, compute $R^k X$:** To do this, we compute

$$R^k X = R^k (c_1 A_1 + \cdots + c_m A_m) = c_1 r_1^k A_1 + \cdots + c_m r_m^k A_m.$$

REMARK: Construct the matrix

$$M = A_1 \mid A_2 \mid \cdots \mid A_m,$$

that is, the jth column of M is A_j. Also let

$$C = \begin{pmatrix} c_1 \\ \vdots \\ c_m \end{pmatrix}.$$

Then step 3 is to solve the system of equations

$$MC = X.$$

The solution is

$$C = M^{-1}X.$$ □

Let's demonstrate these steps by computing $R^3 X$, where

$$R = \begin{pmatrix} 1 & 2 \\ -1 & 4 \end{pmatrix} \quad \text{and} \quad X = \begin{pmatrix} 5 \\ 1 \end{pmatrix}.$$

First we find the characteristic values, that is, we solve

$$|R - rI| = (1 - r)(4 - r) - (2)(-1)$$
$$= r^2 - 5r + 6 = (r - 3)(r - 2) = 0,$$

for r. Therefore, the characteristic values are $r = 3$ and $r = 2$.

Next, we find the characteristic vectors starting with a characteristic vector corresponding to $r = 3$. To do this we solve the system of equations $(R - 3I)A_1 = 0$, that is,

$$\begin{pmatrix} -2 & 2 \\ -1 & 1 \end{pmatrix} \begin{pmatrix} x \\ y \end{pmatrix} = \begin{pmatrix} 0 \\ 0 \end{pmatrix}.$$

This is in reality only a single equation, that is,

$$-x + y = 0.$$

Since we can let y be anything we want, let's pick $y = 1$. Then $x = 1$, and a characteristic vector corresponding to $r = 3$ is

$$A_1 = \begin{pmatrix} 1 \\ 1 \end{pmatrix}.$$

We could have let $x = 1$ in the above and then solved for y.

To find a characteristic vector corresponding to $r = 2$, solve the system of equations $(R - 2I)A_2 = 0$, that is,

$$\begin{pmatrix} -1 & 2 \\ -1 & 2 \end{pmatrix} \begin{pmatrix} x \\ y \end{pmatrix} = \begin{pmatrix} 0 \\ 0 \end{pmatrix}.$$

This is in reality only a single equation,

$$-x + 2y = 0.$$

Since we can let y be anything we want, let's pick $y = 1$. Then $x = 2$, and a characteristic vector corresponding to $r = 2$ is

$$A_2 = \begin{pmatrix} 2 \\ 1 \end{pmatrix}.$$

Our third step is to rewrite X in terms of A_1 and A_2, that is, solve

$$c_1 \begin{pmatrix} 1 \\ 1 \end{pmatrix} + c_2 \begin{pmatrix} 2 \\ 1 \end{pmatrix} = \begin{pmatrix} 5 \\ 1 \end{pmatrix},$$

or

$$c_1 + 2c_2 = 5 \qquad \text{and} \qquad c_1 + c_2 = 1.$$

You should find that the solution to these equations is $c_1 = -3$ and $c_2 = 4$. Thus,

$$X = -3A_1 + 4A_2.$$

Finally we can compute the answer. From the above calculations, we get

$$R^3 X = R^3(-3A_1 + 4A_2) = -3(3^3)A_1 + 4(2^3)A_2 = -81A_1 + 32A_2$$

$$= -81 \begin{pmatrix} 1 \\ 1 \end{pmatrix} + 32 \begin{pmatrix} 2 \\ 1 \end{pmatrix} = \begin{pmatrix} -17 \\ -49 \end{pmatrix}.$$

You should compute $R^3 X$ using matrix multiplication to check the answer.

Matrix multiplication is easier in this example. But if you were asked to compute $R^{20} X$, this procedure would easily give the answer

$$R^{20} X = R^{20}(-3A_1 + 4A_2) = -3(3^{20})A_1 + 4(2^{20})A_2$$

$$= \begin{pmatrix} -3(3^{20}) + 8(2^{20}) \\ -3(3^{20}) + 4(2^{20}) \end{pmatrix},$$

while matrix multiplication would require long and tedious computations.

Notice that our procedure gives a relatively simple method for performing certain matrix multiplications without ever having to multiply any matrices.

PROBLEMS

1. Consider the matrix R and the vector X given as

$$R = \begin{pmatrix} -2 & 3 \\ 1 & 0 \end{pmatrix} \qquad \text{and} \qquad X = \begin{pmatrix} -2 \\ 6 \end{pmatrix}.$$

   a. Find the characteristic values for R.

   b. Find corresponding characteristic vectors.

   c. Write X as a sum of these characteristic vectors.

   d. Compute $R^4 X$.

2. Consider the matrix R and the vector X given as

$$R = \begin{pmatrix} 3 & 4 \\ 2 & 1 \end{pmatrix} \quad \text{and} \quad X = \begin{pmatrix} 7 \\ 8 \end{pmatrix}.$$

   a. Find the characteristic values for R.

   b. Find corresponding characteristic vectors.

   c. Write X as a sum of these characteristic vectors.

   d. Compute $R^3 X$.

3. Consider the matrix R and the vector X given as

$$R = \begin{pmatrix} -4 & -4 & -23 \\ 2 & 2 & 11 \\ 0 & 0 & 2 \end{pmatrix} \quad \text{and} \quad X = \begin{pmatrix} -25 \\ 13 \\ 4 \end{pmatrix}.$$

   a. Find the characteristic values for R.

   b. Find corresponding characteristic vectors.

   c. Write X as a sum of these characteristic vectors.

   d. Compute $R^5 X$.

4. Consider the matrix R and the vector X given as

$$R = \begin{pmatrix} 14 & -2 & 5 \\ 9 & -1 & 3 \\ -30 & 4 & -11 \end{pmatrix} \quad \text{and} \quad X = \begin{pmatrix} -3 \\ 1 \\ 9 \end{pmatrix}.$$

   a. Find the characteristic values for R.

   b. Find corresponding characteristic vectors.

   c. Write X as a sum of these characteristic vectors.

   d. Compute $R^5 X$.

8.3   First-order dynamical systems of several equations

In section 5.4 we derived two equations which described the process of sex-linked traits in genetics. Letting $p(n)$ be the proportion of A-alleles among women in generation n, and $p'(n)$ be the proportion of A-alleles among men in generation n, we derived the two equations

$$p(n+1) = 0.5p(n) + 0.5p'(n) \quad \text{and} \quad p'(n+1) = p(n).$$

Defining the vector $P(n)$ and the matrix R as

$$P(n) = \begin{pmatrix} p(n) \\ p'(n) \end{pmatrix} \quad \text{and} \quad R = \begin{pmatrix} 0.5 & 0.5 \\ 1 & 0 \end{pmatrix},$$

we can rewrite the above two equations as

$$P(n+1) = RP(n).$$

The general solution to this first-order system of equations is

$$P(k) = R^k P,$$

where P is a fixed vector that depends on the initial values $p(0)$ and $p'(0)$.

To compute $R^k P$, we use the four steps discussed in the previous section. First, we find the characteristic values for R, that is, we solve

$$|R - rI| = (0.5 - r)(-r) - 0.5(1) = r^2 - 0.5r - 0.5 = 0.$$

Factoring, we get

$$|R - rI| = (r - 1)(r + 0.5) = 0,$$

so the characteristic values are $r = 1$ and $r = -0.5$ and $\|R\| = 1$.

Second, we compute characteristic vectors corresponding to $r = 1$ and to $r = -0.5$. To compute the characteristic vector associated with $r = 1$, we solve the system of equations

$$(R - 1I)P_1 = 0$$

for the vector P_1, that is, solve

$$\begin{pmatrix} -0.5 & 0.5 \\ 1 & -1 \end{pmatrix} \begin{pmatrix} x \\ y \end{pmatrix} = \begin{pmatrix} 0 \\ 0 \end{pmatrix}.$$

Both of these equations reduce to

$$x - y = 0.$$

Since y can be anything, let $y = 1$. Then $x = 1$ and

$$P_1 = \begin{pmatrix} 1 \\ 1 \end{pmatrix}.$$

You should find that a characteristic vector corresponding to $r = -0.5$ is

$$P_2 = \begin{pmatrix} -1 \\ 2 \end{pmatrix}.$$

Third, since we wish to compute $R^k P$, we would write P in terms of P_1 and P_2, that is, we would find numbers c_1 and c_2 such that

$$P = c_1 P_1 + c_2 P_2.$$

Fourth, we would find that the general solution is

$$P(k) = R^k P = c_1(1)^k P_1 + c_2(-0.5)^k P_2.$$

If we were given $P(0) = P$, we could compute the numbers c_1 and c_2. But, in this example, we assumed that genetics had been operating for many generations, and that presently we are in generation k where k is large. Since $|-0.5| < 1$, it follows that $(-0.5)^k$ goes to zero so that

$$\lim_{k \to \infty} P(k) = c_1 P_1 = \begin{pmatrix} c_1 \\ c_1 \end{pmatrix}.$$

From this, we see that $p(k) = c_1$ and $p'(k) = c_1$ and we can draw the same conclusion that we did in section 5.4, that is, the proportion of A-alleles is approximately the same in men as in women. Note that c_1 is the proportion of men with the trait determined by the A-allele, which can be easily determined through observation.

We note here that we actually have two methods for solving first-order linear systems of several equations. The first method is to rewrite the equations as one higher-order dynamical system and then solve by factoring the characteristic equation. The second is to solve the system by finding the roots of the determinant, that is, solving $|R - rI| = 0$. These two methods must give **the same results**, since they both give solutions to the same dynamical system, and there is one and only one particular solution to a first-order linear dynamical system of several equations.

The powers of the roots of the characteristic equation control the solution to the higher-order dynamical system. The powers of the roots of the determinant of $|R - rI|$ control the solution to the dynamical system of several equations. But both solutions are in some sense the same, since they describe the same process. Thus the roots of the characteristic equation and the roots of the determinant $|R - rI| = 0$ must be the same. Therefore, the characteristic equation for the higher-order system and the determinant of $|R - rI|$ **must be the same**. Because of this, we will call the equation

$$|R - rI| = 0,$$

the characteristic equation to the first-order dynamical system

$$A(n+1) = RA(n).$$

You might ask which method is better. In some sense, they are the same. On the other hand, the first-order system is aesthetically more pleasing because systems of several equations arise naturally (as in sex-linked genes and populations with age structure) and they have a simple matrix notation.

We will see later that systems of equations are easier to deal with when we have a large number of equations, say a system of four or more equations. In these cases, it is often difficult to factor the characteristic equation, but we can use the system and our computer to estimate the roots (or at least the most important root) of the characteristic equation. This will be enough to determine the long-term behavior of the system.

Let's work some examples.

EXAMPLE 8.5   Find the particular solution to the system of equations

$$a(n+1) = a(n) + 2b(n) \qquad \text{and} \qquad b(n+1) = -a(n) + 4b(n)$$

where $a(0) = 1$ and $b(0) = 2$.

Define the vector $A(n)$ and the matrix R as

$$A(n) = \begin{pmatrix} a(n) \\ b(n) \end{pmatrix} \qquad \text{and} \qquad R = \begin{pmatrix} 1 & 2 \\ -1 & 4 \end{pmatrix}.$$

Note that

$$A(0) = \begin{pmatrix} 1 \\ 2 \end{pmatrix}.$$

The above system of equations can now be written as

$$A(n+1) = RA(n).$$

The general solution is $A(k) = R^k A$ and the particular solution is $A(k) = R^k A(0)$. The first step is to compute the roots of the characteristic equation, that is, the characteristic values for $|R - rI| = 0$. The characteristic equation is

$$(1-r)(4-r) - 2(-1) = 0, \qquad \text{or} \qquad (r-2)(r-3) = 0$$

after simplification. The roots are $r = 3$ and $r = 2$.

If A_1 and A_2 are characteristic vectors corresponding to $r = 3$ and $r = 2$, respectively, then the general solution is

$$A(k) = R^k A = R^k(c_1 A_1 + c_2 A_2) = c_1 3^k A_1 + c_2 2^k A_2,$$

where c_1 and c_2 are two unknown constants which can be determined from the initial conditions, $a(0)$ and $b(0)$.

The second step is to compute characteristic vectors corresponding to $r = 3$ and $r = 2$, respectively. For $r = 3$, solving the system of equations

$$(R - 2I)A_1 = \begin{pmatrix} -2 & 2 \\ -1 & 1 \end{pmatrix} \begin{pmatrix} x \\ y \end{pmatrix} = \begin{pmatrix} 0 \\ 0 \end{pmatrix}$$

(which reduces to the single equation $-x + y = 0$) gives the characteristic vector

$$A_1 = \begin{pmatrix} 1 \\ 1 \end{pmatrix}$$

(or any multiple of this vector).

Likewise, you should find that a characteristic vector corresponding to $r = 2$ is

$$A_2 = \begin{pmatrix} 2 \\ 1 \end{pmatrix}.$$

The third step is to write $A(0)$ in terms of A_1 and A_2. Thus, we need to solve

$$c_1 A_1 + c_2 A_2 = A(0),$$

that is, solve the two equations,

$$c_1 + 2c_2 = 1 \quad \text{and} \quad c_1 + c_2 = 2.$$

You should get $c_1 = 3$ and $c_2 = -1$ as the solution. Thus,

$$A(0) = 3A_1 - A_2.$$

The fourth step is to compute the particular solution. This is,

$$A(k) = R^k A(0) = R^k(3A_1 - A_2) = 3(3^k)A_1 - (2^k)A_2$$
$$= \begin{pmatrix} 3(3^k) - 2(2^k) \\ 3(3^k) - (2^k) \end{pmatrix}. \qquad \blacksquare$$

EXAMPLE 8.6   Let's find the general solution and then the particular solution to the dynamical system

$$A(n+1) = RA(n),$$

given that

$$R = \begin{pmatrix} 0.5 & -0.25 \\ 1 & 0.5 \end{pmatrix} \quad \text{and} \quad A(0) = \begin{pmatrix} 1 \\ 2 \end{pmatrix}.$$

First, we find the roots of the characteristic equation

$$(0.5 - r)(0.5 - r) - (-0.25)(1) = r^2 - r + 0.5 = 0.$$

Using the quadratic formula, the roots are the complex numbers

$$r = 0.5 + 0.5i \quad \text{and} \quad r = 0.5 - 0.5i.$$

Thus, the general solution is

$$A(k) = c_1 (0.5 + 0.5i)^k A_1 + c_2 (0.5 - 0.5i)^k A_2,$$

where c_1 and c_2 are two unknown constants and A_1 and A_2 are the corresponding characteristic vectors.

Here,

$$|0.5 \pm 0.5i| = \sqrt{0.5^2 + 0.5^2} = \sqrt{0.5} = \frac{1}{\sqrt{2}} = \|R\|.$$

To solve for the characteristic vector corresponding to $r = 0.5 + 0.5i$, we need to find a vector that satisfies the equations

$$(R - (0.5 + 0.5i)I)A_1 = \begin{pmatrix} -0.5i & -0.25 \\ 1 & -0.5i \end{pmatrix} \begin{pmatrix} x \\ y \end{pmatrix} = \begin{pmatrix} 0 \\ 0 \end{pmatrix},$$

which means solving the two equations

$$-0.5ix = 0.25y \quad \text{and} \quad x = 0.5iy.$$

After multiplying both sides of the second equation by $-0.5i$ it is seen that the second equation is a multiple of the first. Thus, we need to solve one of the equations, say

$$x = 0.5iy.$$

Since y can be any real (or complex) number, let $y = 2$. Then we get $x = i$, and a characteristic vector is

$$A_1 = \begin{pmatrix} i \\ 2 \end{pmatrix}.$$

A similar calculation gives a characteristic vector corresponding to $r = 0.5 - 0.5i$ as

$$A_2 = \begin{pmatrix} -i \\ 2 \end{pmatrix}.$$

Thus, the general solution to the real dynamical system is given by

$$A(k) = R^k A = R^k (c_1 A_1 + c_2 A_2)$$

$$= c_1 (0.5 + 0.5i)^k \begin{pmatrix} i \\ 2 \end{pmatrix} + c_2 (0.5 - 0.5i)^k \begin{pmatrix} -i \\ 2 \end{pmatrix},$$

where c_1 and c_2 are unknown (complex) constants that can be determined from the initial value $A(0)$.

If we wish to find the particular solution, given $A(0)$ above, we need to solve for c_1 and c_2 in the equation

$$A(0) = c_1 A_1 + c_2 A_2,$$

that is, solve the equations

$$ic_1 - ic_2 = 1 \qquad \text{and} \qquad 2c_1 + 2c_2 = 2.$$

Using Gauss–Jordan elimination, the augmented matrix is

$$M' = \begin{pmatrix} i & -i & 1 \\ 2 & 2 & 2 \end{pmatrix}.$$

Multiply the first row by $-i$, giving

$$M' = \begin{pmatrix} 1 & -1 & -i \\ 2 & 2 & 2 \end{pmatrix}.$$

Subtract twice the first row from the second, giving

$$M' = \begin{pmatrix} 1 & -1 & -i \\ 0 & 4 & 2 + 2i \end{pmatrix}.$$

Divide the second row by 4, giving

$$M' = \begin{pmatrix} 1 & -1 & -i \\ 0 & 1 & 0.5 + 0.5i \end{pmatrix}.$$

Add the second row to the first to get

$$M' = \begin{pmatrix} 1 & 0 & 0.5 - 0.5i \\ 0 & 1 & 0.5 + 0.5i \end{pmatrix}.$$

Thus, $c_1 = 0.5 - 0.5i$ and $c_2 = 0.5 + 0.5i$.

The particular solution is then

$$A(k) = (0.5 - 0.5i)(0.5 + 0.5i)^k A_1 + (0.5 + 0.5i)(0.5 - 0.5i)^k A_2.$$

Noting that $(0.5 - 0.5i)(0.5 + 0.5i) = 0.5$, this solution can be simplified to

$$A(k) = 0.5(0.5 + 0.5i)^{k-1} A_1 + 0.5(0.5 - 0.5i)^{k-1} A_2.$$

It seems amazing that such a complicated solution, when multiplied out for a given k, gives real numbers for answers, the complex numbers all canceling out. This must happen, though, since this is the solution, and when we observe the original dynamical system, it is clear that we always get real numbers for the underlying $a(k)$ and $b(k)$.   ∎

PROBLEMS

1. Find a general solution, by finding the characteristic values and their corresponding characteristic vectors, for the dynamical system

$$A(n+1) = RA(n),$$

where R equals each of the following matrices.

    a. $\begin{pmatrix} -2 & 2 \\ 5 & 1 \end{pmatrix}$,   **b.** $\begin{pmatrix} 1.5 & 1 \\ 0.5 & 1 \end{pmatrix}$,

    c. $\begin{pmatrix} 2 & -5 \\ 1 & -2 \end{pmatrix}$,   **d.** $\begin{pmatrix} 2 & 5 & 1 \\ 0 & 1 & 4 \\ 0 & 0 & 3 \end{pmatrix}$.

2. Find the particular solution to the corresponding dynamical system in problem 1, when $A(0)$ equals the following vectors.

    a. $\begin{pmatrix} -3 \\ 10 \end{pmatrix}$,   **b.** $\begin{pmatrix} 0 \\ -3 \end{pmatrix}$,

    c.   **i.** $\begin{pmatrix} 4 \\ 2 \end{pmatrix}$, **ii.** $\begin{pmatrix} 10 \\ 4 \end{pmatrix}$,   **d.** $\begin{pmatrix} 7 \\ 3 \\ 1 \end{pmatrix}$.

3. Find the particular solution to the first-order linear system of equations

$$a(n+1) = -0.5a(n) + 0.5b(n)$$
$$b(n+1) = -0.5a(n) + 0.75b(n)$$

given that $a(0) = 6$ and $b(0) = 6$.

8.4   Regular Markov chains

In this section, we continue the discussion of Markov chains started in section 3.4. Suppose we conduct an experiment that has m possible results or **states**. Also suppose we keep repeating this experiment, but that the probability of each of the states occurring on the $(n+1)$th repetition of the experiment depends only on the result of the nth repetition of the experiment. This process is called a Markov chain.

One example of a Markov chain is our marble-drawing experiment of section 3.4. Recall that there were two bags, the red bag and the blue bag. The red bag contained 3 red marbles and 2 blue marbles, while the blue bag contained 3 red marbles and 7 blue marbles. A marble is drawn from a bag, its color is recorded, and the marble is returned to the bag from which it was drawn. The next marble is then chosen from the bag that is the same color as the last marble drawn. There are two states or results for each draw, a red marble or a blue marble. The probability of getting a red marble on the $(n+1)$th draw depends only on what color marble we got on the nth draw, since that determines the bag from which we will draw.

Suppose we have m states, $e_1, e_2, \ldots, e_m$. State e_1 is red, state e_2 is blue, and $m = 2$ in the marble-drawing example. Let $p_1(n)$ be the probability of state e_1 occurring on the nth repetition of our experiment. Likewise let $p_2(n), \ldots, p_m(n)$ be the probabilities of states $e_2, \ldots, e_m$, respectively, occurring on the nth repetition of the experiment. Since one of the m states must occur on the nth repetition, it follows that

$$p_1(n) + p_2(n) + \cdots + p_m(n) = 1. \tag{8.3}$$

Given that state e_1 was the result of the last repetition of the experiment, we must be given the probabilities that $e_1, e_2, \ldots,$ and e_m will occur on the next repetition of the experiment. Reading $P(e_2 \mid e_1)$ as the probability of e_2 occurring on the next repetition given that e_1 occurred on the last repetition, we denote these probabilities as

$$P(e_1 \mid e_1) = p_{11}, \quad P(e_2 \mid e_1) = p_{21}, \quad \ldots, \quad P(e_m \mid e_1) = p_{m1}.$$

Notice that

$$p_{11} + p_{21} + \cdots + p_{m1} = 1,$$

since, given that e_1 has happened on the last repetition, one of $e_1, e_2, \ldots,$ or e_m must occur on the next repetition.

REMARK: The probability that e_2 occurs given that e_1 occurred is denoted p_{21}. Observe that event e_1 occurred **before** event e_2, that is, the number that occurs first in the subscript corresponds to the event that happened last.   □

In the marble-drawing example of section 3.4,

$$P(e_1 \mid e_1) = p_{11} = 0.6 \qquad \text{and} \qquad P(e_2 \mid e_1) = p_{21} = 0.4$$

since there are 3 red and 2 blue marbles in the red bag, the bag from which you draw given that the previous draw was $e_1 =$ red. Note that $0.6 + 0.4 = 1$.

Likewise, given that e_2 occurred on the last repetition of the experiment, we must know the probabilities of $e_1, \ldots, e_m$ occurring on the next repetition, that is, we must know

$$P(e_1 \mid e_2) = p_{12}, \quad P(e_2 \mid e_2) = p_{22}, \quad \ldots, \quad P(e_m \mid e_2) = p_{m2}.$$

Again in the marble-drawing example, we have

$$P(e_1 \mid e_2) = p_{12} = 0.3 \qquad \text{and} \qquad P(e_2 \mid e_2) = p_{22} = 0.7$$

since there are 3 red and 7 blue marbles in the blue bag, the bag from which you draw given that the previous draw was $e_2 =$ blue. Note that $0.3 + 0.7 = 1$.

Similarly, if $m > 2$, we must know

$$P(e_1 \mid e_3) = p_{13}, \quad \ldots, \quad P(e_m \mid e_3) = p_{m3},$$
$$\vdots \qquad \qquad \ddots \qquad \qquad \vdots$$
$$P(e_1 \mid e_m) = p_{1m}, \quad \ldots, \quad P(e_m \mid e_m) = p_{mm}.$$

Again note that

$$p_{12} + p_{22} + \quad \cdots \quad + p_{m2} = 1,$$
$$\vdots$$
$$p_{1m} + p_{2m} + \quad \cdots \quad + p_{mm} = 1.$$

EXAMPLE 8.7 Suppose we have three bags of marbles.

- The red bag contains 2 red, 3 blue, and 5 yellow marbles.
- The blue bag contains 1 red, 5 blue, and 4 yellow marbles.
- The yellow bag contains 3 red, 1 blue, and 6 yellow marbles.

A marble is drawn from a bag, its color is noted, and the marble is returned. Red is e_1, blue is e_2, and yellow is e_3. The next marble is drawn from the bag that is the same color as the marble just drawn.

Given that a red marble e_1 was just drawn, we know that the next marble will be drawn from the red bag so that the probabilities that the next marble drawn is red, blue, and yellow are 0.2, 0.3, and 0.5, respectively, that is,

$$p_{11} = 0.2, \qquad p_{21} = 0.3, \qquad p_{31} = 0.5.$$

Note that $0.2 + 0.3 + 0.5 = 1$, since we must get one of the marbles from the red bag. Likewise,

$$p_{12} = 0.1, p_{22} = 0.5, p_{32} = 0.4$$
$$p_{13} = 0.3, p_{23} = 0.1, p_{33} = 0.6. \qquad \blacksquare$$

Suppose we wish to compute $p_1(n+1)$, that is, the probability that the $(n+1)$th repetition of the experiment will result in state e_1. There are m ways this can happen. The first case is that we get e_1 on repetitions n and $n+1$; the second case is that we get e_2 on repetition n and e_1 on repetition $n+1$; $\ldots$; and the mth case is that we get e_m on repetition n and e_1 on repetition $n+1$.

Case 1 is a two-stage experiment. The probability of the first stage, getting e_1 on the nth repetition, is $p_1(n)$. The probability of the second stage, getting e_1 given that we just got e_1 is p_{11}. By the multiplication principle,

$$p(\text{case } 1) = p_{11} p_1(n).$$

Likewise, the probability of the second, third, $\ldots$, and mth cases are

$$p(\text{case } 2) = p_{12} p_2(n), \quad \ldots, \quad p(\text{case } m) = p_{1m} p_m(n).$$

Recall that p_{jm} is the probability of e_j occurring on one repetition given that e_m occurred on the previous repetition.

Adding up the probabilities of these cases gives us

$$p_1(n+1) = p_{11} p_1(n) + p_{12} p_2(n) + \cdots + p_{1m} p_m(n).$$

Similarly,

$$p_2(n+1) = p_{21} p_1(n) + p_{22} p_2(n) + \cdots + p_{2m} p_m(n)$$
$$\vdots$$
$$p_m(n+1) = p_{m1} p_1(n) + p_{m2} p_2(n) + \cdots + p_{mm} p_m(n).$$

This gives us a dynamical system of m equations.

Let's compute $p_1(n+1)$ for example 8.7, that is, the probability that we get a red marble on the $(n+1)$th draw. We can get a red marble in three ways or cases: from the red bag, from the blue bag, or from the yellow bag. The probability we are drawing from the red bag on the $(n+1)$th draw is the probability that we got a red marble on our nth draw, $p_1(n)$. The probability that we then draw a red marble (second stage) is p_{11}. Thus, the probability of the first case is

$$p_{11}p_1(n) = 0.2p_1(n).$$

The probability that we get a red marble from the blue bag on the $(n+1)$th draw is the probability that the nth marble was blue, $p_2(n)$, and that we then got a red marble from the blue bag, $p_{12} = 0.1$. By the multiplication principle, this probability is

$$p_{12}p_2(n) = 0.1p_2(n).$$

Similarly, the probability that we get a red marble from the yellow bag is

$$p_{13}p_3(n) = 0.3p_3(n).$$

Thus,

$$p_1(n+1) = 0.2p_1(n) + 0.1p_2(n) + 0.3p_3(n).$$

Similarly,

$$p_2(n+1) = 0.3p_1(n) + 0.5p_2(n) + 0.1p_3(n)$$
$$p_3(n+1) = 0.5p_1(n) + 0.4p_2(n) + 0.6p_3(n).$$

We can rewrite the m equations as the first-order dynamical system

$$P(n+1) = RP(n),$$

by defining the vector $P(n)$ and the matrix R as

$$P(n) = \begin{pmatrix} p_1(n) \\ \vdots \\ p_m(n) \end{pmatrix} \quad \text{and} \quad R = \begin{pmatrix} p_{11} & \cdots & p_{1m} \\ \vdots & \ddots & \vdots \\ p_{m1} & \cdots & p_{mm} \end{pmatrix}.$$

The general solution is

$$P(k) = R^k P(0).$$

The vector $P(k)$ is called the **probability vector** associated with the Markov chain, and the m-by-m matrix R is called the **transition matrix**.

Since each of the components $p_j(k)$ of $P(k)$ is the probability of event e_j occurring on the kth draw, we must have

$$0 \le p_j(k) \le 1,$$

that is, each component of $P(k)$ must be between 0 and 1, inclusive. Also, since one of the events e_j must occur, **the sum of the components of $P(k)$ must equal 1.**

Since each of the components of R is a probability, every number in the transition matrix R must be between 0 and 1, inclusive. Also, from the way R was constructed, **the sum of the numbers in each column of R must equal 1.**

In example 8.7

$$P(n) = \begin{pmatrix} p_1(n) \\ p_2(n) \\ p_3(n) \end{pmatrix} \quad \text{and} \quad R = \begin{pmatrix} 0.2 & 0.1 & 0.3 \\ 0.3 & 0.5 & 0.1 \\ 0.5 & 0.4 & 0.6 \end{pmatrix}.$$

Notice that the numbers in each column of R add up to 1. This is because the numbers in the first column are p_{11}, p_{21}, and p_{31}, and it was given above that they add up to 1, **since they are just the proportions of red, blue, and yellow marbles in the red bag.** Likewise, the numbers in the second and third columns add up to 1 since they are just the proportions of red, blue, and yellow marbles in the blue and yellow bags, respectively.

We can actually compute that

$$R^{10} = \begin{pmatrix} 0.229 & 0.229 & 0.229 \\ 0.243 & 0.243 & 0.243 \\ 0.529 & 0.529 & 0.529 \end{pmatrix},$$

where the components are rounded to 3 decimal places. (Each column adds to 1, but the rounding makes each column appear to add to 1.001.) This seems interesting. Even more interesting is that we find that

$$R^{20} = \begin{pmatrix} 0.229 & 0.229 & 0.229 \\ 0.243 & 0.243 & 0.243 \\ 0.529 & 0.529 & 0.529 \end{pmatrix}.$$

One goal of this section is to understand what we have just observed.

Let's find the general solution to the dynamical system

$$P(n+1) = RP(n)$$

of example 8.7. Since we have an aversion to computing R^k, we find the characteristic values for R. Thus,

$$|R - rI| = (0.2 - r)(0.5 - r)(0.6 - r) + 0.005 + 0.036$$
$$- 0.15(0.5 - r) - 0.03(0.6 - r) - 0.04(0.2 - r).$$

Simplifying this, after a good bit of careful computation, you should get

$$|R - rI| = -r^3 + 1.3r^2 - 0.3r = -r(r-1)(r-0.3) = 0.$$

Thus, the three characteristic values are $r = 1$, 0, and 0.3, and the general solution is

$$P(k) = c_1 1^k P_1 + c_2 0^k P_2 + c_3 0.3^k P_3 = c_1 P_1 + c_3 0.3^k P_3,$$

where P_1, P_2, and P_3 are characteristic vectors corresponding to $r = 1$, 0, and 0.3, respectively. Since 0.3^k goes to zero as k goes to infinity,

$$\lim_{k \to \infty} P(k) = c_1 P_1.$$

Let's find a characteristic vector corresponding to $r = 1$. Note that we have saved ourselves some work in that we do not need to know the characteristic vectors corresponding to $r = 0$ and $r = 0.3$ to approximate $P(k)$ for large k. To find P_1, we compute

$$(R - I)P_1 = \begin{pmatrix} -0.8 & 0.1 & 0.3 \\ 0.3 & -0.5 & 0.1 \\ 0.5 & 0.4 & -0.4 \end{pmatrix} \begin{pmatrix} x \\ y \\ z \end{pmatrix} = \begin{pmatrix} 0 \\ 0 \\ 0 \end{pmatrix}.$$

At this point you should be able to find a solution to these equations. One solution (characteristic vector) is

$$P_1 = \begin{pmatrix} 16 \\ 17 \\ 37 \end{pmatrix}.$$

Since $P(k) \to c_1 P_1$ and the components $(p_1(k), p_2(k),$ and $p_3(k))$ must add up to 1, it follows that the components of $c_1 P_1$ must add up to 1, that is,

$$16c_1 + 17c_1 + 37c_1 = 1, \qquad \text{or} \qquad c_1 = 1/70.$$

Thus,

$$\lim_{k \to \infty} P(k) = \begin{pmatrix} 16/70 \\ 17/70 \\ 37/70 \end{pmatrix} = \begin{pmatrix} 0.229 \\ 0.243 \\ 0.529 \end{pmatrix}$$

when rounded. Note that this vector equals each column of the matrix R^{20} computed earlier, that is, R^k appears to converge to a matrix in which each column equals a characteristic vector corresponding to the characteristic value $r = 1$.

The answer to our problem is then

$p_1(k) = $ the probability the kth draw is red $\to 16/70$,

$p_2(k) = $ the probability the kth draw is blue $\to 17/70$,

$p_3(k) = $ the probability the kth draw is yellow $\to 37/70$.

Notice that we did not need $P(0)$ in order to compute the long term probabilities, that is, we did not need to know from which bag we drew our first marble in order to compute the probability that our 100th marble was red.

REMARK: Many expositions on Markov chains multiply row vectors by matrices PR instead of matrices by column vectors RP. We did not use that convention, because it does not seem aesthetically pleasing to write the dynamical system as

$$P(n+1) = P(n)R.$$

As a word of caution when reading other expositions on Markov chains, writing the vector on the left, $P(n)R$, the matrix R is not the same matrix that we have been considering but is its transpose, that is, columns of our matrix become rows of their matrix.   □

The general solution to the dynamical system

$$P(n+1) = RP(n)$$

is

$$P(k) = c_1(r_1)^k P_1 + \cdots + c_m(r_m)^k P_m$$

where $r_1, \ldots, r_m$ are the characteristic values (assuming that there are no double roots). Since the components of $P(k)$ are the probabilities of the m states happening on the kth repetition, the components are all positive and add up to 1. Thus, $P(k)$ cannot go to infinity and so $|r_j| \leq 1$ for $j = 1, \ldots, m$, that is, $\|R\| \leq 1$.

Likewise, since the sum of the components equals 1, $P(k)$ does not go to zero, so there must be at least one characteristic value, say r_1, whose absolute value equals 1, that is, $|r_1| = 1$ and $\|R\| = 1$. There may be more than one characteristic value whose absolute value equals 1.

The simplest type of Markov chain is one in which exactly one characteristic value of R, say r_1, equals one (and is not a double root), and all other characteristic values satisfy

$$|r_j| < 1 \qquad \text{for} \quad j = 2, \ldots, m.$$

In such a case,

$$\lim_{k \to \infty} P(k) = c_1 P_1.$$

The vector P_1 is a characteristic vector corresponding to $r_1 = 1$, and thus it satisfies

$$(R - I)P_1 = 0.$$

The number c_1 must be chosen so that the components of $P(k)$ add up to 1, that is, if

$$P_1 = \begin{pmatrix} p_1 \\ \vdots \\ p_m \end{pmatrix},$$

then c_1 is chosen so that $c_1 p_1 + \cdots + c_1 p_m = 1$, or

$$c_1 = \frac{1}{p_1 + \cdots + p_m}.$$

This was the case for our marble-drawing example of section 3.4 and in example 8.7. All that was necessary in these examples to find the probabilities $P(k)$ for large k was to find the characteristic vector corresponding to $r_1 = 1$, and solve a simple equation for the number c_1.

Such matrices as these are easy to work with, but how does one identify them? A **regular Markov chain** is one in which the matrix, R^k, has **no zeros** in any position, for **some** value of k. The 3-by-3 matrix R in example 8.7, has no zeros in any position, so this example is a regular Markov chain. In the next section, we will see transition matrices R in which R^k has some zeros for every power k. It can be shown that the matrix corresponding to a regular Markov chain has one characteristic value equal to 1 and the rest are less than 1 in absolute value.

EXAMPLE 8.8   Consider two tennis players, Ace and Spike. We will assume that when someone wins a tennis match, it gives him confidence and the other player loses confidence. Because of this, if a player wins a match, he is more likely to win the next match, and if he loses a match he is more likely to lose the next match. Specifically, suppose if Spike beats Ace in a match, then the likelihood Spike will win the next match is 0.6, while if Spike loses to Ace, the likelihood Spike will win the next match is 0.3.

Let the probability that Spike wins match n be $p(n)$, and the probability that Ace wins match n be $q(n)$. From the above, you should be able to derive

$$p(n+1) = 0.6p(n) + 0.3q(n)$$

as the equation that gives the probability that Spike will win match $n + 1$. From the above, we also know that if Spike wins a match, the probability that Ace wins the next match (Spike loses) is $1 - 0.6 = 0.4$, and if Ace wins a match, the probability that Ace wins the next match is 0.7. Thus,

$$q(n+1) = 0.4p(n) + 0.7q(n).$$

We now have the Markov chain

$$\begin{pmatrix} p(n+1) \\ q(n+1) \end{pmatrix} = \begin{pmatrix} 0.6 & 0.3 \\ 0.4 & 0.7 \end{pmatrix} \begin{pmatrix} p(n) \\ q(n) \end{pmatrix}.$$

Since the matrix R has no zeros, this is a regular Markov chain. Therefore, we solve

$$(R - I)P_1 = \begin{pmatrix} -0.4 & 0.3 \\ 0.4 & -0.3 \end{pmatrix} \begin{pmatrix} x \\ y \end{pmatrix} = \begin{pmatrix} 0 \\ 0 \end{pmatrix}.$$

One solution is that $y = 4$ and $x = 3$. Thus,

$$\lim_{k \to \infty} P(k) = c_1 \begin{pmatrix} 3 \\ 4 \end{pmatrix} = \begin{pmatrix} 3c_1 \\ 4c_1 \end{pmatrix}.$$

Since $3c_1 + 4c_1 = 1$, we get $c_1 = 1/7$. We then have

$$\lim_{k \to \infty} p(k) = \frac{3}{7} \quad \text{and} \quad \lim_{k \to \infty} q(k) = \frac{4}{7}.$$

Thus, Spike will win (approximately) $3/7$ of the games while Ace will win $4/7$.

Note that this system could have been rewritten as a first-order affine dynamical system and solved similarly using the techniques developed in section 3.4. ∎

You may wonder why a regular Markov chain has one root equal to one and all others less than one (in absolute value). While the answer to this requires a bit more mathematics than we presently have at our disposal, we can get some insight into why this is true by considering the converse. Suppose we have a Markov chain in which the transition matrix R has $r = 1$ as a single root of the characteristic equation and all other roots are less than one in absolute value. Also assume that all the components of P_1, the characteristic vector corresponding to $r = 1$, are nonzero. Then this Markov chain is regular, that is, for some value of k, the matrix R^k has no zeros.

To see that this is true, let's reconsider the regular Markov chain of example 8.7. The matrix R and the characteristic vector P_1 corresponding to the characteristic value $r = 1$ are

$$R = \begin{pmatrix} 0.2 & 0.1 & 0.3 \\ 0.3 & 0.5 & 0.1 \\ 0.5 & 0.4 & 0.6 \end{pmatrix} \quad \text{and} \quad P_1 = \begin{pmatrix} 16/70 \\ 17/70 \\ 37/70 \end{pmatrix}.$$

Suppose that you drew the first marble from the red bag, that is,

$$P(0) = \begin{pmatrix} 1 \\ 0 \\ 0 \end{pmatrix}.$$

Thus, for large values of k we have, approximately,

$$P(k) = R^k P(0) = P_1.$$

Denote the matrix R^k as

$$R^k = \begin{pmatrix} a_{11} & a_{12} & a_{13} \\ a_{21} & a_{22} & a_{23} \\ a_{31} & a_{32} & a_{33} \end{pmatrix}.$$

Then

$$R^k P(0) = \begin{pmatrix} a_{11} & a_{12} & a_{13} \\ a_{21} & a_{22} & a_{23} \\ a_{31} & a_{32} & a_{33} \end{pmatrix} \begin{pmatrix} 1 \\ 0 \\ 0 \end{pmatrix} = \begin{pmatrix} a_{11} \\ a_{21} \\ a_{31} \end{pmatrix} = P_1.$$

Thus, $a_{11} = 16/70$, $a_{21} = 17/70$, and $a_{31} = 37/70$.

By letting

$$P(0) = \begin{pmatrix} 0 \\ 1 \\ 0 \end{pmatrix},$$

we also get

$$R^k P(0) = \begin{pmatrix} a_{11} & a_{12} & a_{13} \\ a_{21} & a_{22} & a_{23} \\ a_{31} & a_{32} & a_{33} \end{pmatrix} \begin{pmatrix} 0 \\ 1 \\ 0 \end{pmatrix} = \begin{pmatrix} a_{12} \\ a_{22} \\ a_{32} \end{pmatrix} = P_1,$$

and the second column of the matrix R^k is the same as P_1.

Likewise, the third column of R^k is P_1. Thus,

$$\lim_{k \to \infty} R^k = \begin{pmatrix} 16/70 & 16/70 & 16/70 \\ 17/70 & 17/70 & 17/70 \\ 37/70 & 37/70 & 37/70 \end{pmatrix}.$$

Thus, if there are no zeros in the components of P_1, and since each column of R^k is approximately equal to P_1, then R^k has no zeros, for some power of k (and for all powers k greater than some value k_0).

REMARK: We have seen that R^k converges to a matrix in which each column equals the characteristic vector corresponding to the characteristic value $r = 1$ and such that the sum of the vector's components equals 1. If you have a computer or appropriate calculator available, it may be easier to compute the characteristic vector by computing R^k for a reasonably large value of k than it is to actually solve a system of equations to find this vector directly. □

It is more difficult to show that if R is the transition matrix for a regular Markov chain, that is, if R^k has no zeros for some k, then $r = 1$ is the only root whose absolute value equals 1, and its characteristic vector P_1 has no zeros.

PROBLEMS

1. Which of the following vectors cannot be a probability vector $P(n)$ and why?

   a. $\begin{pmatrix} 0.4 \\ 0.5 \\ 0.2 \end{pmatrix}$, **b.** $\begin{pmatrix} 0.3 \\ 0.2 \\ 0.5 \end{pmatrix}$, **c.** $\begin{pmatrix} 0.8 \\ -0.1 \\ 0.3 \end{pmatrix}$, **d.** $\begin{pmatrix} 0 \\ 0 \\ 1 \end{pmatrix}$.

2. Which of the following matrices cannot be a transition matrix and why? Which transition matrices are regular?

   a. $\begin{pmatrix} 0.2 & 0.3 \\ 0.9 & 0.7 \end{pmatrix}$, **b.** $\begin{pmatrix} -0.1 & 0.6 \\ 1.1 & 0.4 \end{pmatrix}$,

   c. $\begin{pmatrix} 0.4 & 1 \\ 0.6 & 0 \end{pmatrix}$, **d.** $\begin{pmatrix} 0.2 & 0.3 \\ 0.8 & 0.7 \end{pmatrix}$,

   e. $\begin{pmatrix} 0 & 1 \\ 1 & 0 \end{pmatrix}$, **f.** $\begin{pmatrix} 0.4 & 0.6 \\ 0.3 & 0.7 \end{pmatrix}$.

3. For each of these regular transition matrices, find

$$\lim_{k \to \infty} P(k),$$

that is, find a characteristic vector P_1 corresponding to $r = 1$ and then find the value of a such that the components of aP_1 add up to 1.

$$\textbf{a.} \begin{pmatrix} 0.2 & 0.6 \\ 0.8 & 0.4 \end{pmatrix}, \quad \textbf{b.} \begin{pmatrix} 0.6 & 1 \\ 0.4 & 0 \end{pmatrix}, \quad \textbf{c.} \begin{pmatrix} 0.7 & 0.3 & 0.9 \\ 0.3 & 0.6 & 0 \\ 0 & 0.1 & 0.1 \end{pmatrix}.$$

4. Suppose we have three bags: a red bag containing 2 red, 2 blue, and 2 green marbles; a blue bag containing 2 red and 6 green marbles; and a green bag containing 2 red and 6 blue marbles. A marble is drawn at random from one of the bags and its color is recorded. The next marble is drawn from the bag that is the same color as the marble just drawn. What are the probabilities that the 100th marble drawn is red, is blue, and is green? You do not need to know from which bag you drew the first marble.

5. Two hamburger chains (to avoid controversy, call them A and B) are in competition. Suppose that if a person goes to A for a hamburger, the probability of going back to A for the next hamburger is 0.8, while the probability of going to B for the next hamburger is 0.2. Suppose also that if a person goes to B for a hamburger, the probability of going back to B for the next hamburger is 0.7, while the probability of going to A for the next hamburger is 0.3. Compute chain A's percentage of the hamburger market.

8.5 Absorbing Markov chains

Suppose we have a Markov chain in which there is at least one state, say e_j, such that if this state occurs on the nth repetition of the experiment, then it will occur on the $(n+1)$th repetition as well. This state is then called an **absorbing state**. In other words, a state e_j is an absorbing state if the probability that state e_j occurs on the next repetition of the experiment given that e_j occurred on the last repetition is one, that is, $p_{jj} = 1$. In this case, $p_{ij} = 0$ for $i \neq j$, where p_{ij} is the probability that state e_i will occur on the next repetition of the experiment given state e_j just occurred. Thus, the jth column of the transition matrix R will have a 1 in the jth position and 0's in every other position of that column.

DEFINITION 8.9 *Suppose we have a Markov chain in which there is at least one absorbing state. Suppose that for any starting state, it is possible to end at some absorbing state. This process is called an **absorbing Markov chain**.*

EXAMPLE 8.10 Suppose we have four bags of marbles.

- The red bag contains 1 red marble.
- The blue bag contains 1 blue marble.
- The green bag contains 3 red, 1 blue, 4 green, and 2 yellow marbles.
- The yellow bag contains 1 red, 2 blue, 3 green, and 4 yellow marbles.

The $(n+1)$th marble is drawn from the bag that is the same color as the nth marble drawn.

In this problem, red and blue marbles are absorbing states since once a red marble is drawn, we must get red marbles from that point on, and the same goes for blue marbles. Let $r(n)$, $b(n)$, $g(n)$, and $y(n)$ be the probabilities of red, blue, green, and yellow, respectively, occurring on the nth draw. Then the dynamical system that describes this problem is

$$P(n+1) = RP(n),$$

where

$$P(n) = \begin{pmatrix} r(n) \\ b(n) \\ g(n) \\ y(n) \end{pmatrix} \quad R = \begin{pmatrix} 1 & 0 & 0.3 & 0.1 \\ 0 & 1 & 0.1 & 0.2 \\ 0 & 0 & 0.4 & 0.3 \\ 0 & 0 & 0.2 & 0.4 \end{pmatrix}. \qquad \blacksquare$$

The above example is an absorbing Markov chain. Notice that we don't have to get red or blue on an early draw. In fact, you may have to go through many repetitions before you can possibly reach an absorbing state.

EXAMPLE 8.11 Suppose we have four bags of marbles.

- The red bag contains 1 red marble.
- The blue bag contains 1 blue marble, 1 red marble, and 2 green marbles.

- The green bag contains 4 green and 2 yellow marbles.
- The yellow bag contains 3 green and 4 yellow marbles.

A marble is drawn from a bag, its color is recorded, and the marble is returned to that bag. The $(n+1)$th marble is drawn from the bag that is the same color as the nth marble drawn. This is **not** an absorbing Markov chain nor a regular Markov chain. The reason it is not an absorbing Markov chain is that if the first marble is drawn from the yellow bag, it is impossible ever to get a red marble, and you will keep getting green and yellow marbles forever. It is not a regular Markov chain because there is an absorbing state, red, and R^k will always have some zeros.   ■

The problem with absorbing Markov chains with more than one absorbing state is that $r = 1$ is a multiple root of the characteristic equation $|R - rI| = 0$, a situation which we are not covering in this book. The goal of the rest of this section is to find an alternative method for determining the probabilities of ending in each of the absorbing states. We do this by way of an "application".

Let's consider what is called a **random walk**. Suppose a man leaves a bar after having too much to drink. His home is two blocks to the left, while a lake is one block to his right. He will stumble one block to the left with probability 0.2 or one block to the right with probability 0.8. For simplicity, let's assume it takes him 1 minute to walk one block. After walking one block, he will again walk left with probability 0.2 or right with probability 0.8. He continues in this manner until he arrives home or falls into the lake. If he arrives home or falls into the lake, he remains there. See figure 8.1.

Let home be corner a and the lake be corner d, as indicated in the figure. Let $a(n)$ be the probability that the man will be at corner a when n minutes have passed. Likewise, $b(n)$, $c(n)$, and $d(n)$ are the probabilities that the man will be at corners b, c, and d when n minutes have passed.

To compute $a(n + 1)$, the probability that the man is at corner a when $n + 1$ minutes have passed, we must consider two cases. Case

FIGURE 8.1.   Man leaving bar must arrive at home to the left or fall into the lake to the right.

1 is that the man is at corner a when n minutes have passed, since he then stays at home. The probability of case 1 is $a(n)$. Case 2 is that the man is at corner b when n minutes have passed and he then walks left. Case 2 is a two-stage process. The first stage is that he is at corner b when n minutes have passed and this has probability $b(n)$. The second stage is that he turns left, and the probability of this is 0.2. By the multiplication principle, the probability of case 2 is $0.2b(n)$. Adding the two cases together gives our first equation

$$a(n+1) = a(n) + 0.2b(n).$$

For the man to be at corner b when $n+1$ minutes have passed, he must have been at corner c when n minutes had passed and then turned left. Notice that if he had been at corner a when n minutes had passed, he would have stayed there. Thus, we only have one case. Using the multiplication principle, we get

$$b(n+1) = 0.2c(n).$$

Similarly, we get the equations

$$c(n+1) = 0.8b(n) \qquad \text{and} \qquad d(n+1) = 0.8c(n) + d(n)$$

corresponding to corners c and d, respectively.

Combining these four equations, we get the dynamical system

$$P(n+1) = RP(n),$$

where

$$P(n) = \begin{pmatrix} a(n) \\ b(n) \\ c(n) \\ d(n) \end{pmatrix} \qquad \text{and} \qquad R = \begin{pmatrix} 1 & 0.2 & 0 & 0 \\ 0 & 0 & 0.2 & 0 \\ 0 & 0.8 & 0 & 0 \\ 0 & 0 & 0.8 & 1 \end{pmatrix}.$$

Our first step is to distinguish notationally between the absorbing states and the nonabsorbing states. Let's denote the absorbing states by e_1, e_2, and so forth. Let's denote the nonabsorbing states by f_1, f_2, and so forth. Thus, we will denote corner a by e_1 and corner d by e_2. We will denote corner b by f_1 and corner c by f_2.

Let $p_j(n)$ be the probability of absorbing state e_j occurring on the nth repetition of the experiment, that is, $p_1(n)$ for corner a and $p_2(n)$ for corner d in this case. Also, let $q_j(n)$ be the probability of nonabsorbing state f_j occurring on the nth repetition of the experiment, that is, $q_1(n)$ for corner b and $q_2(n)$ for corner c in this case.

Using this notation, we write our dynamical system as

$$\begin{pmatrix} p_1(n+1) \\ p_2(n+1) \\ q_1(n+1) \\ q_2(n+1) \end{pmatrix} = \begin{pmatrix} 1 & 0 & 0.2 & 0 \\ 0 & 1 & 0 & 0.8 \\ 0 & 0 & 0 & 0.2 \\ 0 & 0 & 0.8 & 0 \end{pmatrix} \begin{pmatrix} p_1(n) \\ p_2(n) \\ q_1(n) \\ q_2(n) \end{pmatrix}. \tag{8.3}$$

Notice that this matrix is a rearrangement of the matrix R given above. This is because the states are given in a different sequence, that is, the corners are listed as a, d, b, and c instead of a, b, c, and d.

The first equation from this system is

$$p_1(n+1) = p_1(n) + 0.2q_1(n).$$

This again says that the probability of ending at corner a after $n+1$ minutes is the probability of being at corner a after n minutes plus the probability of being at corner b after n minutes, $q_1(n)$, and then turning left, 0.2. You should be able to check that the other three equations are also correct.

Let's rewrite dynamical system (8.3) as

$$\left(\begin{array}{c} p_1(n+1) \\ p_2(n+1) \\ \hline q_1(n+1) \\ q_2(n+1) \end{array} \right) = \left(\begin{array}{cc|cc} 1 & 0 & 0.2 & 0 \\ 0 & 1 & 0 & 0.8 \\ \hline 0 & 0 & 0 & 0.2 \\ 0 & 0 & 0.8 & 0 \end{array} \right) \left(\begin{array}{c} p_1(n) \\ p_2(n) \\ \hline q_1(n) \\ q_2(n) \end{array} \right). \tag{8.4}$$

Observe from the way this is written that the probability vector can be broken into the two smaller vectors

$$P(n) = \begin{pmatrix} p_1(n) \\ p_2(n) \end{pmatrix} \quad \text{and} \quad Q(n) = \begin{pmatrix} q_1(n) \\ q_2(n) \end{pmatrix}.$$

Also, the transition matrix can be broken into the four smaller matrices

$$I_2 = \begin{pmatrix} 1 & 0 \\ 0 & 1 \end{pmatrix}, \quad S = \begin{pmatrix} 0.2 & 0 \\ 0 & 0.8 \end{pmatrix},$$

$$0_{22} = \begin{pmatrix} 0 & 0 \\ 0 & 0 \end{pmatrix}, \quad R = \begin{pmatrix} 0 & 0.2 \\ 0.8 & 0 \end{pmatrix}.$$

Using this notation, we can rewrite our absorbing Markov chain as

$$\begin{pmatrix} P(n+1) \\ Q(n+1) \end{pmatrix} = \begin{pmatrix} I_2 & S \\ 0_{22} & R \end{pmatrix} \begin{pmatrix} P(n) \\ Q(n) \end{pmatrix}.$$

While $P(n)$ and $Q(n)$ are vectors and I_2, S, 0_{22}, and R are matrices, let's proceed as if they were all just numbers. Dynamical system (8.4) can be rewritten as the two equations

$$P(n+1) = P(n) + SQ(n) \qquad (8.5)$$

and

$$Q(n+1) = RQ(n). \qquad (8.6)$$

Dynamical system (8.6) is in reality a first-order dynamical system of 2 equations, and its solution is

$$Q(k) = R^k Q(0).$$

We note here that since we can eventually reach some absorbing state from each of the nonabsorbing states, the characteristic values for R all have absolute value less than one, that is, $\|R\| < 1$. You should find the characteristic values of R to see that this is true. Thus, the probability of being at any nonabsorbing state after k tries must go to zero. So

$$\lim_{k \to \infty} R^k Q(0) = \begin{pmatrix} 0 \\ 0 \end{pmatrix}.$$

Thus, the probabilities $q_1(k)$ and $q_2(k)$ must decrease to zero.

Substitution of $Q(n)$ into equation (8.5) gives

$$P(n+1) = P(n) + SR^n Q(0). \qquad (8.7)$$

This is a first-order nonhomogeneous dynamical system.

Recall that the general solution of the first-order nonhomogeneous dynamical system

$$a(n+1) = a(n) + r^n b$$

is of the form

$$a(k) = a + r^k c,$$

where the number a depends on the initial value $a(0)$ and the number c depends on the equation. Thus, we might suspect that the general solution to the nonhomogeneous dynamical system (8.7) could be of the form

$$P(k) = A + SR^k C,$$

where the vectors A and C depend on the system and the initial value $P(0)$.

To see if this is true, let's substitute our "guess" into the dynamical system (8.7) and see if we can find vectors A and C that satisfy the equation. Since

$$P(n) = A + SR^n C \quad \text{and} \quad P(n+1) = A + SR^{n+1} C = A + SR^n RC,$$

substitution into equation (8.7) gives

$$A + SR^n RC = A + SR^n C + SR^n Q(0).$$

Canceling the A's, and bringing all the terms to the left gives

$$SR^n RC - SR^n C - SR^n Q(0) = 0_2,$$

where 0_2 is the vector with two zeros. Factoring out SR^n gives

$$SR^n (RC - C - Q(0)) = 0_2.$$

Thus, the equation is satisfied if

$$RC - C - Q(0) = 0_2, \quad \text{that is, if} \quad (R - I)C = Q(0).$$

Note that since 1 is not a characteristic value for matrix R, $|R - 1I| \neq 0$, so $(R - 1I)^{-1}$ exists. After multiplying both sides above by $(R - I)^{-1}$ on the left, we get

$$C = (R - I)^{-1} Q(0).$$

Thus, the general solution to dynamical system (8.7) is

$$P(k) = A + SR^k (R - I)^{-1} Q(0).$$

Let's now use the initial value $P(0)$ to find the vector A.

$$P(0) = A + SR^0 (R - I)^{-1} Q(0) = A + S(R - I)^{-1} Q(0),$$

since $R^0 = I$. Thus,

$$A = P(0) - S(R - I)^{-1} Q(0) = P(0) + S(I - R)^{-1} Q(0)$$

(note that we used the rule that $-(R - I)^{-1} = (I - R)^{-1}$) and the particular solution is

$$P(k) = P(0) + S(I - R)^{-1} Q(0) - SR^k (I - R)^{-1} Q(0).$$

Since R^k goes to zero as k goes to infinity, it follows that

$$\lim_{k \to \infty} P(k) = P(0) + S(I - R)^{-1} Q(0).$$

Remember that the components of the vector $P(0)$ give the probabilities that we **start** in the absorbing states e_1 and e_2. But if we start in one of these states, we will always remain there, and the problem is easy. Therefore, in many applications we assume that we do not start in an absorbing state, that is, each component of $P(0)$ is zero.

In this case, the probabilities of ending in the absorbing states are given by

$$\lim_{k \to \infty} P(k) = S(I - R)^{-1} Q(0).$$

We have shown the following.

THEOREM 8.12   *Suppose we have an absorbing Markov chain given by the dynamical system*

$$\left(\begin{array}{c} P(n+1) \\ Q(n+1) \end{array} \right) = \left(\begin{array}{cc} I_m & S \\ 0_{\ell m} & R \end{array} \right) \left(\begin{array}{c} P(n) \\ Q(n) \end{array} \right),$$

where $P(n)$ is the m-vector which gives the probabilities of being in the m absorbing states after n repetitions, $Q(n)$ is the ℓ-vector which gives the probabilities of being in the ℓ nonabsorbing states after n repetitions, S is an m-by-ℓ matrix giving the probabilities of going from the nonabsorbing states to the absorbing states, R is the ℓ-by-ℓ matrix giving the probabilities of going from the nonabsorbing states to the nonabsorbing states, I_m is the m-by-m identity matrix, and $0_{\ell m}$ is an ℓ-by-m matrix with all zeros. Then the probabilities of ending in each of the absorbing states is given by the vector

$$P(0) + S(I - R)^{-1} Q(0).$$

Applying this theorem to our bar problem, we note that

$$I - R = \left(\begin{array}{cc} 1 & -0.2 \\ -0.8 & 1 \end{array} \right).$$

You should then be able to compute that

$$(I - R)^{-1} = \frac{5}{21} \left(\begin{array}{cc} 5 & 1 \\ 4 & 5 \end{array} \right).$$

Since you start at corner c which is state f_2, we have that

$$Q(0) = \left(\begin{array}{c} 0 \\ 1 \end{array} \right) \qquad \text{and} \qquad P(0) = \left(\begin{array}{c} 0 \\ 0 \end{array} \right)$$

so that

$$S(I - R)^{-1} Q(0) = \frac{5}{21} \left(\begin{array}{cc} 0.2 & 0 \\ 0 & 0.8 \end{array} \right) \left(\begin{array}{cc} 5 & 1 \\ 4 & 5 \end{array} \right) \left(\begin{array}{c} 0 \\ 1 \end{array} \right)$$

$$= \left(\begin{array}{c} 1/21 \\ 20/21 \end{array} \right).$$

Thus, the probability the man will eventually find his way home is 1/21 and the probability he will be "absorbed" in the lake is 20/21.

EXAMPLE 8.13 Let's apply theorem 8.12 to example 8.10. Recall the composition of the four bags.

- The red bag contains 1 red marble.
- The blue bag contains 1 blue marble.
- The green bag contains 3 red, 1 blue, 4 green, and 2 yellow marbles.
- The yellow bag contains 1 red, 2 blue, 3 green, and 4 yellow marbles.

Suppose we draw our first marble from the yellow bag. What is the probability the 100th marble will be red, that is, what fraction of the cases will end in absorbing state "red"?

Let e_1 be the state that a red marble is drawn. Likewise, e_2, f_1, and f_2 are the states that a blue, green, and yellow marble are drawn, respectively. Notice that e_1 and e_2 are absorbing states and that f_1 and f_2 are nonabsorbing states. Similarly, we define $p_1(n)$, $p_2(n)$, $q_1(n)$, and $q_2(n)$ as the probabilities that the corresponding colored marbles are drawn on the nth draw. Recall that the dynamical system that modeled this marble drawing situation was

$$
\begin{pmatrix} p_1(n+1) \\ p_2(n+1) \\ q_1(n+1) \\ q_2(n+1) \end{pmatrix} = \begin{pmatrix} 1 & 0 & 0.3 & 0.1 \\ 0 & 1 & 0.1 & 0.2 \\ 0 & 0 & 0.4 & 0.3 \\ 0 & 0 & 0.2 & 0.4 \end{pmatrix} \begin{pmatrix} p_1(n) \\ p_2(n) \\ q_1(n) \\ q_2(n) \end{pmatrix}.
$$

Notice that

$$
P(n) = \begin{pmatrix} p_1(n) \\ p_2(n) \end{pmatrix}, \qquad Q(n) = \begin{pmatrix} q_1(n) \\ q_2(n) \end{pmatrix}, \qquad Q(0) = \begin{pmatrix} 0 \\ 1 \end{pmatrix},
$$

$$
S = \begin{pmatrix} 0.3 & 0.1 \\ 0.1 & 0.2 \end{pmatrix}, \qquad R = \begin{pmatrix} 0.4 & 0.3 \\ 0.2 & 0.4 \end{pmatrix}.
$$

Thus the probabilities of eventually ending in the red bag or in the blue bag are given by

$$
S(I-R)^{-1}Q(0)
$$

since $P(0) = 0_2$, the 2-vector with all zeros. We compute that

$$
I - R = \begin{pmatrix} 0.6 & -0.3 \\ -0.2 & 0.6 \end{pmatrix}, \qquad \text{so} \quad (I-R)^{-1} = \frac{1}{3} \begin{pmatrix} 6 & 3 \\ 2 & 6 \end{pmatrix}.
$$

Now

$$S(I - R)^{-1} = \frac{1}{3} \begin{pmatrix} 0.3 & 0.1 \\ 0.1 & 0.2 \end{pmatrix} \begin{pmatrix} 6 & 3 \\ 2 & 6 \end{pmatrix} = \frac{1}{6} \begin{pmatrix} 4 & 3 \\ 2 & 3 \end{pmatrix}.$$

Since the first marble is drawn from the yellow bag, we have

$$S(I - R)^{-1}Q(0) = \frac{1}{6} \begin{pmatrix} 4 & 3 \\ 2 & 3 \end{pmatrix} \begin{pmatrix} 0 \\ 1 \end{pmatrix} = \begin{pmatrix} 0.5 \\ 0.5 \end{pmatrix},$$

so that the probability we end in state e_1 is 0.5 and the probability that we end in state e_2 is 0.5.

Suppose we draw the first marble from the green bag instead of from the yellow bag. What is the probability of ending in state e_1. This is easy, once we have computed the matrix $S(I - R)^{-1}$. The answer is

$$S(I - R)^{-1}Q(0) = \frac{1}{6} \begin{pmatrix} 4 & 3 \\ 2 & 3 \end{pmatrix} \begin{pmatrix} 1 \\ 0 \end{pmatrix} = \begin{pmatrix} 2/3 \\ 1/3 \end{pmatrix},$$

that is, we end with a red marble two-thirds of the time, and with a blue marble one-third of the time.

Suppose we flip a fair coin and if head comes up, we draw our first marble from the yellow bag, while if tails comes up, we draw our first marble from the green bag. What are the probabilities of ending in the absorbing states? In this case, since

$$Q(0) = \begin{pmatrix} 0.5 \\ 0.5 \end{pmatrix},$$

the answer is given by

$$S(I - R)^{-1}Q(0) = \frac{1}{6} \begin{pmatrix} 4 & 3 \\ 2 & 3 \end{pmatrix} \begin{pmatrix} 0.5 \\ 0.5 \end{pmatrix} = \begin{pmatrix} 7/12 \\ 5/12 \end{pmatrix},$$

that is, we end with a red marble seven-twelfths of the time and with a blue marble five-twelfths of the time.   ■

PROBLEMS

1. Suppose we have four bags of marbles. The red bag contains 1 red marble; the blue bag contains 1 blue marble; the green bag contains 2 red, 2 blue, 4 green, and 2 yellow marbles; and the yellow bag contains 2 red, 1 blue, 3 green, and 4 yellow marbles. A marble is drawn from the yellow bag, its color is recorded, and the marble is returned to the yellow bag. The $(n + 1)$th marble is drawn from the bag that is the same color as the nth marble

drawn. What is the probability the 100th marble is red, that is, in what fraction of the cases will we end in the absorbing state, red? What is the probability we end in the absorbing state, blue?

2. What is the answer to problem 1 if the first marble is drawn from the green bag?

3. In problem 1, suppose there is also a white bag containing 2 green marbles, 3 white marbles, and 1 yellow marble.
   a. If the first marble drawn is from the white bag, what are the probabilities of eventually ending in the red bag and eventually ending in the blue bag?
   b. Suppose it is equally likely that the first marble is drawn from any one of the five bags. What are the probabilities of eventually ending in the red bag and eventually ending in the blue bag?

4. Consider the drunk staggering home from the bar. Find the probability that the man arrives home given that the bar is at corner b instead of at corner c.

5. Consider the drunk staggering home from the bar. Suppose he staggers left one block with probability 0.4 and right with probability 0.6. Suppose home is 2 blocks to the left of the bar and the lake is 2 blocks to the right of the bar. What is the probability that the drunk gets home safely? Hint: There are 5 corners, 2 of them are absorbing and 3 of them are nonabsorbing.

6. Suppose that a drunk leaves a bar. His home is one block to the left and a wall is one block to the right. When he is at the bar, he goes left with probability 0.72 and right with probability 0.28. When he is at home, he remains there. When he is at the wall, he goes left with probability 0.5 and remains at the wall for the next minute with probability 0.5. Model this as a Markov chain, that is, find the transition matrix R, then find the particular solution. Use this solution to find $a(10)$, the probability he is at home after 10 minutes.

7. Suppose that a drunk leaves a bar. His home is one block to the left and a wall is one block to the right. When he is at the bar, he goes left with probability 0.2 and right with probability 0.8. When he is at home, he goes right with probability 0.8 and remains home for the next minute with probability 0.2. When he is at the wall, he goes left with probability 0.5 and remains at the wall for the next minute with probability 0.5. Model this

as a Markov chain, that is, find the transition matrix R. This is a regular Markov chain, and therefore to find the long-term probabilities, all that is necessary is to find the characteristic vector corresponding to the characteristic value, $r = 1$. Find this vector.

8.6   Applications of absorbing Markov chains

In this section, we will use our methods to study two problems. The first problem is a bit contrived. The second is a cost-accounting problem.

8.6.1   The good, the bad, and the ugly

In a variation of the movie **The Good, the Bad, and the Ugly**, suppose we have three gunfighters, A, B, and C, who are entering a three-way duel. Gunfighter A is a good shot and hits his target 70 per cent of the time. Gunfighter B is not as good a shot and hits his target 50 per cent of the time. Gunfighter C is wondering how he got into this mess, since he hits his target only 30 per cent of the time. All hits are fatal.

The rules of the gunfight are that on the count of three, all three gunfighters draw and shoot one shot at one of the others, simultaneously. Each gunfighter knows where all three rank in ability. As his strategy, each will shoot at the strongest opponent still remaining. Thus, on the first shot, A shoots at B, while B and C both shoot at A. If more than one gunfighter survives, the process is repeated. Who will most likely win the gunfight?

The states will be the possible survivors of a gunfight. We will list the absorbing states first, that is, states in which the gunfight ends.

- State e_1 is that nobody survives.
- State e_2 is that only A survives.
- State e_3 is that only B survives.
- State e_4 is that only C survives.

We now list the nonabsorbing states.

- State f_1 is the state in which A and C are the only survivors.
- State f_2 is the state in which B and C are the only survivors.
- State f_3 is the state in which all three survive.

Note that when all three gunfighters are alive, nobody shoots at C. Thus, A and/or B must be shot first and, therefore, we cannot have the state in which A and B are the only survivors.

Let $p_1(n)$ be the probability that state e_1 (nobody survives) occurs after n rounds of shots have been fired, and likewise for $p_2(n)$, $p_3(n)$, $p_4(n)$, $q_1(n)$, $q_2(n)$, and $q_3(n)$.

Let's compute $p_1(n+1)$. There are three cases (or ways) in which nobody survives round $n + 1$. These are that nobody survived round n, that only A and C survived round n and then A and C shoot at and hit each other, and that only B and C survived round n and then B and C shoot at and hit each other. The first case occurs with probability $p_1(n)$. The second case is a three-stage process, the first stage being that only A and C survive round n with probability $q_1(n)$, the second stage being that A shoots C with probability 0.7, and the third stage being that C shoots A with probability 0.3. Thus, the probability of the second case is, using the multiplication principle, $0.21q_1(n)$.

Likewise, using three stages, we get that the probability of the third case is $(0.5)(0.3)q_2(n) = 0.15q_2(n)$. Adding the probabilities of these three cases, we get that

$$p_1(n+1) = p_1(n) + 0.21q_1(n) + 0.15q_2(n).$$

To get $p_2(n+1)$, that is, the probability that only A survives round $n + 1$, we consider two cases that (1) only A survives round n which occurs with probability $p_2(n)$, and (2) that only A and C survive round n and then A shoots C but C misses A. This second case is a three-stage process. The probability that only A and C survive round n is $q_1(n)$, the probability A shoots C is 0.7, and the probability C misses A is 0.7. Thus

$$p_2(n+1) = p_2(n) + 0.49q_1(n).$$

In computing the other equations, there is one moderately difficult probability, the probability that B or C or both shoot A. This probability has two cases. The first is that B shoots A, in which case it is irrelevant what C does. The second is that B misses and C hits A. The sum of these cases gives the probability that B and/or C shoots A as $0.5 + (0.5)(0.3) = 0.65$.

You should now be able to find that

$$p_3(n+1) = p_3(n) + 0.35q_2(n),$$

$$p_4(n+1) = p_4(n) + 0.09q_1(n) + 0.15q_2(n) + 0.7(0.65)q_3(n),$$

$$q_1(n+1) = 0.21\,q_1(n) + 0.7(0.5)(0.7)\,q_3(n),$$

$$q_2(n+1) = 0.35\,q_2(n) + 0.3(0.65)\,q_3(n),$$

$$q_3(n+1) = 0.3(0.5)(0.7)\,q_3(n).$$

We now rewrite these equations as the dynamical system

$$
\left(
\begin{array}{c}
p_1(n+1) \\
p_2(n+1) \\
p_3(n+1) \\
p_4(n+1) \\
\hline
q_1(n+1) \\
q_2(n+1) \\
q_3(n+1)
\end{array}
\right)
=
\left(
\begin{array}{cccc|ccc}
1 & 0 & 0 & 0 & 0.21 & 0.15 & 0 \\
0 & 1 & 0 & 0 & 0.49 & 0 & 0 \\
0 & 0 & 1 & 0 & 0 & 0.35 & 0 \\
0 & 0 & 0 & 1 & 0.09 & 0.15 & 0.455 \\
\hline
0 & 0 & 0 & 0 & 0.21 & 0 & 0.245 \\
0 & 0 & 0 & 0 & 0 & 0.35 & 0.195 \\
0 & 0 & 0 & 0 & 0 & 0 & 0.105
\end{array}
\right)
\left(
\begin{array}{c}
p_1(n) \\
p_2(n) \\
p_3(n) \\
p_4(n) \\
\hline
q_1(n) \\
q_2(n) \\
q_3(n)
\end{array}
\right).
$$

In this case, we have the matrices

$$
S=
\begin{pmatrix}
0.21 & 0.15 & 0 \\
0.49 & 0 & 0 \\
0 & 0.35 & 0 \\
0.09 & 0.15 & 0.455
\end{pmatrix}
\quad \text{and} \quad
R=
\begin{pmatrix}
0.21 & 0 & 0.245 \\
0 & 0.35 & 0.195 \\
0 & 0 & 0.105
\end{pmatrix}.
$$

Using Gauss–Jordan elimination we find that

$$
(I-R)^{-1} =
\begin{pmatrix}
1.266 & 0 & 0.347 \\
0 & 1.538 & 0.335 \\
0 & 0 & 1.117
\end{pmatrix},
$$

where the components of $(I-R)^{-1}$ have been rounded to 3 decimal places.

Now we find that

$$
S(I-R)^{-1} =
\begin{pmatrix}
0.266 & 0.231 & 0.123 \\
0.620 & 0 & 0.170 \\
0 & 0.538 & 0.117 \\
0.114 & 0.231 & 0.590
\end{pmatrix}.
$$

Since we start in state f_3, the answer to our problem is

$$
S(I-R)^{-1}Q(0) =
\begin{pmatrix}
0.266 & 0.231 & 0.123 \\
0.620 & 0 & 0.170 \\
0 & 0.538 & 0.117 \\
0.114 & 0.231 & 0.590
\end{pmatrix}
\begin{pmatrix}
0 \\
0 \\
1
\end{pmatrix}
=
\begin{pmatrix}
0.123 \\
0.170 \\
0.117 \\
0.590
\end{pmatrix}.
$$

Therefore, the probability nobody survives the gunfight is 0.123, the probability that A wins is 0.170, the probability that B wins is 0.117, and the probability that C wins is 0.590.

If one of the better gunfighters didn't show up for the fight, say B didn't show, then you are starting in state f_1 and the probabilities of the different results are

$$
\begin{pmatrix}
0.266 & 0.231 & 0.123 \\
0.620 & 0 & 0.170 \\
0 & 0.538 & 0.117 \\
0.114 & 0.231 & 0.590
\end{pmatrix}
\begin{pmatrix}
1 \\
0 \\
0
\end{pmatrix}
=
\begin{pmatrix}
0.266 \\
0.620 \\
0 \\
0.114
\end{pmatrix},
$$

that is, the probability that nobody survives is 0.266, the probability that A wins is 0.620, the probability that B wins is 0 since he wasn't in the fight, and the probability that C wins is 0.114. Obviously, C hopes that B shows up.

8.6.2   Cost-accounting

The next problem we will consider is a problem in cost-accounting. Suppose that a company has four departments, Marketing, Manufacturing, Maintenance, and Accounting. Two of the departments, Marketing and Manufacturing are production departments, and their expenses are kept to themselves. Maintenance and Accounting are service departments and their expenses are distributed to the departments that they serve. In other words, all expenses should be distributed to the two departments that make money.

For example, suppose that Marketing's direct expenses are 50 000 dollars, Manufacturing's direct expenses are 70 000 dollars, Maintenance's direct expenses are 40 000 dollars, and Accounting's direct expenses are 20 000 dollars for a total of 180 000 dollars for the company.

Assume that Maintenance spends 20 per cent of its time working for the Marketing department (cleaning and repairing Marketing's facilities), 40 per cent of its time working for Manufacturing, 30 per cent of its time working for itself, and 10 per cent of its time working for Accounting. Similarly Accounting figures it spends 30, 30, 20, and 20 per cent of its time working for Marketing, Manufacturing, Maintenance, and Accounting, respectively.

The idea is that Maintenance will distribute its 40 000 dollars to the four departments according to the proportion of time it spends working for each department, that is, it will distribute 20 per cent of

40 000 or 8000 dollars to Marketing, 16 000 dollars to Manufacturing, 12 000 to Maintenance, and 4000 dollars to Accounting. Likewise, Accounting will distribute its 20 000 dollars in expenses to the other departments in the amounts of 6000 dollars to Marketing, 6000 dollars to Manufacturing, 4000 dollars to Maintenance, and 4000 dollars to Accounting.

After the redistribution, the costs for each department are 64 000 dollars for Marketing, 92 000 dollars for Manufacturing, 16 000 dollars for Maintenance, and 8000 dollars for Accounting for the same total of 180 000 dollars.

The problem is that Maintenance and Accounting still have costs. The solution is to redistribute these new costs according to the same formula as for the previous costs. This redistribution will continue until all of the 180 000 dollars in costs are given to Marketing and Manufacturing. What are the total costs for Marketing and Manufacturing?

To solve this problem, we let the costs to Marketing and Manufacturing after the nth redistribution be represented by $p_1(n)$ and $p_2(n)$, respectively. Also, we let the costs to Maintenance and Accounting after the nth redistribution be represented by $q_1(n)$ and $q_2(n)$, respectively.

The costs to Marketing after the $(n + 1)$th redistribution are the costs after the nth redistribution plus the new costs from Maintenance and Accounting, that is,

$$p_1(n+1) = p_1(n) + 0.2q_1(n) + 0.3q_2(n).$$

Similarly, we get

$$p_2(n+1) = p_2(n) + 0.4q_1(n) + 0.3q_2(n),$$

$$q_1(n+1) = 0.3q_1(n) + 0.2q_2(n),$$

$$q_2(n+1) = 0.1q_1(n) + 0.2q_2(n).$$

This gives us the dynamical system

$$\left(\begin{array}{c} p_1(n+1) \\ p_2(n+1) \\ \hline q_1(n+1) \\ q_2(n+1) \end{array}\right) = \left(\begin{array}{cc|cc} 1 & 0 & 0.2 & 0.3 \\ 0 & 1 & 0.4 & 0.3 \\ \hline 0 & 0 & 0.3 & 0.2 \\ 0 & 0 & 0.1 & 0.2 \end{array}\right) \left(\begin{array}{c} p_1(n) \\ p_2(n) \\ \hline q_1(n) \\ q_2(n) \end{array}\right).$$

But this is just an absorbing Markov chain in which

$$S = \begin{pmatrix} 0.2 & 0.3 \\ 0.4 & 0.3 \end{pmatrix}, \quad R = \begin{pmatrix} 0.3 & 0.2 \\ 0.1 & 0.2 \end{pmatrix},$$

$$P(0) = \begin{pmatrix} 50\,000 \\ 70\,000 \end{pmatrix}, \quad Q(0) = \begin{pmatrix} 40\,000 \\ 20\,000 \end{pmatrix}.$$

Although $P(0)$ and $Q(0)$ are not probability vectors, that is irrelevant as far as the mathematical analysis goes. (You could use the fraction of the total costs for each component of $P(0)$ and $Q(0)$.) Therefore, since

$$(I - R)^{-1} = \frac{5}{27} \begin{pmatrix} 8 & 2 \\ 1 & 7 \end{pmatrix},$$

the answer to our problem is

$$\lim_{k \to \infty} P(k) = P(0) + S(I - R)^{-1}Q(0)$$

$$= \begin{pmatrix} 50\,000 \\ 70\,000 \end{pmatrix} + \frac{5}{27} \begin{pmatrix} 0.2 & 0.3 \\ 0.4 & 0.3 \end{pmatrix} \begin{pmatrix} 8 & 2 \\ 1 & 7 \end{pmatrix} \begin{pmatrix} 40\,000 \\ 20\,000 \end{pmatrix}$$

$$= \begin{pmatrix} 73\,333 \\ 106\,667 \end{pmatrix}$$

rounded to the nearest dollar.

Thus, of the 180\,000 dollars expenses for the company, 73\,333 dollars are attributed to Marketing and 106\,667 dollars are attributed to Manufacturing.

PROBLEMS

1. Suppose that, in our gunfight, A hits his target with probability 0.4, B hits his target with probability 0.3, and C hits his target with probability 0.2. What are the probabilities that each gunfighter survives and that nobody survives this gunfight?

2. Suppose in problem 1 that there is a fourth gunfighter D who hits his target with probability 0.1. What are the different non-absorbing states in a gunfight with four gunfighters, A, B, C, and D?

3. In the cost-accounting problem of this section, what are the costs attributed to Marketing and Manufacturing if the initial costs attributed to each department are 80\,000 for Marketing, 120\,000 dollars for Manufacturing, 70\,000 dollars for Maintenance, and 40\,000 dollars for Accounting.

4. Suppose a company has five departments, Marketing, Toy Manufacturing, Machine Manufacturing, Maintenance, and Accounting, with initial costs of 40 000, 60 000, 70 000, 60 000, and 30 000 dollars, respectively. Suppose that the amount of time spent servicing Marketing, Toy Manufacturing, Machine Manufacturing, Maintenance, and Accounting by the Maintenance department is 10, 20, 30, 30, and 10 per cent, and by the Accounting department is 10, 30, 30, 10, and 20 per cent, respectively. Of the 260 000 dollars total costs for the company, what amount is attributed to Marketing, Toy Manufacturing, and Machine Manufacturing?

5. Suppose we have a pair of dogs. They mate and produce two offspring, one of each sex. The two offspring mate and produce another pair of offspring, one of each sex, and so forth. The possible states are that both offspring are dominant homozygotes denoted e_1, both offspring are recessive homozygotes denoted e_2, both offspring are heterozygotes denoted f_1, one of the offspring is a dominant and the other a recessive homozygote denoted f_2, one of the offspring is a dominant homozygote and the other is a heterozygote denoted f_3, and one of the offspring is a recessive homozygote and the other is a heterozygote denoted f_4.
   a. Construct the transition matrix for this problem, and identify the matrices S and R.
   b. Find $(I - R)^{-1}$ and use this to compute the probabilities of ending in the states e_1 and e_2 given that the original pair of dogs were both heterozygotes.
   c. Repeat part b) for the case when one of the original pair of dogs was a dominant homozygote and the other was a heterozygote.

6. In the gunfight earlier in this section, we might wonder if A and B would enter this gunfight with such low probabilities of surviving. Might they form a coalition to first eliminate C? Might C bargain with one of them to break-up the coalition? Could they possibly discuss the situation until they get tired and go home?
   a. Find the probability of A, B, C, and nobody winning if A and B both shoot at C while C survives.
   b. Find the probability of A, B, C, and nobody winning if A and C both shoot at B while B survives.

   c. List any other possibly approaches to the gunfight that A, B, and C might take and discuss the implications for each surviving the gunfight.

7. Suppose for the original gunfight discussed in this section, we let $\tilde{f}_1(n)$ be the expected number of times we are in state f_1 during the first n rounds of the gunfight. Likewise for $\tilde{f}_2(n)$ and $\tilde{f}_3(n)$. Also let $F(n)$ be the 3-vector whose components are $\tilde{f}_1(n)$, $\tilde{f}_2(n)$, and $\tilde{f}_3(n)$. Since we start in state f_3, we have $\tilde{f}_1(0) = 0$, $\tilde{f}_2(0) = 0$, and $\tilde{f}_3(0) = 1$, that is, $F(0) = Q(0)$. The expected number of times for being in state f_1 after $n+1$ rounds is the expected number of times of being in state f_1 after n rounds plus the probability of being in state f_1 on round $n+1$. This gives the equation

$$\tilde{f}_1(n+1) = \tilde{f}_1(n) + q_1(n+1).$$

We can derive similar equations for $\tilde{f}_2(n+1)$ and $\tilde{f}_3(n+1)$, giving the dynamical system

$$F(n+1) = F(n) + Q(n+1) = F(n) + R^{n+1}Q(0). \qquad (8.10)$$

   a. Find the particular solution to dynamical system (8.10).
   b. By computing the limit as k goes to infinity, find the expected number of times for being in each nonabsorbing state before the gunfight ends.
   c. Find the expected total number of rounds of shots it will take for the gunfight to end.
   d. Find the expected total number of rounds of shots it will take for the gunfight to end if B does not show up for the fight.
   e. Find the expected total number of rounds of shots it will take for the gunfight to end if A does not show up for the fight.

8.7   Long-term behavior of systems

Let's review our study of linear dynamical systems. We know that the solution to a first-order linear dynamical system of m equations

$$A(n+1) = RA(n), \qquad (8.11)$$

is

$$A(k) = R^k A(0).$$

For small values of k, this form of the solution is useful, but for studying the **behavior** of the system for large values of k, this form of the long-term solution is almost useless. To get a more useful form for the solution, we found the characteristic values for R, that is, the roots to the equation,

$$|R - rI| = 0.$$

We then found the corresponding characteristic vectors, A_1, A_2, and so forth. The solution to the dynamical system was then given as

$$A(k) = c_1 r_1^k A_1 + \cdots + c_m r_m^k A_m.$$

For systems of three or more equations it is often difficult to find the characteristic values and the corresponding characteristic vectors. There are methods such as Newton's method from calculus that would theoretically allow us to approximate the roots, but even computing the determinant of $|R - rI|$ may not be easy.

This was the case for many examples of Markov chains. Thus, we learned additional techniques that simplified computations in certain special cases. For regular Markov chains we only needed to find the characteristic vector corresponding to the characteristic value $r = 1$ in order to determine the long-term behavior (probabilities). In the case of absorbing Markov chains, $r = 1$ is a multiple root, so this method did not apply. We then learned a clever technique to determine long-term behavior in section 8.5. The result of solving a simple dynamical system gave the long-term behavior as

$$\lim_{k \to \infty} P(k) = P(0) + S(I - R)^{-1} Q(0).$$

But these were special cases. What can we do when we are given a dynamical system in which (1) we cannot find the roots of the characteristic equation, and (2) the largest characteristic value, which we called the norm of R, $\|R\|$, is not $r = 1$ so that we cannot use the techniques that we applied to Markov chains? Our goal in this section is to find a relatively easy computational method for studying these types of dynamical systems. Let's make a moderately restrictive assumption.

ASSUMPTION: Suppose that there is one characteristic value, say r_1, such that $|r_1|$ is **larger** than the absolute value of each of the other characteristic values, that is,

$$|r_1| > |r_j|, \qquad \text{for} \quad j = 2, \ldots, m. \qquad \square$$

This assumption implies that r_1 is not a double root, and that r_1 is a real number, since complex characteristic values occur in conjugate pairs which have the same absolute value.

Recall that the solution to a dynamical system of m equations is

$$A(k) = c_1 r_1^k A_1 + c_2 r_2^k A_2 + \cdots + c_m r_m^k A_m.$$

The trick to this section is, instead of studying the solution $A(k)$, study

$$B(k) = r_1^{-k} A(k).$$

Notice that by dividing the solution $A(k)$ by r_1^k, we have

$$B(k) = c_1 \left(\frac{r_1}{r_1}\right)^k A_1 + c_2 \left(\frac{r_2}{r_1}\right)^k A_2 + \cdots + c_m \left(\frac{r_m}{r_1}\right)^k A_m. \qquad (8.12)$$

Notice that

$$\frac{r_1}{r_1} = 1, \qquad \text{so that} \qquad \left(\frac{r_1}{r_1}\right)^k = 1.$$

Letting

$$s_2 = \frac{r_2}{r_1}, \ldots, s_m = \frac{r_m}{r_1},$$

we can rewrite expression (8.12) as

$$B(k) = c_1 A_1 + c_2 s_2^k A_2 + \cdots + c_m s_m^k A_m.$$

From our assumption, we know that $|r_j| < |r_1|$, so $|s_j| < 1$ for $j = 2,$ ..., m, so that s_j^k goes to zero as k goes to infinity. We now have that

$$\lim_{k \to \infty} B(k) = \lim_{k \to \infty} \frac{A(k)}{r_1^k} = c_1 A_1,$$

where A_1 is the original characteristic vector corresponding to r_1.

Note that if we know the characteristic value r_1, we can then approximate a characteristic vector $c_1 A_1$ corresponding to r_1 with the vector

$$B(k) = \frac{A(k)}{r_1^k}$$

where k is a large integer. Note that this characteristic vector is a multiple of the original characteristic vector, but that will be no problem.

The question remains, how do we find r_1? Let's denote the first component of the vector $A(k)$ by $a_1(k)$ and the first component of the vector A_j by a_{1j} for $j = 1, \ldots, m$. Then the first component of the solution to the original dynamical system (8.11) is given by

$$a_1(k) = c_1 r_1^k a_{11} + c_2 r_2^k a_{12} + \cdots + c_m r_m^k a_{1m}.$$

(The notation is a little messy, so try not to let it confuse you.) Now the trick to answering our question is to study

$$\frac{a_1(n+1)}{a_1(n)}.$$

First, we observe that

$$r_1^{-k} a_1(k) = c_1 a_{11} + c_2 s_2^k a_{12} + \cdots + c_m s_m^k a_{1m} \to c_1 a_{11}$$

and

$$r_1^{-k} a_1(k+1) = c_1 r_1 a_{11} + c_2 r_2 s_2^k a_{12} + \cdots + c_m r_m s_m^k a_{1m} \to c_1 r_1 a_{11}$$

as k goes to infinity. Now we see that

$$\lim_{k \to \infty} \frac{a_1(k+1)}{a_1(k)} = \lim_{k \to \infty} \frac{r_1^{-k} a_1(k+1)}{r_1^{-k} a_1(k)} = \frac{c_1 r_1 a_{11}}{c_1 a_{11}} = r_1.$$

provided $a_{11} \neq 0$.

To summarize, consider the dynamical system (8.11)

$$A(n+1) = RA(n).$$

If you are not given an $A(0)$ value, just make one up. Then compute $A(1), A(2)$, and so forth, preferably with the aid of a computer or calculator. Let $a_1(k)$ be the first component of $A(k)$. Then approximate r_1 as

$$r_1 = \frac{a_1(k+1)}{a_1(k)}$$

where k is a "large" integer. We then approximate a characteristic vector A_1 with

$$A_1 = \frac{A(k)}{r_1^k},$$

where k is again a "large" integer.

Let's use this technique on a simple example in which we know r_1 and A_1, so that we can compare results.

EXAMPLE 8.14   In example 8.5, we considered the dynamical system
$A(n+1) = RA(n)$ where

$$A(n) = \left(\begin{array}{c} a_1(n) \\ a_2(n) \end{array} \right) \quad \text{and} \quad R = \left(\begin{array}{cc} 1 & 2 \\ -1 & 4 \end{array} \right).$$

Recall that the characteristic equation was

$$|R - rI| = (r-2)(r-3) = 0$$

and the largest characteristic value was $r_1 = 3$. We also computed the characteristic vector

$$A_1 = \left(\begin{array}{c} 1 \\ 1 \end{array} \right)$$

corresponding to $r_1 = 3$, or any multiple of this vector. Thus, we have

$$A(k) = c_1 3^k A_1 + c_2 2^k A_2$$

and

$$\lim_{k \to \infty} \frac{A(k)}{3^k} = c_1 A_1,$$

meaning that for large values of k, $a_1(k)$ and $a_2(k)$ are in the same ratio as the components of vector A_1, that is,

$$\frac{a_1(k)}{a_2(k)} = \frac{3^k c_1 1}{3^k c_1 1} = 1.$$

Suppose we couldn't find the roots of the characteristic equation. Let's use the techniques of this section to approximate r_1 and A_1. First, we need to make up an initial vector, say

$$A(0) = \left(\begin{array}{c} 1 \\ 0 \end{array} \right).$$

Then using a computer, we compute $A(1)$, $A(2)$, and so forth. The vectors $A(0)$, $A(1)$, $A(2)$, and $A(3)$ are

$$\left(\begin{array}{c} 1 \\ 0 \end{array} \right), \quad \left(\begin{array}{c} 1 \\ -1 \end{array} \right), \quad \left(\begin{array}{c} -1 \\ -5 \end{array} \right), \quad \left(\begin{array}{c} -11 \\ -19 \end{array} \right),$$

respectively. Omitting the intermediate calculations, we get

$$A(9) = \left(\begin{array}{c} -18\,659 \\ -19\,171 \end{array} \right) \quad \text{and} \quad A(10) = \left(\begin{array}{c} -57\,001 \\ -58\,025 \end{array} \right).$$

Our first three approximations for r_1 are

$$\frac{a_1(1)}{a_1(0)} = \frac{1}{1} = 1, \qquad \frac{a_1(2)}{a_1(1)} = -1, \qquad \frac{a_1(3)}{a_1(2)} = 11.$$

Continuing, but omitting the computations, we get the approximation

$$\frac{a_1(10)}{a_1(9)} = \frac{-57\,001}{-18\,659} = 3.05.$$

As you can see, if we approximated r_1 with 3.05, we would only be off by 0.05. Continuing these computations gives

$$\frac{a_1(20)}{a_1(19)} = 3.000\,90,$$

the moral being that the larger the value of k that you use in the approximation $a_1(k+1)/a_1(k)$, the better the approximation. The study of how large k needs to be to have a predetermined degree of accuracy is studied in the mathematical field of numerical analysis.

We now approximate a characteristic vector for $r = 3$ with

$$\frac{A(10)}{3^{10}} = \begin{pmatrix} -0.965 \\ -0.983 \end{pmatrix}.$$

Our approximations predict that the long term behavior of this system is

$$\lim_{k \to \infty} \frac{A(k)}{r_1^k} \approx c_1 \begin{pmatrix} -0.965 \\ -0.983 \end{pmatrix}$$

for some value of c_1, and, after dividing each component of this vector by $-0.983c_1$, we find that the two components are in the ratio of $(0.982 : 1)$ which is fairly close to the actual ratio of $(1 : 1)$.

Technically, we should use our approximation of $r_1 = 3.05$ instead of $r_1 = 3$ in the above calculation of A_1. But this would give us a vector that is just a multiple of the one we computed, and the ratio would be exactly the same.

If we used $3^{-20}A(20)$ or $(3.000\,90)^{-20}A(20)$ for the approximation, the ratio would be $(0.9997 : 1)$.   ■

There are a few problems with this approach. First, if the first component of A_1 is zero, then the quotient $a_1(k+1)/a_1(k)$ may not go to r_1. One solution would be to approximate r_1 with $a_2(k+1)/a_2(k)$ if it appears that $\lim_{k\to\infty} a_1(k) = 0$. A better approach would be to approximate η_1 with

$$\frac{\sqrt{a_1^2(k+1) + \cdots + a_m^2(k+1)}}{\sqrt{a_1^2(k) + \cdots + a_m^2(k)}}.$$

In the above example, we would have

$$\frac{\sqrt{a_1^2(3) + a_2^2(3)}}{\sqrt{a_1^2(2) + a_2^2(2)}} = \frac{\sqrt{482}}{\sqrt{26}} = 4.31,$$

and

$$\frac{\sqrt{a_1^2(10) + a_2^2(10)}}{\sqrt{a_1^2(9) + a_2^2(9)}} = 3.04.$$

By taking the square root of the sum of the squares of the components, we are finding how far vector $A(k)$ is from the origin. This is just the Pythagorean theorem. In fact, we can define the absolute value of an m-vector

$$A = \begin{pmatrix} a_1 \\ \vdots \\ a_m \end{pmatrix}$$

as

$$|A| = \sqrt{a_1^2 + \cdots + a_m^2}.$$

In this case, we are saying that

$$\lim_{k \to \infty} \frac{|A(k+1)|}{|A(k)|} = |\eta_1|.$$

This seems to be algebraically more complicated than using the first components. But it is easier to write a program using $|A(k)|$ than it is to use $a_1(k)$.

REMARK: We give the following program for computing $|\eta_1|$ and A_1 on the TI-81 calculator. Before running this program, store matrix R under $\boxed{\text{[B]}}$ and store vector $A(0)$ under $\boxed{\text{[A]}}$.

: $\boxed{\text{Lbl}}$ 1
: $\boxed{\text{[A]}}$ $\boxed{\text{T}}$ $\boxed{\text{[A]}}$ (where $\boxed{\text{T}}$ is on the $\boxed{\text{MATRIX}}$ menu.)
: $\boxed{\sqrt{}}$ $\boxed{\text{det}}$ $\boxed{\text{ANS}}$ $\to R$ (where $\boxed{\text{det}}$ is on the $\boxed{\text{MATRIX}}$ menu.)

: [B] [A] → [A]
: [A] [T] [A]
: [√] [det] [ANS] → S
: S/R → T
: [Disp] T
: [Pause]
: S⁻¹ [[A]] → [A]
: [Disp] [[A]]
: [Pause]
: [Goto] 1

When run, this program gives the first estimate for $|r_1|$ as $T = |A(1)|/|A(0)|$. Press [ENTER] and it gives you the first estimate for A_1 by using $A(1)/|A(1)|$. Press [ENTER] again and it gives you an estimate for $|r_1|$ using $|A(2)|/|A(1)|$. Keep pressing [ENTER] and it keeps giving further estimates for $|r_1|$ and A_1. To exit the program, press [ON]. □

Another problem is if the initial vector $A(0)$ is a multiple of one of the other characteristic vectors, say A_2, then $a_1(k+1)/a_1(k)$ will approximate r_2, the characteristic value corresponding to A_2. There are ways to avoid this problem, such as doing the approximation using several different initial vectors and using the largest approximation you get. One good choice would be to use m initial vectors. The jth of these vectors has a 1 in the jth position and 0's in each of the other positions. For the above example, you could use each of the vectors

$$\begin{pmatrix} 1 \\ 0 \end{pmatrix}, \quad \begin{pmatrix} 0 \\ 1 \end{pmatrix}$$

as $A(0)$.

REMARK: Sometimes there is a multiple root r_j. In this case, the solution may contain a term of the form $kr_j^k A_j$. Then $r_1^{-k} A(k)$ contains a term of the form $ks_j^k A_j$. This term also goes to zero as k goes to infinity and so it poses no problem.

If r_1 is a multiple root, then the solution to dynamical system (8.11) may be of the form

$$A(k) = r_1^k(A_1 + kA_2) + r_3^k A_3 + \cdots + r_m^k A_m.$$

In this case,

$$\lim_{k \to \infty} \frac{a_1(k+1)}{a_1(k)} = \lim_{k \to \infty} \frac{r_1(a_1' + ka_2' + a_2')}{a_1' + ka_2'} = r_1,$$

where a'_j is the first component of vector A_j. Thus, we see that this also does not pose a problem in finding r_1. But, in this case, $A(k)/r_1^k$ goes to infinity. If you compute $A(k)/kr_1^k$ you will get an estimate for A_2, which will give the long-term ratios for the components of the solution.

There is a real problem if there are other characteristic values, say r_2, such that $|r_2| = |r_1|$, but $r_2 \neq r_1$. For example, suppose we have a population with m equal age groups which is modeled by a dynamical system of m equations. If the largest characteristic value (in absolute value) for this dynamical system is positive, then we can estimate that value by computer computations, and then we can estimate the characteristic vector A_1. This vector will give the **stable age distribution** for this population, such as was found in section 5.10. But sometimes there are complex characteristic values, $r_2 = a + ib$ and $r_3 = a - ib$, for this dynamical system whose absolute value equals that of the largest positive characteristic value, that is,

$$|r_2| = |r_3| = \sqrt{a^2 + b^2} = r_1.$$

In this case, the numbers

$$\frac{a_1(k+1)}{a_1(k)}$$

will oscillate instead of converging to a fixed number. This tells us that the size of the population of each age group $a_j(k)$ will oscillate, that is, increase, decrease, and then increase again. This is an example of **population waves**, and is thought to explain population behavior for many species. In particular, population waves have been observed among humans. □

This should indicate that, although we can now successfully analyze a large number of linear dynamical systems, there are still many problems that could be considered, such as cases in which several characteristic values have the same absolute value as r_1 or cases in which r_1 is a multiple root. Hopefully, the reader will further these studies on his or her own.

PROBLEMS

1. Compute $a(4)/a(3)$ in example 8.14, where $a_1(0) = 1$ and $a_2(0) = 0$.

2. Consider the dynamical system, $A(n+1) = RA(n)$, where

$$A(n) = \begin{pmatrix} a(n) \\ b(n) \end{pmatrix}, \qquad R = \begin{pmatrix} 4 & -2 \\ 1 & 1 \end{pmatrix}, \qquad A(0) = \begin{pmatrix} 1 \\ -1 \end{pmatrix}.$$

   a. Find the characteristic values for this dynamical system. Also find a characteristic vector

$$A_1 = \begin{pmatrix} a \\ b \end{pmatrix},$$

   corresponding to the characteristic value $\eta_1 = \|R\|$.
   b. Using the given value for $A(0)$, compute $A(1)$ through $A(6)$. Using these vectors, compute

$$\frac{a(1)}{a(0)}, \quad \cdots, \quad \frac{a(6)}{a(5)}$$

   and let the last number be your estimate of η_1. Compare your results with part a).
   c. Estimate $|\eta_1|$ using $|A(6)|/|A(5)|$. Compare your answer to parts a) and b).
   d. Estimate A_1 with the vector $A(6)/|A(6)|$.
   e. Find the ratio of the two components of A_1 from part (a) and compare that to the ratio of the two components of your estimate for A_1 in part (d).

3. Repeat problem 2 with

$$R = \begin{pmatrix} 1 & 0.5 \\ 3 & 0.5 \end{pmatrix} \qquad \text{and} \qquad A(0) = \begin{pmatrix} 1 \\ 1 \end{pmatrix}.$$

4. Consider a population with three equal age groups, 0–1 year, 1–2 years, and 2–3 years old, denoted a_1, a_2, and a_3, respectively. Suppose in one time period all of age group a_1 survives to age group a_2, half of age group a_2 survives to age group a_3, and all of age group a_3 dies by the end of their third year. Suppose the birth rates for the three age groups are 1, 10, and 16 per individual in that age group, respectively.

   a. Develop a dynamical system of three equations that models this population.
   b. Find a characteristic vector A_1 corresponding to the characteristic value $\eta_1 = \|R\|$. This vector gives the stable age distribution of the age groups. Compute the actual vector, not an estimate.

c. Estimate r_1 and A_1 by using

$$A(0) = \begin{pmatrix} 1 \\ 0 \\ 0 \end{pmatrix}$$

and computing $A(1)$ through $A(5)$.

5. For some species, the reproductive rate and survival rate for the young of the species, say 0–1 year and 1–2 years, differ from the rates for the adults, but the reproductive rate and survival rate for the adults is approximately the same for every year, 2–3, 3–4, ... Instead of developing a dynamical system with a large number of equations, one for each possible age, the population can be approximated with three equations, one for those that are under 1 year old (called calves), one for those that are between 1 and 2 years old (called yearlings), and one for those that are over 2 years old (called adults). Let $a_1(n)$, $a_2(n)$, and $a_3(n)$ be the number of calves, yearlings, and adults, respectively, at the beginning of year n. Let s_1, s_2, and s_3 be the survival rates and b_1, b_2, and b_3 be the reproductive rates for the calves, yearlings, and adults, respectively.
    a. Write a dynamical system to model this population growth.
    b. For American bison, it is estimated that $b_1 = b_2 = 0$, $b_3 = 0.42$, $s_1 = 0.6$, $s_2 = 0.75$, and $s_3 = 0.95$. Estimate $\|R\|$ with $a_1(5)/a_1(4)$ where $a_1(0) = a_2(0) = a_3(0) = 1$.
    c. Estimate the characteristic vector A_1 for the characteristic value $r_1 = \|R\|$ by computing $A(5)/|A(5)|$. Then give the long-term population distribution.

8.8   The heat equation

We now apply our techniques to another type of problem, the distribution of heat through a long rod or bar. We will approximate the movement of heat through a bar by picking several evenly spaced points on our bar and approximating the temperature at those points. We assume that points close to each other will have about the same temperature. For this discussion, let's pick three points on a bar, say a, b, and c. See figure 8.2.

We let $a(n)$, $b(n)$, and $c(n)$ be the temperature at time n at the points a, b, and c, respectively. Let's assume the bar is in a vacuum, so that the only thing that affects the temperature at each point is

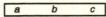

FIGURE 8.2.   A study of the temperature of a bar at three points at time n.

the temperature of the points next to it, that is, b affects a, b affects c, and a and c both affect b.

Let's consider how b is affected by a. At time n, if the temperature at point a is higher than at point b, $a(n) > b(n)$, then point a will cause the temperature at point b to increase, $b(n+1) > b(n)$. The amount of increase, $b(n+1) - b(n)$, will be proportional to the amount that a is hotter than b, $p[a(n) - b(n)]$, where p is the constant of proportionality. Similarly if a is cooler than b, the change in temperature, $b(n+1) - b(n)$, is negative since $p[a(n) - b(n)]$ is negative. Thus, the net effect of point a on point b is summarized by the equation

$$b(n+1) - b(n) = p[a(n) - b(n)].$$

This is known as **Newton's law of cooling**.

The number p depends on the following: the particular properties of the material from which the bar is made; the length of one unit of time (the smaller the unit of time, the less affect a has on b in that unit of time, and thus the number p is smaller); and the distance between the points a and b (the closer the points are together, the more affect a has on b, and thus p will be larger).

The point c has a similar affect on point b, that is, the affect of c on b is given by

$$b(n+1) - b(n) = p[c(n) - b(n)],$$

where the number p is the same as before. Thus, the **net** change in temperature at point b is the **sum of the affects of points a and c**, and the temperature at point b is completely described by the equation

$$b(n+1) - b(n) = p[a(n) - b(n)] + p[c(n) - b(n)],$$

or, after simplification,

$$b(n+1) = p[a(n) + c(n)] + (1 - 2p)b(n).$$

Similarly, after simplification, the equations that describe the temperature at points a and c are

$$a(n+1) = (1 - p)a(n) + pb(n)$$

and

$$c(n+1) = (1-p)c(n) + pb(n).$$

Combining these equations, we have the dynamical system

$$A(n+1) = RA(n)$$

where

$$A(n) = \begin{pmatrix} a(n) \\ b(n) \\ c(n) \end{pmatrix} \quad \text{and} \quad R = \begin{pmatrix} 1-p & p & 0 \\ p & 1-2p & p \\ 0 & p & 1-p \end{pmatrix}.$$

The characteristic equation is

$$|R - rI| = (1-p-r)^2(1-2p-r) - 2p^2(1-p-r)$$
$$= (1-p-r)(1-r)(1-3p-r),$$

and the three characteristic values are $r = 1$, $r = 1-p$, and $r = 1-3p$. Assuming that p is reasonably small, $(1-p)^k$ and $(1-3p)^k$ go to zero and the eventual temperature in the bar is given by

$$\lim_{k \to \infty} A(k) = \lim_{k \to \infty} [c_1 A_1 + (1-p)^k c_2 A_2 + (1-3p)^k c_3 A_3] = c_1 A_1,$$

where A_1 is a characteristic vector corresponding to the characteristic value $r_1 = 1$. Solving $(R-I)A_1$ for A_1 gives

$$A_1 = \begin{pmatrix} 1 \\ 1 \\ 1 \end{pmatrix}$$

or some multiple of this vector. Thus, eventually all points on the bar are the same temperature.

This was not very illuminating, so we will add a new component to our problem. We now assume that we have a bar with several (equidistant) points on it, say six, counting the end points. Let's assume that the left end of the bar is being chilled, that is, the point at the left of the bar is kept at a constant temperature of 0 degrees centigrade. Let's also assume that the point at the right of the bar is kept at a constant temperature of 20 degrees centigrade. See figure 8.3.

Here we have a total of six points, counting the ends which are kept at constant temperatures. For the four interim points of the bar, let $a_j(n)$ be the temperature at point a_j at time n, for $j = 1, 2, 3$, and 4, counting the points from left to right. For simplicity of

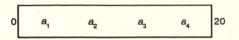

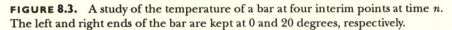

FIGURE 8.3. A study of the temperature of a bar at four interim points at time n. The left and right ends of the bar are kept at 0 and 20 degrees, respectively.

exposition, denote the temperatures at time n at the left and right end-points by $a_0(n)$ and $a_5(n)$, respectively. Note that $a_0(n) = 0$ and $a_5(n) = 20$ for all n.

As before, we assume that the temperature change at each point a_j, with $j = 1, 2, 3$, and 4 during one time-period is proportional to the difference in temperature between that point and the points immediately to the left and right, that is,

$$a_j(n+1) - a_j(n) = p[a_{j-1}(n) - a_j(n)] + p[a_{j+1}(n) - a_j(n)],$$

for $j = 1, 2, 3$, and 4. After simplification, this $a_j(n)$ satisfies an equation similar to the one for point b in the discussion above, that is,

$$a_j(n+1) = p[a_{j-1}(n) + a_{j+1}(n)] + (1 - 2p)a_j(n).$$

For points a_1 and a_4, this becomes

$$a_1(n+1) = p[0 + a_2(n)] + (1 - 2p)a_1(n),$$

$$a_4(n+1) = p[a_3(n) + 20] + (1 - 2p)a_4(n)$$

since $a_0(n) = 0$ and $a_5(n) = 20$.

For simplicity, let's assume that $p = 0.5$, so that the four equations become

$$a_1(n+1) = 0.5a_2(n),$$

$$a_2(n+1) = 0.5a_1(n) + 0.5a_3(n),$$

$$a_3(n+1) = 0.5a_2(n) + 0.5a_4(n),$$

$$a_4(n+1) = 0.5a_3(n) + 10.$$

This becomes the first-order affine dynamical system

$$A(n+1) = RA(n) + B, \tag{8.13}$$

where

$$A(n) = \begin{pmatrix} a_1(n) \\ a_2(n) \\ a_3(n) \\ a_4(n) \end{pmatrix}, R = \begin{pmatrix} 0 & 0.5 & 0 & 0 \\ 0.5 & 0 & 0.5 & 0 \\ 0 & 0.5 & 0 & 0.5 \\ 0 & 0 & 0.5 & 0 \end{pmatrix}, B = \begin{pmatrix} 0 \\ 0 \\ 0 \\ 10 \end{pmatrix}.$$

We suspect that the solution to this dynamical system might be of the form

$$A(k) = R^k C + A,$$

where A and C are two yet-to-be-determined vectors. The vector C depends on the initial value, that is, the initial temperature distribution of the 4 points in the bar. The vector A will be the **equilibrium vector** for the dynamical system, that is, the (unknown) vector A that satisfies

$$A = RA + B, \qquad \text{or} \qquad A = (I - R)^{-1} B,$$

if $(I - R)^{-1}$ exists. You can check that

$$A(k) = R^k C + (I - R)^{-1} B$$

satisfies dynamical system (8.13). This is a system of four equations and four unknowns (the four unknown components of A). The solution is then

$$A = (I - R)^{-1} B = \frac{1}{5} \begin{pmatrix} 8 & 6 & 4 & 2 \\ 6 & 12 & 8 & 4 \\ 4 & 8 & 12 & 6 \\ 2 & 4 & 6 & 8 \end{pmatrix} \begin{pmatrix} 0 \\ 0 \\ 0 \\ 10 \end{pmatrix} = \begin{pmatrix} 4 \\ 8 \\ 12 \\ 16 \end{pmatrix}.$$

We now have the solution

$$A(k) = R^k C + A,$$

and if $\|R\| < 1$, then $R^k C$ goes to zero and the temperature of the bar goes to its equilibrium A, which is "stable" in the sense that this happens regardless of $A(0)$.

In order to avoid finding and factoring the characteristic equation, we will estimate $\|R\|$ using the techniques of the previous section. To do this, we first make up an initial vector, say

$$A(0) = \begin{pmatrix} 1 \\ 1 \\ 1 \\ 1 \end{pmatrix}.$$

We then compute $A(1)$, $A(2)$, and so forth, for the **linear dynamical system**

$$A(n+1) = RA(n),$$

not the first-order affine dynamical system (8.13) we have been studying. The vectors $A(1)$ through $A(5)$ are

$$
\begin{pmatrix} 0.5 \\ 1 \\ 1 \\ 0.5 \end{pmatrix},\quad
\begin{pmatrix} 0.5 \\ 0.75 \\ 0.75 \\ 0.5 \end{pmatrix},\quad
\begin{pmatrix} 0.375 \\ 0.625 \\ 0.625 \\ 0.375 \end{pmatrix},\quad
\begin{pmatrix} 0.313 \\ 0.5 \\ 0.5 \\ 0.313 \end{pmatrix},\quad
\begin{pmatrix} 0.25 \\ 0.406 \\ 0.406 \\ 0.25 \end{pmatrix},
$$

respectively. Our estimates, $a_1(n+1)/a_1(n)$, for r_1 are 0.5, 1, 0.75, 0.833, and 0.800 for $n = 0, 1, 2, 3$, and 4, respectively. If you continued, you would find that

$$\frac{a_1(n+1)}{a_1(n)} = 0.8090, \qquad \text{for} \quad n = 11, 12, \ldots,$$

to 4 decimal place accuracy. Therefore, $\|R\| = 0.8090$, approximately, so

$$\|R\| < 1.$$

Hence, $A(k)$ gets closer to A. This also implies that 1 is not a characteristic value for R, so that we would know that $(I - R)^{-1}$ exists, if we hadn't already computed it.

For your information, we give the temperature distribution for dynamical system (8.13) at times $n = 0, 1, 5, 10$, and 30, respectively.

$$
A(0) = \begin{pmatrix} 1 \\ 1 \\ 1 \\ 1 \end{pmatrix},\quad
A(1) = \begin{pmatrix} 0.5 \\ 1 \\ 1 \\ 10.5 \end{pmatrix},\quad
A(5) = \begin{pmatrix} 1.5 \\ 4.78 \\ 7.91 \\ 14 \end{pmatrix},
$$

$$
A(10) = \begin{pmatrix} 3.31 \\ 6.59 \\ 10.88 \\ 15.13 \end{pmatrix},\quad \text{and} \quad
A(30) = \begin{pmatrix} 3.99 \\ 7.98 \\ 11.98 \\ 15.99 \end{pmatrix}.
$$

Thus, you can see $A(k)$ converging to the equilibrium vector

$$
A = \begin{pmatrix} 4 \\ 8 \\ 12 \\ 16 \end{pmatrix}.
$$

REMARK: Using calculus and differential equations, it is possible to model the continuous change in temperature throughout a bar. In fact, the continuous version of our dynamical system can be derived by first constructing a dynamical system (much like the one we derived) and then letting the time interval and the distance between the points both go to zero. This shows the close relationship that often exists between discrete and continuous models, that is, between dynamical systems and calculus.   □

In many applications, a dynamical system is derived, much like

$$A(n+1) = RA(n) + B,$$

in which there is an equilibrium vector A. The goal is to show that the equilibrium vector is **stable** by showing that the matrix R satisfies $\|R\| < 1$. If this is true, then the equilibrium vector is stable and the solution $A(k)$ goes to the equilibrium vector A as k goes to infinity. We have seen this time and again in our discussions. In such cases, mathematicians call the matrix R a **contraction mapping**. Because of this, one major area of mathematical research is the study of **mappings** such as the matrix R, to determine which mappings are contractions, that is, which mappings satisfy

$$\|R\| < 1.$$

PROBLEMS

1. Consider the dynamical system

$$a_1(n+1) = p[c_1 + a_2(n)] + (1 - 2p)a_1(n),$$

$$a_2(n+1) = p[a_1(n) + a_3(n)] + (1 - 2p)a_2(n),$$

$$a_3(n+1) = p[a_2(n) + a_4(n)] + (1 - 2p)a_3(n),$$

$$a_4(n+1) = p[a_3(n) + c_2)] + (1 - 2p)a_4(n),$$

which describes the transfer of heat between the four points in the bar, where c_1 is the constant temperature at the left of the bar and c_2 is the constant temperature at the right of the bar.
   a. Rewrite these as a system of matrices, that is, give the matrix R and the vector B, when $p = 0.25$, $c_1 = 0$, and $c_2 = 20$.
   b. Find the equilibrium vector for this dynamical system.

    c. Compute $A(1)$, $A(2)$, $A(3)$, and $A(4)$, where $A(n+1) = RA(n)$, and

$$A(0) = \begin{pmatrix} 16 \\ 16 \\ 16 \\ 16 \end{pmatrix}.$$

    d. Find $a_1(4)/a_1(3)$. Does $\|R\|$ appear to be less than 1? (Note that, to 4-decimal-place accuracy, $\|R\| = 0.9045$. This would be found if you computed $a_1(10)/a_1(9)$.)

    e. Find $A(4)$ using the dynamical system $A(n+1) = RA(n) + B$ and $A(0)$ from part (c).

2. Repeat problem 1 with $p = 0.75$, $c_1 = 10$, and $c_2 = 14$. It must be pointed out that $a_1(4)/a_1(3)$ is not a good estimate for $\|R\|$. To 4-decimal-place accuracy, the largest characteristic value (in absolute value) is $r_1 = -0.9635$, which is not found until $a_1(41)/a_1(40)$. Thus, $\|R\| = |r_1| = 0.9635$.

3. Consider a bar, such that part of the bar is in air that is kept at a constant temperature of 20 degrees, and part of the bar is submersed in water that is kept at a constant temperature of 0 degrees. See figure 8.4. We will assume that the temperature at each point on the bar, a_1, a_2, and a_3, depends on the temperature of the four nearest points, that is, the points above, below, to the left, and to the right. The points that affect the temperature of the point a_1 are the point above which is 20 degrees, the points to the left and to the right which are both 20 degrees, and the point below which is $a_2(n)$ degrees. Thus, the change

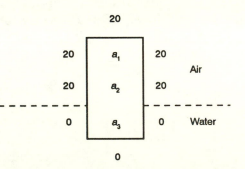

FIGURE 8.4. A study of the temperature of a bar at three points at time n. One-third of the bar is immersed in water at 0 degrees and the remainder is in air at 20 degrees.

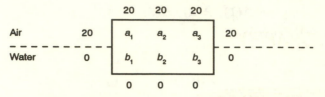

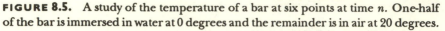

FIGURE 8.5. A study of the temperature of a bar at six points at time n. One-half of the bar is immersed in water at 0 degrees and the remainder is in air at 20 degrees.

in temperature is given by the equation

$$a_1(n+1) = a_1(n) + 3p[20 - a_1(n)] + p[a_2(n) - a_1(n)]$$
$$= (1 - 4p)a_1(n) + pa_2(n) + 60p.$$

  a. Derive the equations for the temperature at points a_2 and a_3.

  b. Give the first-order affine dynamical system describing the transfer of heat at the three points a_1, a_2, and a_3, that is, find the matrix R and the vector B in terms of p.

  c. Give the system when $p = 0.25$.

  d. Solve for the equilibrium vector.

  e. Approximate $\|R\|$.

  f. Find $A(4)$ using the dynamical system $A(n+1) = RA(n) + B$ with $A(0)$ having $a_1(0) = a_2(0) = a_3(0) = 8$.

4. Suppose we have a grid of six points on our bar as shown in figure 8.5.

  a. Write a dynamical system of six equations that describes the flow of heat in this bar, using the same assumptions as in problem 3.

  b. Rewrite these equations as a first-order affine dynamical system and find the equilibrium vector for this system.

  c. Approximate $\|R\|$.

BIBLIOGRAPHY

Barnsley, M. (1988). *Fractals Everywhere.* Academic Press, Boston.

Baumol, W. J. (1961). Pitfalls on contracyclical policies: some tools and results. *Review of Economics and Statistics*, **43**, 21–6.

Cavalli-Sforza, L. L. and **Bodmer, W. F.** (1971). *The Genetics of Human Populations.* W. H. Freeman, San Francisco.

Devaney, R. L. (1986). *An Introduction to Chaotic Dynamical Systems.* Benjamin/Cummings, Menlo Park, California.

Devaney, R. L. (1989). *Chaos, Fractals, and Dynamics: Computer Experiments in Mathematics.* Addison-Wesley, Menlo Park.

Gleick, J. (1987). *Chaos: Making a New Science.* Viking, New York.

Hoppensteadt, F. C. (1982). *Mathematical Methods of Population Biology.* Cambridge University Press.

Malkevitch, J. and **Meyer, W.** (1974) *Graphs, Models, and Finite Mathematics.* Prentice Hall, Englewood Cliffs, New Jersey.

Peitgen, H.-O. and **Richter, P.** (1986). *The Beauty of Fractals.* Springer-Verlag, New York.

Richardson, L. F. (1960). *Arms and Insecurity: A Mathematical Study of the Causes and Origins of War.* Boxwood Press, Pittsburgh.

Samuelson, P. A. (1939). Interactions between the multiplier analysis and the principle of acceleration. *Review of Economics and Statistics*, **21**, 75–8.

Smith, J. M. (1968). *Mathematical Ideas in Biology*. Cambridge University Press.

ANSWERS

Section 1.1

1. **a.** $a(n+1) = 0.8a(n) + 10$
   b. 34.64 ml
   c. 50 ml

3. 8 ml

Section 1.2

1. **a.** nonlinear,
   b. affine,
   c. not a dynamical system,
   d. not a dynamical system
   e. nonlinear,
   f. affine (when transposed),
   g. not a dynamical system,
   h. linear nonhomogeneous.

3. **a.** $a(n) = 4a(n-1)$, for $n = 3, 4, \ldots$
   b. $a(n) = a(n-1) + 2(n+3) = a(n-1) + 2n + 6$, for $n = 1, 2, \ldots$
   c. $a(n) = (n+1)a(n-1)$, for $n = 1, 2, \ldots$

5. $a(6) = -10$

7. $a(3) = 222.5$ and $b(3) = 42.5$

Section 1.3

1. a. -6
    b. $7/5 = 1.4$
    c. none
    d. $a = 2$ and $a = -1$
    e. $a = 0$, $a = 0.5$, and $a = -0.5$

3. $b = 3$

5. $r = 1/4 = 0.25$

7. Equilibrium $a = 2$ is unstable.

9. $a(0) = 2$

11. If $a(0) = 1.1$ then $a(1) = 1.09$, $a(2) = 1.0819$, and $a(3) = 1.0752$. If $a(0) = 0.9$ then $a(1) = 0.89$, $a(2) = 0.8779$, and $a(3) = 0.8630$. Implies $a = 1$ is semistable.

13. $a = -1$ and $a = 3$ are both unstable.

15. stable

Section 1.4

1. $t(n) = [p_i/(p_i - p_s)]t(n-1) + c/p_i$ has the unstable equilibrium value $t = (p_s - p_i)c/(p_i p_s)$.

3. $i(n) = p_i(t(n) - t(n-1)) + qt(n)$. If $q < p_s$, then $t(k)$ goes to infinity. If $q > p_s$, then $t(k)$ goes to zero.

5. $p(n+1) = -(4/3)p(n) + 35/3$ has unstable equilibrium value $p = 5$.

7. a. $p(n+1) = -(3/4)p(n) + 35$, $p = 20$ is stable.
    b. $s(n+1) = -(3/4)s(n) + 3.5$, $s = 2$ is stable.
    c. $d(n+1) = -(3/4)d(n) + 3.5$, $d = 2$ is stable.

Section 1.5

1. a. 35 quarters (50 005.52 dollars)
    b. 56 quarters

3. a. 15 years
    b. 8 years (almost doubled)

5. 41 210.74 dollars

7. 2149.29 dollars

9. 273.02 dollars

11.   **a.** 21 914.91 dollars
          b. 2 130 530.72 dollars
          c. approximately 180 dollars (depending on when the first deposit is made).

13. 14.47 per cent per year

Section 2.1

1.   **a.** $a(1) = -2.2$, $a(2) = 2.42$, $a(3) = -2.662$, $a(4) = 2.9282$, $a(5) = -3.221$, $a(k) = 2(-1.1)^k$, solution oscillates to infinity.
         b. $a(5) = -1.61$, $a(k) = -(1.1)^k$, solution decreases to negative infinity.
         c. $a(5) = -0.32$, $a(k) = 1000(-0.2)^k$, solution oscillates to zero.
         d. $a(5) = -0.32$, $a(k) = -1000(0.2)^k$, solution increases to zero.

3.   **a.** $a(k) = 4 \times 3^k$
         b. $a(k) = 4(-0.5)^k$
         c. $a(k) = 4(-1)^k$

5.   **a.** 34.3 years
         b. The half-life is between 29 and 42 years.

7.   **a.** 9901 years
         b. The half-life is between 9764 and 10 039 years.

9.   **a.** approximately 66 698 years
         b. The half-life is between 59 998 and 73 405 years.

11. **a.** 1326 years
         b. 486 years

Section 2.2

1.   **a.** $a(k) = c - 5k$
         b. $a(k) = c + 3k$

3.   **a.** $a(k) = 10 - 5k$
         b. $a(k) = -6 + 3k$

5. $a(n+1) = a(n) + 100$ with $a(0) = 150$, so $a(k) = 150 + 100k$. If $k = 9$ encyclopedias then $a(9) = 1050$ dollars.

7. $a(n+1) = a(n) + 0.7$ so $a(k) = 330 + 0.7k$ meters per second at k degrees Celsius.

Section 2.3

1. **a.** $u = 0.16$, $v = 0.48$, and $w = 0.36$.
    b. $q = 0.4$ and 36 per cent are dominant homozygotes
    c. 400, 3200, and 6400 respectively. There are 4000 A-alleles and $p = 0.2$.

3. **a.** 50 generations
    b. 83.3 or 84 generations
    c. 50/3 or 17 generations

5. $p(n+1) = (1-\mu-\nu)p(n)+\nu$ with equilibrium value $p = \nu/(\nu+\mu)$.

Section 2.4

1. **a.** $a(1) = 1.4$, $a(2) = 0.12$, $a(3) = -0.904$, $a(4) = -1.723$, $a(5) = -2.379$, $a(k) = 8(0.8)^k - 5$, and solution decreases to negative five.
    b. $a(5) = -2$, $a(k) = 3.5(-1)^k + 1.5$, and -2 and 5 form a 2-cycle for this solution.
    c. $a(5) = -5.594$, $a(k) = (-1.5)^k + 2$ and solution oscillates to infinity.

3. **a.** $a(k) = 4 + 3k$
    b. $a(k) = -6(1.5)^k + 10$
    c. $a(k) = 2(-0.9)^k + 2$

5. $r = 0.5$

7. $a(k) = c(2/3)^k - 5$

9. $t(0) < 200$

11. $t(n + 1) = 0.8t(n) + 20$ and 100 billion dollars is a stable equilibrium.

13. 2/7ths are A-alleles, 5/7ths are a-alleles, and 24/49ths exhibit the dominant trait.

15. **a.** $q(n+1) = q(n)(2 - q(n))/(2 - q^2(n))$, $q = 0$ and $q = 1$ are equilibrium values.
    b. $q(n+1) = [2(1 - \mu)q(n) - q^2(n) + 2\mu]/[2 - q^2(n)]$ with equilibrium values of $q = 1$, $q = \sqrt{2\mu}$, and $q = -\sqrt{2\mu}$.

Section 2.5

1. **a.** $a(n+1) = 1.02a(n)$ with $a(0) = 1000$ gives $a(4) = 1082.43$ dollars.
   b. $a(n+1) = 1.025a(n)$ with $a(0) = 200$ gives $a(10) = 256.02$ dollars.

3. **a.** 99 425.49 dollars
   b. 88 915.96 dollars

5. 2149.29 dollars

7. 273.02 dollars per quarter.

9. 8.45 per cent

11. **a.** 14.21 or 15 years
    b. 8.04 or 9 years

Section 3.1

1. 12

3. **a.** 80
   b. 56
   c. 22

5. **a.** 210
   b. $15!/6! = 1\,816\,214\,400$

7. $2^{10} = 1024$

9. **a.** 15^{15}
   b. $15!$
   c. $15!/10! = 360\,360$

11. $7! = 5040$

13. **a.** $2^6 = 64$
    b. 15
    c. 22
    d. $2^6 - 1 = 63$

Section 3.2

1. **a.** $5 \times 6^4 \times 4/10^5 = 0.259$
   b. $(5 \times 6^4 \times 4 + 6^5)/10^5 = 0.337$
   c. 0.087

3. **a.** $4(1500)^3(2500)/(4000)^4 = 0.132$
   b. 0.330

      c. 0.481

5.   **a.**   **i.** $3^2/7^2 = 9/49$
         ii. $3 \times 2/(7 \times 6) = 1/7$
    b.   **i.** 24/49
         ii. 4/7

7.   **a.** $(5/6)^3 = 0.5787$
    b. $5 \times 4 \times 3/(6 \times 5 \times 4) = 0.5$

9. $1 - (45 \times 44 \times 43)/47 \times 46 \times 45) = 0.125$

11. $10 \times 9/(47 \times 46) = 0.042$

13.   **a.** $(15!/10!)/(20!/15!) = 0.194$
    b. $5 \times 15 \times (5!/1!)/(20!/15!) = 0.0048$
    c. Answer to last part plus $5!/(20!/15!)$ which equals 0.0049.

15.   **a.** $(6 \times 10^2 \times 15^2 + 4 \times 10 \times 15^3 + 15^4)/(25)^4 = 0.821$
    b. 0.841

Section 3.3

1. $(0.5)(0.8) + (0.5)^2(0.8) = 0.6$

3. $(3/10)(3/5) + (7/10)(2/5) = 23/50$

5. $(0.6)(0.2) + (0.4)(0.9) = 0.48$

7. $(0.6)^3[1 + 3(0.4) + 6(0.4)^2] = 0.682\,56$

9. 2/3 (One of three possible heads is showing. In two of those cases, heads is on the other side also.)

Section 3.4

1.   **a.** $p(n+1) = (7/24)p(n) + 3/8$
    b. $p(k) = c(7/24)^k + 9/17$, so $p(k)$ goes to 9/17.
    c. $p(4) = 0.533$
    d. $p(k) = (-1/34)(7/24)^k + 9/17$ and $p(4) = 0.529$.

3.   **a.** 0.305
    b. 2/7

5.   **a.** 0.3052
    b. 2.15

Section 4.1

1.   **a.** $a(k) = c(2)^k + 3^k$
    b. $a(k) = c(-3)^k + 3(2)^k$

    c. $a(k) = c - 0.5(-1)^k$
    d. $a(k) = c(2)^k + 3^k + 0.5(4)^k$
    e. $a(k) = c + 2^k - 0.25(5)^k$
    f. $a(k) = c(3)^k + 2(4)^k + 3$

3. $a(k) = c(2)^k + (k - 8)3^k$

5. $5/16$

7.   **a.** $a(n+1) = 1.1a(n) - 40\,000(1.05)^{n+1}$ or $a(n+1) = 1.1a(n) - 42\,000(1.05)^n$
    b. $a(k) = c(1.1)^k + 840\,000(1.05)^k$
    c. $c = a_0 - 840\,000$
    d. $508\,707.53$ dollars

9.   **a.** $a(n+1) = 1.1a(n) + 10\,000(1.2)^n$ with $a(0) = 20\,000$ gives $a(24) = 7\,161\,706.11$ dollars
    b. $b(n + 1) = b(n) + 2000(1.3)^{n+1}$ with $b(0) = 2000$ gives $b(24) = 4\,697\,606.68$ dollars
    c. $c(n+1) = 1.1c(n) + 10\,000(1.2)^n - 2000(1.3)^{n+1}$ with $c(0) = 18\,000$ gives $c(24) = 213\,643.15$ dollars

11.   **a.** $a(k) = 20(0.95)^k - 20(0.9)^k + 2.5(0.99)^k$, $b(k) = 10(0.9)^k + (0.99)^k$, and $c(k) = 9(0.99)^k$.
    b. $b(k)/c(k)$ goes to $1/9$
    c. $a(k)/c(k)$ goes to $5/18$

13.   **a.** 3
    b. $2/3$
    c. $1/48$
    d. $25/21$

Section 4.2

1.   **a.** $a(k) = (c - 0.5)2^k$
    c. $a(k) = (c - k)(-1)^k$
    e. $a(k) = (c + k/3)3^k - 4$

2.   **a.** $c = 1$
    c. $c = 1$
    e. $c = 5$

3. $a(k) = (c_1 + c_2 k + c_3 k^2)r^k$ with $c_2 = (b_1 - 0.5b_2)/r$ and $c_3 = 0.5b_2/r$.

5. $a_0 = 40\,000 \times 20/1.07 = 747\,663.55$ dollars

Section 4.3

1.   **b.** $p(k) = 6 - 2(0.5)^k$ goes to 6.

    c. $a(k) = 4/3 - (1/3)(0.25)^k$ goes to $4/3$.

3.   **b.** $1/(3 - \sqrt{3}) = 0.5 + \sqrt{3}/6$

    c. $1 + \sqrt{3}/3$

5.   **b.** $1/(2 - 2\cos 36)$

Section 4.4

1.   **a.** $a(k) = c\,2^k + 1 + k$

    c. $a(k) = c(-1)^k - 1 - 2k + 2k^2$

2.   **a.** $c = 0$

    c. $c = 2$

3. $67\,956.35$ dollars

Section 4.5

1.   **a.** $a(k) = c - 2k + 3k^2$

    c. $a(k) = c + 2k - 2k^2$

2.   **a.** $c = 2$

    c. $c = 2$

3.   **a.** $c = 1$

    c. $c = 2$

5.   **a.** $(k + 1)(k + 2)/2$

    b. $a(k) = (1/3)k + (1/2)k^2 + (1/6)k^3$

7. $a(k) = (-0.25k + 0.25k^2)2^k$

9.   **a.** $a(n + 1) = a(n) + 2\pi + 0.004n\pi$

    b. $a(k) = 1.998\pi k + 0.002\pi k^2$

    c. 4709.25 inches

    d. 1.15 inches

Section 5.1

1.   **a.** $x^2 = x + 2$ has roots $x = 2$ and $x = -1$ so the general solution is $a(k) = c_1(-1)^k + c_2(2)^k$. Particular solution when $c_1 = 1$ and $c_2 = 2$.

    c. $a(k) = c_1 3^k + c_2(-5)^k$ with $c_1 = 2$ and $c_2 = -1$.

    e. $a(k) = c_1(2 + i)^k + c_2(2 - i)^k$ with $c_1 = c_2 = 1$.

2.   **a.** $a(n + 2) = 2a(n + 1) + a(n)$

    b. $a(k) = c_1(1 + \sqrt{2})^k + c_2(1 - \sqrt{2})^k$

    c. $a(k) = (\sqrt{2}/4)(1 + \sqrt{2})^{k+1} - (\sqrt{2}/4)(1 - \sqrt{2})^{k+1}$

3. **b.** $c_1 = (a(0)s^N - a(N))/(s^N - r^N)$ and $c_2 = (a(0)r^N - a(N))/(r^N - s^N)$

Section 5.2

1. **a.** $a(k) = 5 - k$
   d. $a(k) = (1 - 3k)(-0.5)^k$
2. **a.** $a(k) = -2 + 3k$
   d. $a(k) = (-20 + 18k)(-0.5)^k$
3. **a.** $a(k) = 2 + (10/3)k$
   d. $a(k) = [2 - (98/3)k](-0.5)^k$
4. **b.** $c_1 = a(0)$ and $c_2 = [a(N) - r^N a(0)]/[Nr^N]$

Section 5.3

1. **a.** 0.881
   b. 0.599
   c. 0.550
   d. 0.520
   e. 0.51
3. **a.** $p(0) = 1$, $p(1) = \frac{30}{41}$, $p(2) = \frac{18}{41}$, $p(3) = p(4) = 0$
   b. $p(n + 3) = -3p(n + 2) + 10p(n + 1) - 6p(n)$, $p(0) = 1$, $p(N) = 0$, and $p(N + 1) = 0$

Section 5.4

1. 9 per cent of women and 30 per cent of men exhibit recessive trait; 42 per cent of women are heterozygotes.
3. **a.** 0.85
   b. $p(k) = (13/15) + (1/30)(-0.5)^k$
   c. 13/15
5. $p(n + 2) = 0.5(1 - \mu)p(n + 1) + 0.5(1 - \mu)^2 p(n)$, and $p(k) = c_1(1 - \mu)^k + c_2(0.5\mu - 0.5)^k$

Section 5.5

1. **a.** $20 + 17i$
   b. $-17 + 11i$
   c. $-45 - 28i$
   d. $\sqrt{53}$
   e. $\sqrt{5}$

3.   **a.** $a(k) = c_1(-0.5 + 0.5i)^k + c_2(-0.5 - 0.5i)^k$
      b. $|-0.5 \pm 0.5i| = \sqrt{0.5} < 1$ so zero is attracting.
      c. $c_1 = i$ and $c_2 = -i$
      d. $a(2) = 1$, $a(3) = -0.5$, $a(11) = -0.03125$, and $a(12) = 0$

5.   **a.** $a(k) = c_1 4^k + c_2 2^k + 2$
      b. $a = 2$ is a repelling fixed point because $|4| > 1$.

7.   **a.** $a(k) = c_1(2 + i)^k + c_2(2 - i)^k + 3$
      b. $|2 \pm i| = \sqrt{5} > 1$ so $a = 3$ is repelling.

Section 5.6

1. $d(k) = c_1[(2 + 0.4i)/2.08]^k + c_2[(2 - 0.4i)/2.08]^k$. Since $|(2 \pm 0.4i)/2.08| = 0.9806 < 1$ the spring oscillates to zero.

3.   **a.** $|(1.99 \pm \sqrt{0.0799}i)/2| = 1/1.01 < 1$ so solution oscillates to zero.
      b. $|(2 - p + q \pm \sqrt{4p - (p - q)^2}i)/(2 + 2q)| = 1/(1 + q) < 1$ so $d(k)$ oscillates to zero.

5.   **a.** $d(n + 2) - 1.6d(n + 1) + 0.63d(n) = 0$
      b. $d(k) = c_1(0.9)^k + c_2(0.7)^k$
      c. $d(k) = (0.9)^k + 3(0.7)^k$, $d(2) = 2.28$, $d(3) = 1.758$, $d(10) = 0.433$, and $d(k)$ decreases to zero.
      d. $d(k) = -4(0.9)^k + 8(0.7)^k$, $d(2) = 0.68$, $d(3) = -0.172$, $d(10) = -1.169$, and $d(k)$ decreases to -1.261 at $k = 8$, then increases to zero.

Section 5.7

1.   **a.** $a(k) = c_1 3^k + c_2(-2)^k$
      b. $a(k) = c_1 3^k + c_2(-2)^k - 5^k$
      c. $a(k) = c_1 3^k + c_2(-2)^k + 2k - 1$
      d. $a(k) = c_1 3^k + c_2(-2)^k + 0.5(2)^k$
      e. $a(k) = c_1 3^k + c_2(-2)^k - 2k3^k$

4.   **a.** $a(k) = c_1 + c_2 k$
      b. $a(k) = c_1 + c_2 k - 3k^2$
      c. $a(k) = c_1 + c_2 k + 2^k$
      d. $a(k) = c_1 + c_2 k + k^3 - k^2$

5.   **a.** $a(k) = 3^{k+1} - 5^k$
      d. $a(k) = 7 + 2k - 3k^2$

6.   **a.** $a(k) = (c_1 + k^2)3^k + c_2(-2)^k$

7. $c_1 = a(0)$ and $c_2 = -c_3 - a(0) + a(1)/r$

Section 5.8

1. 1904 to the nearest integer.
3.   **a.** $d(n+2) = 2d(n+1) - d(n) - 32$
    b. $d(k) = c_1 + c_2 k - 16k^2$
    c. $d(k) = 144 - 16k^2$ so at $k = 3$ seconds, it reaches the ground.

Section 5.9

1. $t = 20$
3. $t(n+2) = 1.8t(n+1) - 0.9t(n) + (1.05)^{n+2}$ with general solution
$t(k) = c_1(0.9+0.3i)^k + c_2(0.9-0.3i)^k + (9/80)(1.05)^{k+2}$
5.   **a.** $t = 3/(1+a-ab)$
    b. $(ab \pm \sqrt{a^2 b^2 - 4a})/2$
    c. $a < 1$
7.   **a.** $p > 0.35$
    b. $|x| = 1.16$ as long as $p < 4.5$.
    c. $|x| = \sqrt{0.9+p}$ which is unstable for $p > 0.1$.

Section 5.10

1. $a(n+2) = 2a(n+1) + 3a(n)$ gives $(a(k) : b(k)) \to (3.75 : 1)$.
3. $b(n+2) = 2.6b(n+1) + 0.5b(n)$ gives $(a(k) : b(k)) \to (5.2 : 1)$.
5. $(a(k) : b(k) : c(k)) \to (0.38 : 0.206 : 1)$
7.   **a.** $p(n+3) = -3p(n+2) + 9p(n+1) - 5p(n)$
    b. $p(k) = c_1 + c_2 k + c_3(-5)^k$
    c. $p(k) = [119(-5)^{100} + 6k(-5)^{99} + (-5)^k]/[1 + 119(-5)^{100}]$
    d. 0.496

Section 5.11

1.   **a.** $r = 2\sqrt{2}$ and $\theta = 7\pi/4$
    b. $r = 2$ and $\theta = 5\pi/6$
    c. $r = 15$ and $\theta = 3\pi/2$
3.   **a.** $a(k) = (\sqrt{2}/4)^k [c_1 \sin(3\pi k/4) + c_2 \cos(3\pi k/4)]$
    b. attracting
    c. $c_1 = -4$ and $c_2 = 0$
5.   **a.** $a(k) = (3\sqrt{2})^k [c_1 \sin(\pi k/4) + c_2 \cos(\pi k/4)] + 1$
    b. repelling

c. $c_1 = 1/3$ and $c_2 = 2$

Section 6.1

1. **a.** $x = 0$ is repelling. $x = 5$ is attracting and $a(k)$ goes to 5 if $0 < a(0) < 85/7$.
    c. $x = 0$ and $x = 1$ are both repelling.
    d. $x = 0$ is stable and $a(k)$ goes to 0 if $-2.4 < a(0) < 2.4$. It is difficult to determine the largest interval of stability for this problem.

Section 6.2

1. **a.** For $a = -1$, $\tilde{e}(n+1) = 5\tilde{e}(n)$ so it is repelling.
    b. For $a = 3$, $\tilde{e}(n+1) = -3\tilde{e}(n)$ so it is repelling.
    c. For $a = -1$, $2ra + b = 5 > 1$ so it is repelling. For $a = 3$, $2ra + b = -3 < -1$ so it is repelling.

3. **a.** $\tilde{e}(n+1) = \tilde{e}(n)$ for $a = 1$.
    b. $2ra + b = 1$ so theorem 6.12 does not apply.

5. **a.** $\tilde{e}(n+1) = 5\tilde{e}(n)$
    b. $\tilde{e}(n+1) = \tilde{e}(n)$

Section 6.3

1. **a.** $a = 2 \pm \sqrt{4 - h}$
    b. $2 - \sqrt{4 - h}$ is unstable for $h < 4$ and is never stable. $2 + \sqrt{4 - h}$ is stable for $0 < h < 4$ and is unstable for $h < 0$.
    c. $a(k) \to -\infty$ when $h > 4$.
    d. When $h = 4$, $a = 2$ is semistable (from above).

3. **a.** $a = 2 - 4p$ is stable for $-1.5 < p < 0.5$.
    b. When $p = 0.25$, $h = 0.25$ which is the maximum harvest.

Section 7.1

1. **a.** $\begin{pmatrix} -18 \\ 4 \\ 16 \end{pmatrix}$
    c. Not possible
    e. $\begin{pmatrix} -17 \\ 49 \end{pmatrix}$

g. $\begin{pmatrix} -3 \\ 35 \\ 16 \end{pmatrix}$

i. Not possible

k. $\begin{pmatrix} -5 & 5 & -18 \\ -4 & 15 & 45 \\ 4 & 0 & 36 \end{pmatrix}$

m. $\begin{pmatrix} 184 \\ 141 \end{pmatrix}$

o. $\begin{pmatrix} -8 \\ 3 \\ -21 \end{pmatrix}$

q. $\begin{pmatrix} 11 & -5 & 34 \\ 8 & -17 & -45 \\ 6 & 14 & -40 \end{pmatrix}$

Section 7.2

1.   a. $\begin{pmatrix} 13 & 7 \\ 23 & 13 \end{pmatrix}, \begin{pmatrix} 10 & 4 \\ 11 & 6 \end{pmatrix}, \begin{pmatrix} 14 \\ 13 \end{pmatrix}$

   c. $\begin{pmatrix} 67 \\ 121 \end{pmatrix}$

   d. $MR = \begin{pmatrix} 13 & 7 \\ 23 & 13 \end{pmatrix}$, and $RM = \begin{pmatrix} 12 & 16 \\ 10 & 14 \end{pmatrix}$

Section 7.3

1.   a. $x = 1$ and $y = 1$
   c. $x = 1, y = -2, z = 0$, and $w = 3$.
   e. No solution

Section 7.4

1.   a. 6
   c. -30

Section 7.5

1.   a. $\begin{pmatrix} 0.5 & -2 \\ 0 & 1 \end{pmatrix}$

$$\mathbf{c.} \ \tfrac{1}{30} \begin{pmatrix} 13 & 11 & -4 \\ -8 & -16 & 14 \\ -11 & -7 & 80 \end{pmatrix}$$

3. $x = 4$, $y = -1$, and $z = 9$

Section 8.1

1. **a.** $A(n+1) = RA(n)$ where $A(n) = \begin{pmatrix} p(n) \\ a(n) \end{pmatrix}$ and $R =$

$$\begin{pmatrix} 0.5 & 0.5 \\ 1 & 0 \end{pmatrix}$$

   b. $R^3 = \begin{pmatrix} 0.625 & 0.375 \\ 0.75 & 0.25 \end{pmatrix}$

   c. $\begin{pmatrix} 2/3 & 1/3 \\ 2/3 & 1/3 \end{pmatrix}$

   d. $\begin{pmatrix} 0.643\,75 \\ 0.6125 \end{pmatrix}$

Section 8.2

1. **a.** $r = 1$ and $r = -3$

   b. $c_1 A_1 = c_1 \begin{pmatrix} 1 \\ 1 \end{pmatrix}$ for $r = 1$ and $c_2 A_2 = c_2 \begin{pmatrix} -3 \\ 1 \end{pmatrix}$ for $r = -3$.

   c. $c_1 = 4$ and $c_2 = 2$.

   d. $R^4 B = 4R^4 A_1 + 2R^4 A_2 = 4A_1 + 2(-3)^4 A_2 = \begin{pmatrix} -482 \\ 166 \end{pmatrix}$

3. **a.** $r = 0$, -2, and 2

   b. $c_1 A_1 = c_1 \begin{pmatrix} 1 \\ -1 \\ 0 \end{pmatrix}$ for $r = 0$; $c_2 A_2 = c_2 \begin{pmatrix} -2 \\ 1 \\ 0 \end{pmatrix}$ for $r = -2$;

   and $c_3 A_3 = c_3 \begin{pmatrix} -11 \\ 5 \\ 2 \end{pmatrix}$ for $r = 2$.

   c. $c_1 = -3$, $c_2 = 0$, and $c_3 = 2$.

   d. $R^5 B = -3(0)^5 A_1 + 0(-2)^5 A_2 + 2(2)^5 A_3 = 2^6 A_3$

Section 8.3

1. **a.** $A(k) = c_1 (-4)^k \begin{pmatrix} 1 \\ -1 \end{pmatrix} + c_2 (3)^k \begin{pmatrix} 2 \\ 5 \end{pmatrix}$

   c. $A(k) = c_1 (i)^k \begin{pmatrix} 5 \\ 2-i \end{pmatrix} + c_2 (-i)^k \begin{pmatrix} 5 \\ 2+i \end{pmatrix}$

2. **a.** $c_1 = -5$ and $c_2 = 1$ for our choice of characteristic vectors.

For any choice, $A(k) = \begin{pmatrix} -5(-4)^k + 2(3)^k \\ 5(-4)^k + 5(3)^k \end{pmatrix}$

   c. **i.** $A(k) = \begin{pmatrix} (2 - 9i)(i)^k + (2 + 9i)(-i)^k \\ -(1 + 4i)(i)^k - (1 - 4i)(-i)^k \end{pmatrix}$

   ii. $A(k) = \begin{pmatrix} 5(i)^k + 5(-i)^k \\ (2 - i)(i)^k + (2 + i)(-i)^k \end{pmatrix}$

3. $A(k) = \begin{pmatrix} a(k) \\ b(k) \end{pmatrix} = \begin{pmatrix} 2(0.5)^k + 4(-0.25)^k \\ 4(0.5)^k + 2(-0.25)^k \end{pmatrix}$

Section 8.4

1. **a.** Not a probability vector since column does not sum to one.
   b. Probability vector.
   c. Not a probability vector since it has a negative component.
   d. Probability vector.
2. **a.** Not a transition matrix since first column does not sum to one.
   c. It is a regular transition matrix since squaring it gives a matrix with all nonzero components.
   e. It is a transition matrix, but is not regular. All even powers of this matrix have zeros on the main diagonal and all odd powers have zeros off the main diagonal.
3. **a.** $\begin{pmatrix} 3/7 \\ 4/7 \end{pmatrix}$

   b. $\begin{pmatrix} 5/8 \\ 3/8 \end{pmatrix}$

   c. $\begin{pmatrix} 6/11 \\ 9/22 \\ 1/22 \end{pmatrix}$

5. 60 per cent

Section 8.5

1. $r(k) \to 0.6$ and $b(k) \to 0.4$, that is, 60 per cent of the time we end with red and 40 per cent of the time we end with blue.
3. **a.** $r(k) \to \frac{5}{9}$ and $b(k) \to \frac{4}{9}$
   b. $r(k) \to 0.537\,78$ and $b(k) \to 0.462\,22$
5. Probability of ending at home is 0.3077 and probability of ending in the lake is 0.6923.

7. $A(n+1) = RA(n)$ where

$$A(k) = \begin{pmatrix} h(k) \\ b(k) \\ w(k) \end{pmatrix}, \qquad R = \begin{pmatrix} 0.2 & 0.2 & 0 \\ 0.8 & 0 & 0.5 \\ 0 & 0.8 & 0.5 \end{pmatrix},$$

$$A_1 = \begin{pmatrix} 5/57 \\ 20/57 \\ 32/57 \end{pmatrix} = \begin{pmatrix} 0.088 \\ 0.351 \\ 0.561 \end{pmatrix}.$$

Here, $h(k)$, $b(k)$, and $w(k)$ are the probabilities of being at home, the bar, and the wall, respectively, after k minutes. A_1 gives the long-term probabilities.

Section 8.6

1. prob(A wins) $= 0.208$, prob(B wins) $= 0.217$, prob(C wins) $= 0.469$, and prob(nobody survives) $= 0.106$

3. 123 148 dollars for Marketing and 186 852 dollars for Manufacturing

5.   a. $P(n+1) = MP(n)$ where

$$M = \frac{1}{16} \begin{pmatrix} 16 & 0 & 1 & 0 & 4 & 0 \\ 0 & 16 & 1 & 0 & 0 & 4 \\ 0 & 0 & 4 & 16 & 4 & 4 \\ 0 & 0 & 2 & 0 & 0 & 0 \\ 0 & 0 & 4 & 0 & 8 & 0 \\ 0 & 0 & 4 & 0 & 0 & 8 \end{pmatrix}, \qquad P(k) = \begin{pmatrix} e_1(k) \\ e_2(k) \\ f_1(k) \\ f_2(k) \\ f_3(k) \\ f_4(k) \end{pmatrix},$$

$$S = \frac{1}{16} \begin{pmatrix} 1 & 0 & 4 & 0 \\ 1 & 0 & 0 & 4 \end{pmatrix}, \qquad R = \frac{1}{16} \begin{pmatrix} 4 & 16 & 4 & 4 \\ 2 & 0 & 0 & 0 \\ 4 & 0 & 8 & 0 \\ 4 & 0 & 0 & 8 \end{pmatrix}.$$

   b. $(I - R)^{-1} = \frac{1}{6} \begin{pmatrix} 16 & 16 & 8 & 8 \\ 2 & 8 & 1 & 1 \\ 8 & 8 & 16 & 4 \\ 8 & 8 & 4 & 16 \end{pmatrix}$, and prob(both being dominant homozygotes) $=$ prob(both being recessive homozygotes) $= 0.5$.

   c. prob(both being dominant homozygotes) $= 0.75$ and prob(both being recessive homozygotes) $= 0.25$.

7.   a. $F(k) = (R^{k+1} - I)(R - I)^{-1}Q(0)$. To get this, you can use the fact that $R(I - R)^{-1} = (I - R)^{-1}R$.

b. $F(k) \rightarrow (I - R)^{-1} Q(0) = \begin{pmatrix} 0.347 \\ 0.335 \\ 1.117 \end{pmatrix}$ since $R^{k+1} \rightarrow 0_3$

where 0_3 is the 3-vector with all zeros.

c. $0.347 + 0.335 + 1.117 = 1.799$ rounds on the average.

d. 1.266 rounds

e. 1.538 rounds

Section 8.7

1. 11

3. **a.** $r = 2$ and $r = -0.5$ with characteristic vectors $A_1 = \begin{pmatrix} 1 \\ 2 \end{pmatrix}$

and $A_2 = \begin{pmatrix} 1 \\ -3 \end{pmatrix}$, respectively.

b. $a(1)/a(0) = 1.5$, $a(2)/a(1) = 2.167$, ..., $a(6)/a(5) = 2.00057$.

c. $|A(6)|/|A(5)| = 1.99939$

d. $A(6)/|A(6)| = \begin{pmatrix} 0.447 \\ 0.894 \end{pmatrix}$

e. Ratio of a_1 to a_2 of vector A_1 of part (a) is $(1 : 2)$. Ratio $a_1(6)$ to $a_2(6)$ from vector $A(6)/|A(6)|$ of part (d) is also $(1 : 2)$.

5. **a.** $A(n + 1) = RA(n)$ where $\begin{pmatrix} b_1 & b_2 & b_3 \\ s_1 & 0 & 0 \\ 0 & s_2 & s_3 \end{pmatrix}$.

b. $a_1(5)/a_1(4) = 1.099$

c. $A(5)/|A(5)| = \begin{pmatrix} 0.346 \\ 0.189 \\ 0.919 \end{pmatrix}$ and $\lim_{k \rightarrow \infty}(a_1(k) : a_2(k) : a_3(k)) \approx (1.83 : 1 : 1.486)$.

Section 8.8

1. **a.** $R = \begin{pmatrix} 0.5 & 0.25 & 0 & 0 \\ 0.25 & 0.5 & 0.25 & 0 \\ 0 & 0.25 & 0.5 & 0.25 \\ 0 & 0 & 0.25 & 0.5 \end{pmatrix}$, and $B = \begin{pmatrix} 0 \\ 0 \\ 0 \\ 5 \end{pmatrix}$.

b. $\begin{pmatrix} 4 \\ 8 \\ 12 \\ 16 \end{pmatrix}$

c. $\begin{pmatrix} 7.81 \\ 12.5 \\ 12.5 \\ 7.81 \end{pmatrix}$

d. 0.893

e. $\begin{pmatrix} 7.89 \\ 13.28 \\ 16.09 \\ 17.97 \end{pmatrix}$

3. a. $a_2(n+1) = pa_1(n) + (1 - 4p)a_2(n) + pa_3(n) + 40p$, and
$a_3(n+1) = pa_2(n) + (1 - 4p)a_3(n)$

b. $R = \begin{pmatrix} 1 - 4p & p & 0 \\ p & 1 - 4p & p \\ 0 & p & 1 - 4p \end{pmatrix}$ and $B = \begin{pmatrix} 60p \\ 40p \\ 0 \end{pmatrix}$.

c. $R = \begin{pmatrix} 0 & 0.25 & 0 \\ 0.25 & 0 & 0.25 \\ 0 & 0.25 & 0 \end{pmatrix}$ and $B = \begin{pmatrix} 15 \\ 10 \\ 0 \end{pmatrix}$.

d. $\begin{pmatrix} 18.93 \\ 15.71 \\ 3.93 \end{pmatrix}$,

e. $|R| = 0.3535$

f. $\begin{pmatrix} 18.875 \\ 15.594 \\ 3.875 \end{pmatrix}$

INDEX